ORGANIC AGRICULTURAL PRACTICES

ALTERNATIVES TO CONVENTIONAL AGRICULTURAL SYSTEMS

ORGANIC AGRICULTURAL PRACTICES

ALTERNATIVES TO CONVENTIONAL AGRICULTURAL SYSTEMS

Edited by
Kimberly Etingoff

Apple Academic Press

TORONTO   NEW JERSEY

Apple Academic Press Inc.	Apple Academic Press Inc.
3333 Mistwell Crescent	9 Spinnaker Way
Oakville, ON L6L 0A2	Waretown, NJ 08758
Canada	USA

©2015 by Apple Academic Press, Inc.

First issued in paperback 2021

Exclusive worldwide distribution by CRC Press, a member of Taylor & Francis Group
No claim to original U.S. Government works

ISBN 13: 978-1-77463-214-7 (pbk)
ISBN 13: 978-1-77188-082-4 (hbk)

Library of Congress Control Number: 2014944606

Library and Archives Canada Cataloguing in Publication

Organic agricultural practices: alternatives to conventional agricultural systems/edited by Kimberly Etingoff.

Includes bibliographical references and index.
ISBN 978-1-77188-082-4 (bound)
1. Organic farming. I. Etingoff, Kim, editor

S605.5.O63 2014   631.5'84   C2014-904813-0

Apple Academic Press also publishes its books in a variety of electronic formats. Some content that appears in print may not be available in electronic format. For information about Apple Academic Press products, visit our website at **www.appleacademicpress.com** and the CRC Press website at **www.crcpress.com**

ABOUT THE EDITOR

KIMBERLY ETINGOFF

Kimberly Etingoff's background includes city and regional planning, farming, food systems programming, sociology, and urban geography. She studied at the University of Rochester, New York, and Tufts University, Boston, Massachusetts, and has done extensive field work with food systems and agricultural sociology. She has been writing and editing academic and educational books on topics such as nutrition, farming, and aspects of sociology for more than ten years.

CONTENTS

ACKNOWLEDGMENT AND HOW TO CITE

The editor and publisher thank each of the authors who contributed to this book, whether by granting their permission individually or by releasing their research as open source articles or under a license that permits free use, provided that attribution is made. The chapters in this book were previously published in various places in various formats. To cite the work contained in this book and to view the individual permissions, please refer to the citation at the beginning of each chapter. Each chapter was read individually and carefully selected by the editor; the result is a book that provides a nuanced study of organic agriculture and its environmental effects. The chapters included examine the following topics:

- In Chapter 1, the authors use energy accounting and Life Cycle Accounting to assess the environmental impacts of an organic agricultural system, further comparing the system to high- and low-yielding produce systems.
- The comparison of organic and conventional practices' effects on fungi, bacteria, and other soil organisms in Chapter 2 shows a number of similarities and differences, primarily a larger number of microartropods in organic plots.
- In response to nitrate contamination of the Seine watershed, Chapter 3 theorizes two economically and environmentally sustainable responses based on organic and local agriculture.
- Chapter 4 investigates the benefits of the application of vermicompost to China aster, finding it superior to chemical fertilizers.
- Chapter 5 contains a comparison between organic and conventional agriculture, including carbon sequestration, soil fertility, costs, and labor structure.
- Combinations of organic materials were applied to mung bean crops in Chapter 6 to determine the best specific applications for increasing crop yields.
- Chapter 7 compares various biofertilizers and fertilizer rates, in order to assess the appropriateness of their use in rice growing contexts.
- In Chapter 8, experiments combining biofertilizer variations and plant density variations determined the best applications for a number of factors including canopy cover, grain yield, and oil yield.

- To promote the benefits of vermicompost, including soil fertility, plant disease management, and pest control, the authors of Chapter 9 provide research on the specific attributes of earthworms, soil microbes, and vermicomposting processes.
- The authors of Chapter 10 introduce a variety of technologies and developmental results in selective plant breeding for use in organic agriculture.
- Chapter 11 explores possibilities and results of collaborative work with researchers and farmers can create context-specific plant breeds for organic agricultural systems.
- Spinach varieties are specifically looked at as an example of efforts to breed cultivars appropriate for withstanding the adaptational needs of organic farming in Chapter 12.
- Chapter 13 finds that variations in agricultural management strongly effect soil enzymes and nutrients, but have a less pronounced effect on soil mircobes.
- In Chapter 14, the authors challenge the assumption that organic agriculture is necessarily less polluting, with a study on nitrogen contamination in an intensive organic system.
- Chapter 15 discusses various innovations in crop nitrogen use efficiency, a key part of reducing fertilizer use and making organic agriculture viable.

LIST OF CONTRIBUTORS

Veronica Acosta-Martínez
USDA-ARS, Cropping Systems Research Laboratory, Wind Erosion and Water Conservation Unit, Lubbock, TX 79415, USA

Edna A. Aguilar
Crop Science Cluster, College of Agriculture (CA), University of the Philippines at Los Baños (UPLB), Laguna 4031, Philippines

A. Babad
Department of Environmental Hydrology&Microbiology, Zuckerberg Institute for Water Research, Blaustein Institutes for Desert Research, Ben-Gurion University of the Negev, Midreshet Ben-Gurion, Israel

Rodrigo B. Badayos
Agricultural Systems Cluster, College of Agriculture (CA), University of the Philippines at Los Baños (UPLB), Laguna 4031, Philippines

Niño Paul Meynard Banayo
International Rice Research Institute, Los Baños, Laguna 4030, Philippi

S. Barles
Geo-cité, Université Paris 1-Sorbonne, Paris, France

Jean-François Berthellot
Réseau Semences Paysannes Cazalens, 81600 Brens, France

Wagner Bettiol
Embrapa Meio Ambiente, C.P. 69 - CEP: 13820-000 - Jaguariúna, SP, CNPq Fellow

G. Billen
UMR Sisyphe, UPMC/CNRS, Paris, France

Timothy M. Bowles
Department of Land, Air and Water Resources, University of California Davis, Davis, CA 95616, USA

Francisco Calderón
USDA-ARS, Central Great Plains Research Station, 40335 County Road GG, Akron, CO 80720, USA

Véronique Chable
INRA, Unité SAD Paysage, Institut National de la Recherche Agronomique, 65 rue de Saint-Brieuc, F-35042 Rennes, France

P. Chatzimpiros
LEESU/Génie Urbain, UPEMLV – cité Descartes, 77454 Marne-la-Vallée, France

Pompe C. Sta. Cruz
Crop Science Cluster, College of Agriculture (CA), University of the Philippines at Los Baños (UPLB), Laguna 4031, Philippines

Dejan Cvikic
Institute for Vegetable Crops, Smederevska Palanka, Serbia

O. Dahan
Department of Environmental Hydrology&Microbiology, Zuckerberg Institute for Water Research, Blaustein Institutes for Desert Research, Ben-Gurion University of the Negev, Midreshet Ben-Gurion, Israel

Dulal Chandra Das
Department of Botany, Raja Narendra Lal Khan Women's College, Gope Palace, Paschim Medinipur, West Bengal, India, 721 102

Julie C. Dawson
INRA, UMR 320 Génétique Végétale, Institut National de la Recherche Agronomique, Ferme du Moulon, F-91190 Gif-sur-Yvette, France, current address 422 Bradfield Hall, Department of Plant Breeding and Genetics, Cornell University, Ithaca, NY 14853, USA

Mohammadreza Davari
Department of Agriculture, Payame Noor University, Arak, Iran

Patrick de Kochko
Réseau Semences Paysannes Cazalens, 81600 Brens, France

Frédéric Dubois
Agrophysiologie, Ecophysiologie et Biologie Intégrative, A3900-AEB, Université de Picardie, 33 rue Saint Leu, F-80039 Amiens, France

Mahdi Faravani
Khorasan Agricultural Research Center, P.O.Box 91735-488, Mashhad, Iran

Nathalie Galic
UMR de Génétique Vegetale, Ferme du Moulon, 91190 Gif-sur-Yvette, France

José Abrahão Haddad Galvão
Embrapa Meio Ambiente, C.P. 69 - CEP: 13820-000 - Jaguariúna, SP

J. Garnier
UMR Sisyphe, UPMC/CNRS, Paris, France

Raquel Ghini
Embrapa Meio Ambiente, C.P. 69 - CEP: 13820-000 - Jaguariúna, SP, CNPq Fellow

Barat Ali Gholami
Khorasan Agricultural Research Center, P.O.Box 91735-488, Mashhad, Iran

Zdenka Girek
Institute for Vegetable Crops, Smederevska Palanka, Serbia

Simon Giuliano
Ecole d'Ingénieurs de Purpan, 75 voie du Toec, 31076 Toulouse, France

Isabelle Goldringer
INRA, UMR 320 Génétique Végétale, Institut National de la Recherche Agronomique, Ferme du Moulon, F-91190 Gif-sur-Yvette, France

Stephan M. Haefele
International Rice Research Institute, Los Baños, Laguna 4030, Philippines

Mostafa Heidari
University of Zabol, Zabol, Iran

Bertrand Hirel
Adaptation des Plantes à leur Environnement. Unité de Recherche 511, Institut Jean-Pierre Bourgin, Institut National de la Recherche Agronomique, Centre de Versailles-Grignon, R.D. 10, F-78026 Versailles Cedex, France

Louise E. Jackson
Department of Land, Air and Water Resources, University of California Davis, Davis, CA 95616, USA

Mohammad Taghi Kashki
Khorasan Agricultural Research Center, P.O.Box 91735-488, Mashhad, Iran

Masakazu Komatsuzaki
College of Agriculture, Ibaraki University, 3-21-1 Ami, Inashiki, Ibaraki 300-0393, Japan

Michal Kulak
Life Cycle Assessment group, Institute for Sustainability Sciences, Agroscope Reckenholzstrasse 191, CH-8046 Zurich, Switzerland

D. Kurtzman
Institute of Soil, Water and Environmental Sciences, Agricultural Research Organization, The Volcani Center, Bet Dagan, Israel

N. Lazarovitch
Wyler Department of Dryland Agriculture, French Associates Institute for Agriculture and Biotechnology of Drylands, Jacob Blaustein Institutes for Desert Research, Ben-Gurion University of the Negev, Midreshet Ben-Gurion, Israel

Peter J. Lea
Lancaster Environment Centre, Lancaster University, Lancaster, LA1 4YQ, UK

Marcos Antônio Vieira Ligo
Embrapa Meio Ambiente, C.P. 69 - CEP: 13820-000 - Jaguariúna, SP

Mads V. Markussen
Center for BioProcess Engineering, Department of Chemical and Biochemical Engineering, Technical University of Denmark DTU, DK-2800 Kgs. Lyngby, Denmark

Florent Mercier
Réseau Semences Paysannes Cazalens, 81600 Brens, France

Jeferson Luiz de Carvalho Mineiro
Embrapa Meio Ambiente, C.P. 69 - CEP: 13820-000 - Jaguariúna, SP

Mohammad Mirzakhani
Department of Agriculture, Islamic Azad University, Farahan branch, Farahan, Iran

Thomas Nemecek
Life Cycle Assessment group, Institute for Sustainability Sciences, Agroscope Reckenholzstrasse 191, CH-8046 Zurich, Switzerland

Hanne Østergård
Center for BioProcess Engineering, Department of Chemical and Biochemical Engineering, Technical University of Denmark DTU, DK-2800 Kgs. Lyngby, Denmark

Jayakumar Pathma
Department of Biotechnology School of Life Sciences, Pondicherry University, Kalapet, Puducherry, 605014, India

Nenad Pavlovic
Institute for Vegetable Crops, Smederevska Palanka, Serbia

Sophie Pin
UMR de Génétique Vegetale, Ferme du Moulon, 91190 Gif-sur-Yvette, France

Pierre Rivière
UMR de Génétique Vegetale, Ferme du Moulon, 91190 Gif-sur-Yvette, France

E. E. Russak
Geological and Environmental Sciences, Ben-Gurion University of the Negev, Beer Sheva, Israel

Natarajan Sakthivel
Department of Biotechnology School of Life Sciences, Pondicherry University, Kalapet, Puducherry, 605014, India

Behjat Salari
University of Zabol, Zabol, Iran

Nicolas Schermann
INRA, Unité SAD Paysage, Institut National de la Recherche Agronomique, 65 rue de Saint-Brieuc, F-35042 Rennes, France

Estelle Serpolay
INRA, Unité SAD Paysage, Institut National de la Recherche Agronomique, 65 rue de Saint-Brieuc, F-35042 Rennes, France

Shri Niwas Sharma
Division of Agronomy, Indian Agricultural Research Institute, New Delhi, India

M. Silvestre
UMR Sisyphe, UPMC/CNRS, Paris, France

Laurence G. Smith
The Organic Research Centre, Elm Farm, Hamstead Marshall, Newbury, Berkshire RG20 0HR, UK

M. Faiz Syuaib
Department of Agricultural Engineering, Bogor Agricultural University, JI Meranti, Kampus IPB Darmaga, Bogor, 16680, Indonesia

Thierry Tétu
Agrophysiologie, Ecophysiologie et Biologie Intégrative, A3900-AEB, Université de Picardie, 33 rue Saint Leu, F-80039 Amiens, France

V. Thieu
IES, IRC, Ispra, Italy

Mathieu Thomas
UMR de Génétique Vegetale, Ferme du Moulon, 91190 Gif-sur-Yvette, France

Edith T. Lammerts van Bueren
Louis Bolk Institute, Hoofdstraat 24, NL-3972 LA Driebergen, The Netherlands

Jasmina Zdravkovic
Institute for Vegetable Crops, Smederevska Palanka, Serbia

Milan Zdravkovic
Institute for Vegetable Crops, Smederevska Palanka, Serbia

INTRODUCTION

Organic practices are redefining world agriculture. As researchers and others begin to understand the detrimental effects conventional agriculture can have on local and global environments, more studies focusing on organic agriculture are appearing, including studies that compare organic practices to conventional ones. Innovative organic agricultural practices impact soil fertility, microbial growth, soil organism numbers, greenhouse gas release, and overall crop performance. Among the strategies available to farmers who pursue organic practices are a range of biofertilizers and natural pesticides, along with the breeding of new plant cultivars which can work with other organic practices to increase productivity.

Kimberly Etingoff

Resource use and environmental impacts of a small-scale low-input organic vegetable supply system in the United Kingdom were assessed by emergy accounting and Life Cycle Assessment (LCA). The system consisted of a farm with high crop diversity and a related box-scheme distribution system. In Chapter 1, Markussen and colleagues compared empirical data from this case system with two modeled organic food supply systems representing high- and low-yielding practices for organic vegetable production. Further, these systems were embedded in a supermarket distribution system and they provided the same amount of comparable vegetables at the consumers' door as the case system. The on-farm resource use measured in solar equivalent Joules (seJ) was similar for the case system and the high-yielding model system and higher for the low-yielding model system. The distribution phase of the case system was at least three times as resource efficient as the models and had substantially less environmental impacts when assessed using LCA. The three systems ranked differently for emissions with the high-yielding model system being the worst for terrestrial ecotoxicity and the case system the worst for global warming potential. As a consequence of being embedded in an industrial economy, about 90% of resources (seJ) were used for supporting labor and service.

Despite the recent interest in organic agriculture, little research has been carried out in this area. Thus, the objective of Bettiol and colleagues in Chapter 2 was to compare, in a dystrophic Ultisol, the effects of organic and conventional agricultures on soil organism populations, for the tomato (*Lycopersicum esculentum*) and corn (*Zea mays*) crops. In general, it was found that fungus, bacterium and actinomycet populations counted by the number of colonies in the media, were similar for the two cropping systems. CO_2 evolution during the cropping season was higher, up to the double for the organic agriculture system as compared to the conventional. The number of earthworms was about ten times higher in the organic system. There was no difference in the decomposition rate of organic matter of the two systems. In general, the number of microartropods was always higher in the organic plots in relation to the conventional ones, reflectining on the Shannon index diversity. The higher insect population belonged to the *Collembola* order, and in the case of mites, to the superfamily *Oribatuloidea*. Individuals of the groups *Aranae, Chilopoda, Dyplopoda, Pauropoda, Protura* and *Symphyla* were occasionally collected in similar number in both cropping systems.

The Seine watershed has long been the food-supplying hinterland of Paris, providing most of the animal and vegetal protein consumed in the city. Nowadays, the shift from manure-based to synthetic nitrogen fertilisation, has made possible a strong land specialisation of agriculture in the Seine watershed: it still provides most of the cereal consumed by the Paris agglomeration, but exports 80% of its huge cereal production. On the other hand the meat and milk supply originates mainly from regions in the North and West of France, specialised in animal farming and importing about 30% of their feed from South America. As it works today, this system is responsible for a severe nitrate contamination of surface and groundwater resources. In Chapter 3, Garnier and colleagues explore two scenarios of re-localising Paris's food supply, based on organic farming and local provision of animal feed. The authors show that for the Seine watershed it is technically possible to design an agricultural system able to provide all the plant- and animal-based food required by the population, to deliver sub-root water meeting the drinking water standards and still to export a significant proportion of its production to areas less suitable for cereal cultivation. Decreasing the share of animal products in the human diet has a strong impact on the nitrogen imprint of urban food supply.

The chemical fertilizers led to a loss of faith in agriculture throughout the globe today; conversely, the organic amendments are gradually becoming more reliable components in the field of agriculture. The objective of Das in the present work was to investigate the relative growth analysis of China aster (*Callistephus chinensis* L.), applying the vermi compost and chemical fertilizer (10:26:26) and to observe the growth efficiency of both the amendments. The present investigation also aimed to increase the faith and awareness in the mindset of common people for the use of organic manure in the field of agriculture. Five experimental sets (T1, T2, T3, T4 and T5) with replicas were prepared and nine parameters were selected to study the growth efficiency. The results obtained from the experiments have proved the superiority of the vermicompost over the chemical fertilizer in each and every respect of the parameters that were used and it may be conclude that the vermicompost as an organic manure could be used as an alternative of the chemical fertilizer. These results indicate that the demands of organic agriculture for the yield performance would be more in near future.

Organic farming provides many benefits in Indonesia: it can improve soil quality, food quality and soil carbon sequestration. Chapter 5, by Komatsuzaki and Syuaib, was designed to compare soil carbon sequestration levels between conventional and organic rice farming fields in west Java, Indonesia. The results from soil analysis indicate that organic farming leads to soil with significantly higher soil carbon storage capacity than conventional farming. Organic farming can also cut some farming costs, but it requires about twice as much labor. The sharecropping system of rice farming in Indonesia is highly exploitative of workers; therefore, research should be conducted to develop a fairer organic farming system that can enhance both local and global sustainability.

Chapter 6, by Davari and colleagues, was undertaken to assess the residual influence of organic materials and biofertilizers applied to rice and wheat on yield, nutrient status, and economics of succeeding mung bean in an organic cropping system. The field experiments were carried out on the research farm of IARI, New Delhi during crop cycles of 2006 to 2007 and 2007 to 2008 to study the effects of residual organic manures, crop residues, and biofertilizers applied to rice and wheat on the performance of succeeding mung bean. The experiment was laid out in a randomized block design with three replications. Treatments consisted of six combinations of different residual organic materials, and biofertilizers included residual farmyard manure (FYM) and vermicompost (VC) applied on nitrogen basis at 60 kg ha^{-1} to each rice and wheat crops, FYM + wheat and rice residues at 6 t ha^{-1} and mung bean residue at 3 t ha^{-1} in succeeding crops (CR), VC + CR, FYM + CR + biofertilizers (B), VC + CR + B, and control (no fertilizer applied). For biofertilizers, cellulolytic culture, phosphate-solubilizing bacteria and Rhizobium applied in mung bean. Incorporation of crop residue significantly increased the grain yield of mung bean over residual of FYM and VC by 25.5% and 26.5%, respectively. The combinations of FYM + CR + B and VC + RR + B resulted in the highest increase growth and yield attributing characters of mung bean and increased grain yield of mung bean over the control by 47% and net return by 27%. The present study thus indicates that a combination of FYM + CR + B and VC + CR + B were economical for the nutrient need of mung bean in organic farming of rice-based cropping system.

Biofertilizers are becoming increasingly popular in many countries and for many crops, but very few studies on their effect on grain yield have been conducted in rice. Therefore, in Chapter 7, Banayo and colleagues evaluated three different biofertilizers (based on *Azospirillum, Trichoderma*, or unidentified rhizobacteria) in the Philippines during four cropping seasons between 2009 and 2011, using four different fertilizer rates (100% of the recommended rate [RR], 50% RR, 25% RR, and no fertilizer as Control). The experiments were conducted under fully irrigated conditions in a typical lowland rice environment. Significant yield increases due to biofertilizer use were observed in all experimental seasons with the exception of the 2008/09 DS. However, the effect on rice grain yield varied between biofertilizers, seasons, and fertilizer treatments. In relative terms, the seasonal yield increase across fertilizer treatments was between 5% and 18% for the best biofertilizer (*Azospirillum*-based), but went up to 24% in individual treatments. Absolute grain yield increases due to biofertilizer were usually below 0.5 t·ha^{-1}, corresponding to an estimated additional N uptake of less than 7.5 kg N ha^{-1}. The biofertilizer effect on yield did not significantly interact with the inorganic fertilizer rate used but the best effects on grain yield were achieved at low to medium fertilizer rates. Nevertheless, positive effects of the biofertilizers even occurred at grain yields up to 5 t·ha^{-1}. However, the trends in the results seem to indicate that biofertilizers might be most helpful in rainfed environments with limited inorganic fertilizer input. However, for use in these target environments, biofertilizers need to be evaluated under conditions with abiotic stresses typical of such systems such as drought, soil acidity, or low soil fertility.

In order to understand the effect of organic fertilizer on yield of anise, in Chapter 8 Faravani and colleagues conducted an experiment in the form of split-plot in randomized complete block design with three replications in Mashhad, Khorasan Agriculture and Natural Resource Research Center. Four treatments of fertilization: the control, vermicompost – 5 t/ha, cow manure – 25 t/ha, and mineral fertilizer (NPK) – 60 kg/ha (the same rate of each nutrient) were applied as the main factor. The second factor was plant density, applied at three levels: 17, 25, and 50 plants/m^2. The results showed a significant effect of fertilizer on the number of umbels per plant, number of umbellets per umbel and canopy cover. Plant density had a significant effect on grain yield, biological yield, the number of lateral branches, essential oil percentage and yield of essential oil. Seed and essential oil yield were the highest in the case of the application of vermicompost and plant densities of 50 and 25 plants/m^2 respectively.

Vermicomposting is a non-thermophilic, boioxidative process that involves earthworms and associated microbes. This biological organic waste decomposition process yields the biofertilizer namely the vermicompost. Vermicompost is a finely divided, peat like material with high porosity, good aeration, drainage, water holding capacity, microbial activity, excellent nutrient status and buffering capacity thereby resulting the required physiochemical characters congenial for soil fertility and plant growth. Vermicompost enhances soil biodiversity by promoting the beneficial microbes which inturn enhances plant growth directly by production of plant growth-regulating hormones and enzymes and indirectly by controlling plant pathogens, nematodes and other pests, thereby enhancing plant health and minimizing the yield loss. In Chapter 9, Pathma and Sakthivel argue that due to its innate biological, biochemical and physiochemical properties, vermicompost may be used to promote sustainable agriculture and also for the safe management of agricultural, industrial, domestic and hospital wastes which may otherwise pose serious threat to life and environment.

In Chapter 10, Zdravkovic and colleagues describe the Institute for Vegetable Crops and the rich germplasm collection of vegetables it possesses, utilized as gene resource for breeding specific traits. Onion and garlic breeding programs are based on chemical composition improvement. There are programs for identification and use of genotypes characterized by high tolerance to economically important diseases. Special attention is paid to breeding cucumber and tomato lines tolerant to late blight. As a result, late blight tolerant pickling cucumber line, as well as late blight tolerant tomato lines and hybrids are realized. Research on bean drought stress tolerance is initiated. Lettuce breeding program including research on spontaneous flora is started and interspecies hybrids were observed as possible genetic variability source. It is important to have access to a broad range of vegetable genotypes in order to meet the needs of organic agriculture production. Appreciating the concept of sustainable agriculture, it is important to introduce organic agriculture programs in breeding institutions.

Because organic systems present complex environmental stress, plant breeders may either target very focused regions for different varieties, or create heterogeneous populations which can then evolve specific adaptation through on-farm cultivation and selection. This often leads to participatory plant breeding (PPB) strategies which take advantage of the specific knowledge of farmers. Participatory selection requires increased commitment and engagement on the part of the farmers and researchers. Projects may begin as researcher initiatives with farmer participation or farmer initiatives with researcher participation and over time evolve into true collaborations. These projects are difficult to plan in advance because by nature they change to respond to the priorities and interests of the collaborators. Projects need to provide relevant information and analysis in a time-frame that is meaningful for farmers, while remaining scientifically rigorous and innovative. In Chapter 11, Dawson and colleagues presents two specific studies: the first was a researcher-designed experiment that assessed the potential adaptation of landraces to organic systems through on-farm cultivation and farmer selection. The second is a farmer-led plant breeding project to select bread wheat for organic systems in France. Over the course of these two projects, many discussions among farmers, researchers and farmers associations led to the development of methods that fit the objectives of those involved. This type of project is no longer researcher-led or farmer-led but instead an equal collaboration. Results from the two research projects and the strategy developed for an ongoing collaborative plant breeding project are discussed.

Organic and low-input agriculture needs flexible varieties that can buffer environmental stress and adapt to the needs of farmers. In Chapter 12, Serpolay and colleagues implemented an experiment to investigate the evolutionary capacities of a sample of spinach (*Spinacia oleracea* L.) population varieties for a number of phenotypic traits. Three farmers cultivated, selected and multiplied one or several populations over two years on their farms. The third year, the versions of the varieties cultivated and selected by the different farmers were compared to the original seed lots they had been given. After two cycles of cultivation and on-farm mass selection, all the observed varieties showed significant phenotypic changes (differences between the original version and the version cultivated by farmers) for morphological and phenological traits. When the divergence among versions within varieties was studied, the results show that the varieties conserved their identity, except for one variety, which evolved in such a way that it may now be considered two different varieties. The heterogeneity of the population varieties was assessed in comparison with a commercial F1 hybrid used as control, and we found no specific differences in phenotypic diversity between the hybrid and population varieties. The phenotypic changes shown by the population varieties in response to on-farm cultivation and selection could be useful for the development of specific adaptation. These results call into question the current European seed legislation and the requirements of phenotypic stability for conservation varieties.

Variability in the activity and composition of soil microbial communities may have important implications for the suite of microbially-derived ecosystem functions upon which agricultural systems rely, particularly organic agriculture. In Chapter 13, Bowles and colleagues used an on-farm approach to investigate microbial communities and soil carbon (C) and nitrogen (N) availability on 13 organically-managed fields growing Roma-type tomatoes, but differing in nutrient management, across an intensively-managed agricultural landscape in the Central Valley of California. Soil physicochemical characteristics, potential activities of nine soil enzymes involved in C, N, phosphorus (P), and sulfur (S) cycling, and fatty acid methyl esters (FAMEs) were measured during the growing season and evaluated with multivariate approaches. Soil texture and pH in the 0–15 cm surface layer were similar across the 13 fields, but there was a three-fold range of soil C and N as well as substantial variation in inorganic N and available P that reflected current and historical management practices. Redundancy analysis showed distinct profiles of enzyme activities across the fields, such that C-cycling enzyme potential activities increased with inorganic N availability while those of N-cycling enzymes increased with C availability. Although FAMEs suggested that microbial community composition was less variable across fields than enzyme activities, there were slight community differences that were related to organic amendments (manure vs. composted green waste). Overall, however, the general similarity among fields for particular taxonomic indicators, especially saprophytic fungi, likely reflects the high disturbance and low complexity in this landscape. Variation in potential enzyme activities was better accounted for with soil physicochemical characteristics than microbial community composition, suggesting high plasticity of the resident microbial community to environmental conditions. These patterns suggest that, in this landscape, differences in organic agroecosystem management have strongly influenced soil nutrients and enzyme activity, but without a major effect on soil microbial communities. The on-farm approach provided a wide range of farming practices and soil characteristics to reveal how microbially-derived ecosystem functions can be effectively manipulated to enhance nutrient cycling capacity.

It is commonly presumed that organic agriculture causes only minimal environmental pollution. In Chapter 14, Dahan and colleagues measured the quality of percolating water in the vadose zone, underlying both organic and conventional intensive greenhouses. Their study was conducted in newly established farms where the subsurface underlying the greenhouses has been monitored continuously from their establishment. Surprisingly, intensive organic agriculture relying on solid organic matter, such as composted manure that is implemented in the soil prior to planting as the sole fertilizer, resulted in significant down-leaching of nitrate through the vadose zone to the groundwater. On the other hand, similar intensive agriculture that implemented liquid fertilizer through drip irrigation, as commonly practiced in conventional agriculture, resulted in much lower rates of pollution of the vadose zone and groundwater. It has been shown that accurate fertilization methods that distribute the fertilizers through the irrigation system, according to plant demand, during the growing season dramatically reduce the potential for groundwater contamination from both organic and conventional greenhouses.

In Chapter 15, Hirel and colleagues present the recent developments and future prospects of improving nitrogen use efficiency (NUE) in crops using various complementary approaches. These include conventional breeding and molecular genetics, in addition to alternative farming techniques based on no-till continuous cover cropping cultures and/or organic nitrogen (N) nutrition. Whatever the mode of N fertilization, an increased knowledge of the mechanisms controlling plant N economy is essential for improving NUE and for reducing excessive input of fertilizers, while maintaining an acceptable yield and sufficient profit margin for the farmers. Using plants grown under agronomic conditions, with different tillage conditions, in pure or associated cultures, at low and high N mineral fertilizer input, or using organic fertilization, it is now possible to develop further whole plant agronomic and physiological studies. These can be combined with gene, protein and metabolite profiling to build up a comprehensive picture depicting the different steps of N uptake, assimilation and recycling to produce either biomass in vegetative organs or proteins in storage organs. The authors provide a critical overview as to how our understanding of the agro-ecophysiological, physiological and molecular controls of N assimilation in crops, under varying environmental conditions, has been improved. They have used combined approaches, based on agronomic studies, whole plant physiology, quantitative genetics, forward and reverse genetics and the emerging systems biology. Long-term sustainability may require a gradual transition from synthetic N inputs to legume-based crop rotation, including continuous cover cropping systems, where these may be possible in certain areas of the world, depending on climatic conditions. Current knowledge and prospects for future agronomic development and application for breeding crops adapted to lower mineral fertilizer input and to alternative farming techniques are explored, whilst taking into account the constraints of both the current world economic situation and the environment.

PART I

ORGANIC AGRICULTURAL RESULTS COMPARED WITH CONVENTIONAL AGRICULTURE

CHAPTER 1

EVALUATING THE SUSTAINABILITY OF A SMALL-SCALE LOW-INPUT ORGANIC VEGETABLE SUPPLY SYSTEM IN THE UNITED KINGDOM

MADS V. MARKUSSEN, MICHAL KULAK, LAURENCE G. SMITH, THOMAS NEMECEK, AND HANNE ØSTERGÅRD

1.1 INTRODUCTION

Modern food supply systems (production and distribution) are heavily dependent on fossil energy [1] and other non-renewable resources [2]. The global environmental crisis [3,4] and foreseeable constraints on the supply of energy [5] and fertilizer [6,7] clearly show that there is a need to develop food supply systems that conserve biodiversity and natural systems and rely less on non-renewable resources. A similar conclusion is drawn in a report initiated by the Food and Agriculture Organization of the United Nations (FAO) and The World Bank. It emphasizes the need to maintain productivity, while conserving natural resources by improving nutrient, energy, water and land use efficiency, increasing farm diversification, and

This chapter was originally published under the Creative Commons Attribution License. Markussen MV, Kulak M, Smith LG, Nemecek T, and Østergård H. Evaluating the Sustainability of a Small-Scale Low-Input Organic Vegetable Supply System in the United Kingdom. Sustainability 6,4 (2014); doi:10.3390/su6041913.

supporting agro-ecological systems that take advantage of and conserve biodiversity at both field and landscape scale [8].

It has been shown that the food industry in the UK is responsible for 14% of national energy consumption and for 25% of heavy goods vehicle kilometers [9]. The structural development of the food supply system over the past 60 years means that most goods are now distributed through regional distribution centers before being transported to increasingly centralized and concentrated out-of-town supermarkets. This also means that more shopping trips are done by private cars which make up approximately half of the total food vehicle kilometers [10]. In 2002, 9% of UK's total consumption of petroleum products was used for transportation of food [10]. This clearly shows that if the environmental impacts of the food supply system are to be significantly reduced, then it is necessary to view the production and distribution of food together. Direct marketing and local selling of products offers a way for farms to by-pass the energy intensive mass distribution system. Such distribution systems are particularly appropriate for vegetables, which have a relative short lifetime and are most attractive to consumers when they are fresh. On the other hand, depending on the distance travelled and the mode of transport, the local system may be more energy consuming than the mass distribution system [11,12].

The development in food supply systems has also resulted in a push towards producers being more specialized and production being in larger, uniform units [10]. These changes tend to imply reductions in crop diversity at the farm level, which in the long run may cause problems for society. For example, the biodiversity loss associated with these systems has been shown to result in decreased productivity and stability of ecosystems due to loss of ecosystem services [13]. Specifically, biodiversity at the farm level has been shown often to have many ecological benefits (ecosystem services) like supporting pollination, pest and disease control. Therefore, it has been suggested that it is time for a paradigm shift in agriculture by embracing complexity through diversity at all levels, including soil, crops, and consumers [14]. However, high levels of crop diversity may be rather difficult to combine with the supermarket mass distribution system, which at present sell 85% of food in the UK [10]. On the contrary, local based direct marketing has been identified as a driving force for increasing on-farm biodiversity [15].

The sustainability aspects of resource use and environmental impacts of food supply systems can be assessed by Life Cycle Assessment (LCA) [16,17] or emergy assessment [18,19]. Emergy accounting and LCA are largely based on the same type of inventory (i.e., accounting for energy and material flows) but apply different theories of values and system boundaries [20]. In emergy accounting, all flows of energy and materials are added based on the total available energy (exergy) directly and indirectly required to produce the flow. Emergy accounting is particularly suited for assessing agricultural systems since the method accounts for use of freely available natural resources (sun, rain, wind and geothermal heat) as well as purchased resources from the society [18]. LCA draws system boundaries around human dominated processes (resource extraction, refining, transportation, etc.) and includes indirect resources used throughout the supply chain, such as the transport of inputs supplied into the production system. Unlike emergy accounting, LCA disregards energy used by nature and normally also labor. LCA on the other hand considers emissions to the environment in addition to resource use. Due to the differences in system boundaries and scope of analysis, emergy and LCA are complementary methods [21].

We studied the sustainability of a small-scale low-input organic vegetable food supply system by evaluating empirical data on resource use and emissions resulting from production and distribution of vegetables in a box-scheme. This specific case was chosen because the farm is managed with a strong preference to increase crop diversity and to close the production system with regard to external inputs. Combined with the box-scheme distribution system it thus represents a fundamentally different way of producing and distributing food compared to the dominating supermarket based systems. Our hypothesis was that the food supply system of the case study uses fewer resources (especially fewer non-renewable resources) when compared to standard practices. To test this we developed two organic vegetable food supply model systems, low and high yielding. Each system provided the same amount of food as the case study system, and the food produced was distributed via supermarkets rather than through a box-scheme. The case supply system is benchmarked against these model systems based on a combined emergy and LCA evaluation. Therefore, within this study we aimed to evaluate whether it is possible to perform better than the dominating systems with respect to resource use including labor and environmental impacts, and at the same time increase resilience.

1.2 FARM AND FOOD DISTRIBUTION SYSTEM—EMPIRICAL DATA

The case study farm is a small stockless organic unit of 6.36 ha of which 5.58 ha are cropped and a total of 0.78 ha is used for field margins, parking area and buildings. The box-scheme distribution system supplies vegetables to 200–300 customers on a weekly basis.

Data for 2009 and 2010 were collected by two one-day visits at the farm and follow up contacts in the period 2011 to 2013. Data included all purchased goods for crop production and distribution, as well as a complete list of machineries and buildings. The vegetable production was estimated based on sales records of vegetables delivered to consumers for each week during 2009 and 2010 and subsequently averaged to give an average annual production (Table A1). For the years studied, about 20% of the produce was sold to wholesalers. In our analyses, this share was included in the box-scheme sales.

1.2.1 PRODUCTION SYSTEMS

Forty-eight different crops of vegetables are produced (Table A1) and several different varieties are grown for each crop. Crops are grown in three different systems: open field, intensive managed garden and polytunnels, and greenhouses. The open fields are managed with a 7-year crop rotation and make up 5.09 ha of cropping area. The fields are characterized by a low-fertility soil with a shallow top soil and high stone content. The garden is managed with a 9-year crop rotation and the cropped area is 0.38 ha. In the garden only, walk-behind tractors and hand tools are used for the cultivation. The greenhouse and poly-tunnels make up 0.10 ha.

The farm is managed according to the Stockfree Organic Standard [22], which means that no animals are included in the production system and the farm uses no animal manure. The farm is in general designed and managed with a strong focus on reducing external inputs (e.g., fuel and fertilizers). An example of this is that the fertility is maintained by the use of green manures. The only fertility building input comes from woodchips composted on the farm and small amounts of lime and vermiculite, which are used to produce potting compost for the on-farm production of seedlings. All seed is purchased except for 30% of the seed potatoes, which are farm saved.

1.2.2 DISTRIBUTION SYSTEM

The distribution is done by weekly round-trips of 70 km, where multiple bags are delivered to neighborhood representatives. Other customers may then come to the representatives' collection points to collect the bags. Customers are encouraged to collect the bag on foot or on bike, and the bags are designed to make this easier (i.e., a wooden box is more difficult to carry). Potential customers are rejected if they live in a location from where they would need to drive by car to pick up their bags, even though they offer to pick up the bags themselves and pay the same price. The neighborhood representatives have some administrative tasks and are paid by getting boxes for free.

1.3 ASSESSMENT METHODS—EMERGY AND LCA

The system boundary in this study is the farm and its distribution system. Cooking, consumption, human excretion and wastewater treatment are excluded from the scope of the analysis. The functional unit, which defines the service that is provided, is baskets of vegetables produced during one year and delivered at consumer's door as average of the years 2009 and 2010. Resource consumption and environmental impacts associated with consumers' transport is included except for transport by foot or bike, which was assumed negligible.

1.3.1 EMERGY ACCOUNTING

Emergy accounting quantifies direct input of energy and materials to the system and multiplies these with suitable conversion factors for the solar equivalent joules required per unit input. These are called unit emergy values (UEV) and given in seJ/unit, e.g., seJ/g or seJ/J. Emergy used by a system is divided into different categories [23] and in the following we describe how they are applied in this study.

Local renewable resources (R). The term "R" includes flows of sun, rain, wind and geothermal heat and is the freely available energy flows that an agricultural system captures and transforms into societal useful products. We include the effect of rainfall as evapotranspiration. To avoid double counting only the largest flow of sun, rain and wind is included.

Local non-renewable recourses (N). This includes all stocks of energy and materials within the system boundaries that are subject to depletion. In agricultural systems, this is typically soil carbon and soil nutrients. In this study we assume that these stocks are maintained.

Feedback from the economy (F) consists of purchased materials (M) and purchased labor and services (L&S) [23]. M includes all materials and assets such as machinery and buildings. Assets are worn down over a number of years and the emergy use takes into account the actual age and expected lifetime of each asset. The materials come with a service or indirect labor component. This represents the emergy used to support the labor needed in the bigger economy to make the products and services available for the studied system. It is reflected in the price of purchased goods.

Labor and service (L&S). In this study, the L&S component is accounted for based on monetary expenses calculated from the sales price of the vegetables. This approach rests on the assumption that all money going into the system is used to pay labor and services (including the services provided in return for government taxes or insurances). This revenue is multiplied with the emergy money ratio, designated em£-ratio (seJ/£), which is the total emergy used by the UK society divided by the gross domestic product (GDP). Thus the em£-ratio is the average emergy used per £ of economic activity. To avoid counting the service component twice, UEVs assigned to purchased materials (M) are without the L&S component.

Total emergy use (U). The sum of all inputs is designated "U". We use three emergy indicators to reveal the characteristics of the food supply system: (1) Emergy Yield Ratio (U/F), a measure of how much the system takes advantage of local resources (in this study only R) for each investment from the society in emergy terms (F), (2) Renewability (R/U), a measure of the share of the total emergy use that comes from local renewable resources, and (3) Unit Emergy Value, UEV (U/output from system) [23].

1.3.2 LIFE CYCLE ASSESSMENT (LCA)

The LCA approach quantifies the environmental impacts associated with a product, service or activity throughout its life cycle [24]. The method looks at the impact of the whole system on the global environment by tracing all material flows from their point of extraction from nature through the technosphere and up to the moment of their release into the environment as emissions. LCA takes into account all direct and indirect manmade inputs to the system and all outputs from the system and quantifies the associated impacts on the environment.

Impact categories that are relevant and representative for the assessment of agricultural systems [16] were considered: non-renewable resource use as derived from fossil and nuclear resources [25], Global Warming Potential over 100 years according to the IPCC method [26] and a selection of other impacts from CML01 methods [27] and EDIP2003 [28], (i.e., eutrophication potential to aquatic and terrestrial ecosystems, acidification, terrestrial and aquatic ecotoxicity potentials, human toxicity potential). In addition the use of fossil phosphorus was assessed.

The inventories for the LCA were constructed with the use of Swiss Agricultural Life Cycle Assessment (SALCA) models [28], Simapro V 7.3.3 [29] and the Ecoinvent database v2.2 [30]. The following inputs and emissions were based on other studies: life cycle inventory for vegetable seedlings [31]; biomulch [32]; nitrous oxide and methane emissions from open field woodchip composting on the case study farm [33]. The Life Cycle Inventory for irrigation pipeline from ecoinvent was adjusted to reflect the irrigation system of the case farm and the Swiss inventory for irrigation was adjusted to reflect the British electricity mix.

The analysis was carried out from cradle to the consumer's door with respect to the ISO14040 [24] and ISO14044 [34] standards for environmental Life Cycle Assessment. Upstream environmental impacts related to the production of woodchips or manure were not considered. This is following a cut-off approach that makes a clear division between the system that produces a by-product or waste and the system using it. The emissions from livestock farming (associated with the production of manure used in the models of standard practice) are fully assigned to the livestock farmer and the gardener is responsible for the production of woodchips. However, environmental impacts from the transport of both type of inputs to the farm, their storage and composting at the farm and all the emissions to soil, air and water that arise from their application were considered in this study. The results of the impact category non-renewable resource use were investigated in more detail by looking at the relative contribution of particular processes to the overall resource use, because of some similarities with the emergy assessment.

1.4 MODELS FOR STANDARD PRACTICE OF VEGETABLE SUPPLY SYSTEM

The overall aim of developing these models is to assess the resource use and environmental impacts of providing the same service as the case system but in the dominating supermarket based system. The two model systems, M-Low and M-High, express the range of standard practice for organic vegetable production as defined from the Organic Farm Management Handbook [35]. Since the information in this handbook is independent of scale, i.e., all numbers are given per ha or per kg, then the model systems are also independent of scale. Both model systems provide vegetables in the same quantity at the consumer's door (in food energy) and of comparable quality as the case study. The mix of vegetables provided is identical to the case system for the eight crops (two types of potatoes, carrots, parsnips, beetroots, onions, leeks and squash) constituting 75% of the food energy provided (Table 1). For the remaining 25% representing 40 crops at the case farm, four crops (white cabbage, cauliflower, zucchini and lettuce) have been chosen based on the assumption that they provide a similar utility for the consumer.

TABLE 1: Characteristics of vegetables produced annually in the case system and their counterparts in the model systems.

Case farm crops	Model farm crops	Food energy at consumers (MJ)	Share of total food energy
Storable crops			
Potatoes, main crop	Potatoes, main crop	25,597	34.4%
Potatoes, early	Potatoes, early	8532	11.5%
Carrots (stored and fresh)	Carrots	4635	6.2%
Beetroots (stored and fresh)	Beetroots	4271	5.7%
Onions (stored, fresh and spring)	Onions	3688	5.0%
Parsnips	Parsnips	3555	4.8%
Leeks	Leeks	2902	3.9%
Squash	Squash	2697	3.6%
Cabbages (red-, black-, green-, sprouts, kale, pak choi)	Cabbages, white	5390	7.3%
Cauliflower, broccoli and minor crops (celeriac, fennel, turnips, kohlrabi, rutabaga, daikon, garlic)	Cauliflower	3344	4.5%
Storable crops, total		64,610	86.9%
Fresh crops			
18 different crops (see Table A1 for list of crops)	50% Courgettes	4859	6.5%
	50% Lettuce	4859	6.5%
Fresh crops, total		9717	13.1%
All crops, total (functional unit)		74,328	100.0%

1.4.1 CROP MANAGEMENT FOR M-LOW AND M-HIGH

M-Low and M-High systems were defined from yields per ha using the range in the *Organic Farm Management Handbook* [35]. The M-Low farm represents a standard low-yielding farm, the lowest value in the

Handbook, and the M-High farm represents a standard high yielding farm, the highest value in the *Handbook*. The range is shown in Table 2 for each crop considered. These yield differences, combined with the food chain losses assumed (see Section 4.2), implied that different areas were needed to provide the functional unit, i.e., the average annual amount of vegetables (in food energy) at the consumer's door (Table 2).

TABLE 2: Yields and corresponding areas needed to provide the amount of vegetables sold in the case system for M-Low and M-High. Areas for the case farm are given for comparison.

	Case (ha)	M-Low		M-High	
		Yields[a] (t/ha)	Areas (ha)	Yields[a] (t/ha)	Areas (ha)
Potatoes, early		10	0.42	20	0.21
Potatoes, main crop		15	0.83	40	0.31
Carrots		15	0.41	50	0.12
Beetroots		10	0.35	30	0.12
Onions		10	0.35	25	0.14
Parsnips		10	0.21	30	0.07
Leeks		6	0.48	18	0.16
Squash		15	0.17	40	0.06
Cabbage, white		20	0.26	50	0.11
Cauliflower		16	0.23	24	0.15
Zucchini		7	0.88	13	0.47
Lettuce		6	1.73	9.6	1.08
Vegetables	4.02		6.32		3.01
Green manure	1.56		1.58[b]		0.75[b]
Field margins and infrastructure	0.78		1.12[c]		0.53[c]
Total area	6.36		9.02		4.29

[a] *From* Organic Farm Management Handbook *[35], the lowest and highest yield for each crop;* [b] *20% of cultivated area; c 14% of cultivated area based on the proportion for the case.*

The further definition of the two model systems was based on the assumption that yields are determined by the level of fertilization and irrigation. Therefore, M-Low is defined with a low input of fertilizers and M-High with a higher fertilizer input. The NPK-budgets were calculated based on farm gate inputs and outputs from an average farm with the same crop production and management using a NPK-budget tool from the Organic Research Center [36,37]. Both systems were assumed to have 20% green manure (red clover) in their crop rotation. Input of cattle manure, rock phosphate and rock potash was then modeled such that M-High reached a balance of 90 kgN/ha, 10 kgP/ha and 10 kgK/ha and M-Low a balance of 0 kgP/ha and 0 kgK/ha (Table A2). For the M-Low N balance, the lowest possible value was 54 kgN/ha due to atmospheric deposits and N-fixation.

Further, for M-Low irrigation was only included for the crops for which irrigation is considered essential according to the Handbook [34], whereas for M-High irrigation was also included for crops which "may require irrigation".

Based on the field operations needed for each crop described in *Organic Farm Management Handbook* [35], the resource use in terms of fuel and machinery was modeled according to resource use per unit process [38] (see Supplementary Material for detailed description of the farming model). The approach for determining yields and resource use was similar to a previous study commissioned by Defra [39].

1.4.2 MODEL DISTRIBUTION SYSTEM

The model distribution system from farm gate to consumer's door was modeled on a crop by crop basis based on published LCA reports for supermarket based food distribution chains [40,41,42] (Table 3). The chain is thus assumed to consist of 200 km transport to and storage for 5 days at regional distribution center (RDC), 50 km transport to and storage for 2 days at retailers and 6.4 km transport from the retailer to the customer's home [40] (see Supplementary Material for detailed assumptions.) Transportation from the farm to the RDC is assumed to be in a chilled 32 t truck with an energy consumption of 22.9 mL diesel per euro pallet kilometer

[41]. Throughout the system, food waste is taken into account for each crop [42].

The total expenses to labor, service and materials throughout the supply system were estimated based on 12 month average supermarket prices (from March 2012 to March 2013) for each of the vegetables [43]. The prices were adjusted for inflation to reflect average 2009–2010 prices according to the price index for vegetables including potatoes and tubers [44].

1.5 RESULTS OF SUSTAINABILITY ASSESSMENT

The service provided by the three systems is a comparable "basket" of vegetables produced during one year and delivered to the consumer's door. This service is measured in food energy and is equal to 74,328 MJ/year as an average of 2009 and 2010 (Table 1). This corresponds to the total annual food energy needed for 19–23 people (based on a recommended daily intake of 8.8–11 MJ [45]). The emergy flows are illustrated for the case (Figure 1A) and for the model systems (Figure 1B). The two diagrams demonstrate clearly the different distribution systems and that in the case the full money flow goes to the farm whereas in the model systems part of money flows to the freight companies, supermarkets, and regional distribution centers (RDC).

1.5.1 EMPIRICAL SYSTEM

The basis of any emergy assessment is the emergy table (Table 4) that shows all environmental and societal flows, which support the system. Notably labor and services (L&S) make up 89% of total emergy used by the case (calculated from Table 4). As emergy use for L&S is calculated as a function of the emergy use for the national economy, this reflects the national resource consumption rather than the specific business. To avoid distorting the results of the actual farm with the implications of being embedded in an industrialized economy, we consider the emergy indicators both with and without L&S.

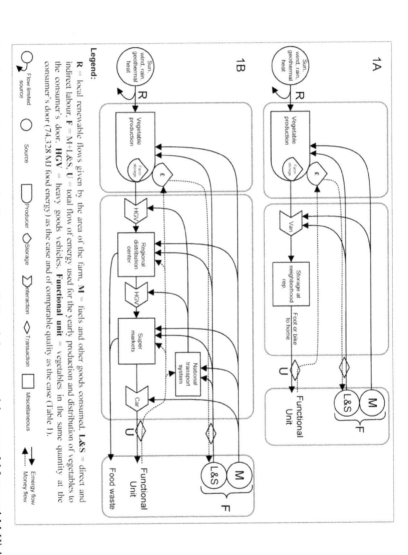

FIGURE 1: Material and emergy flow diagrams for the case system (1A) and the two model systems M-Low and M-High, which have identical distribution systems (1B).

The main result of the emergy evaluation for the case system is the transformity of the vegetables, which amounts to 5.20×10^6 seJ/J with L&S and 5.54×10^5 seJ/J without (Table 5). The Emergy Yield Ratio (EYR) of 1.15 disregarding L&S shows that free local environmental services (R) contribute with only 0.15 seJ per seJ invested from the society. The renewability indicator shows that the system uses 13% local resources when disregarding L&S but only 1% when including L&S. The latter reflects that L&S is considered as non-renewable.

Disregarding L&S, the emergy profiles of the case system are as follows (calculated from Table 4). Purchased miscellaneous materials for the cultivation phase contribute 38% of total emergy used. Fuel used for cultivation and electricity used for production of seedlings are the biggest flows with 18% and 11%, respectively. Notably, irrigation contributes 24% of the total flow with the water used constituting the most important element (17%). Likewise, the woodchips, used as soil enhancement and used to produce potting compost, contribute with 10% and farm assets contribute with 7%. The diesel used on the weekly round-trip was estimated to 465 L/year (1.6×10^{10} J, Table 4) and it is the major component of the emergy used in the distribution phase (7% of the total emergy used).

Analyzing the case system using the LCA perspective, the processes related to the cultivation phase have a much larger environmental impact than the processes involved in the distribution phase, for all nine categories (Figure 2). The LCA impact category non-renewable resource use includes all direct and indirect use of fossil and nuclear fuels converted to MJ. Crude oil in the ground contributes more than 50% of the total raw materials for energy (Figure 3). Crude oil is used to produce diesel for operating tractors and pumping the water for irrigation, and to a smaller extent for the manufacture and transport of other inputs.

1.5.2 BENCHMARKING AGAINST MODEL SYSTEMS

An important difference between the food supply system of the case study and the model systems is the amount of food lost. The long chain in the model systems generates a high percentage of food loss, up to 29% for root vegetables [42]. The direct marketing of the case system implies that the

TABLE 3: Outputs and energy use in model distribution system: from farm to regional distribution center (RDC), to retailer and to consumer's home.

	Farm gate output (t)	Diesel use, transport to RDC (L)[a]	El. use, 5 days storage at RDC (kWh)[b]	RDC gate output (t)[c]	Diesel use, transport to retailer (L)[a]	El. use, storage at retailer (kWh)[d]	NG use, storage at retailer (MJ)[d]	Retail gate output (t)[c]	Gasoline use, transport to home by car (L)[e]	Diesel use, transport to home by bus (L)[e]
Potatoes	16.7	64.6	63.9	11.8	11.5	78.1	346.5	11.5	101.2	2.7
Carrots	6.2	24.0	23.7	4.4	4.3	29.0	128.7	4.3	37.6	1.0
Cabbages	5.3	78.3	77.5	5.3	19.6	34.7	591.7	5.1	45.4	1.2
Cauliflower	3.7	55.4	54.9	3.7	13.9	24.5	419.0	3.6	32.1	0.9
Parsnips	2.1	8.1	8.0	1.5	1.4	9.8	43.5	1.4	12.7	0.3
Beetroots	3.5	13.8	13.6	2.5	2.4	16.6	73.8	2.4	21.6	0.6
Onions	3.5	16.8	16.6	3.2	3.8	21.2	115.6	3.2	28.1	0.8
Leeks	2.9	42.5	42.1	2.9	10.6	18.8	321.7	2.8	24.7	0.7
Squash	2.5	37.7	37.3	2.5	9.4	16.7	285.2	2.5	21.9	0.6
Zucchini	6.1	91.5	90.5	6.1	22.9	40.5	691.4	6.0	53.0	1.4
Lettuce	10.4	219.2	216.9	10.1	53.6	66.8	1620.1	9.9	87.6	2.4
Total	62.9	651.8	645.0	54.0	153.3	356.7	4637.2	52.7	465.9	12.6

[a] The produce is transported 200 km to RDC and 50 km from RDC to retail [40] using 22.9 ml diesel per pallet-km (chilled single drop, 32 t artic) [41]. [b] Electricity consumption in RDC is 0.00059 kWh/l/day [40]. [c] For each crop losses in storage and packaging are taken into account [42]. See Supplementary Material for details. [d] Storage at ambient temperature. Energy use is 0.027 MJ/kg/day (44% electricity for light and 56% natural gas (NG) for heating) [40]. [e] Based on an average UK shopping trip of 6.4 km with an average shopping basket of 28 kg and where 58% of trips made by private car and 8% made by bus [40]. See Supplementary Material for details.

crop loss is smaller due to higher acceptance of less-than-perfect crops. Therefore, the case farm does not need to produce as much to provide the same amount of vegetables at the consumer's door as the model production system.

Land required for providing the food service using standard practices vary between 4.29 ha for the high-yielding model system to 9.02 ha for the low-yielding model system (Table 2). The area required by the case farm (6.36 ha) is within this range. The land use efficiency at the system level may be calculated as the food energy provided to the consumer per hectare of cultivated area (Table 2, vegetables + green manure). This value is 13.3 GJ/ha for the case farm and varies between 9.4 GJ/ha (M-Low) and 19.8 GJ/ha (M-High) for the model systems. This indicates that the case farm has yields within the range of the standard practices.

The consumer price of total output from the case system is £86,800. This is significantly lower than the consumer price for the model systems' output, which is £147,300 (Table 4). That the case farmer is able to sell the products at a significantly lower price may be explained by the fact that the full revenue goes directly to the farm (Figure 1A) whereas in the modeled systems the supermarkets, freight companies and regional distribution centers (RDC) need to make a profit as well (Figure 1B).

1.5.2.1 BENCHMARKING BASED ON EMERGY USE

The emergy use for purchased materials in the model systems is very similar in total but is differently distributed among the different components, e.g., M-Low has twice as much input in the cultivation phase whereas M-High has five times higher input for soil fertility enhancement. By definition, the emergy use in the distribution phase and for L&S is identical for the two systems. The L&S constitute by far the biggest contribution for both M-Low and M-High with about 92% in both systems. The total emergy used for L&S is 5.9×10^{17} seJ for the model systems (Table 4), which is 70% more than for the case system. This directly reflects that consumer price for the vegetables are 70% higher in the supermarket than in the direct marketing scheme.

TABLE 4: Use of emergy per functional unit for the three systems: the case, M-Low and M-High. See Table A3 and Table A4 for notes with details for each item.

	Unit	Case (Unit)	M-Low (Unit)	M-High (Unit)	UEV (seJ/unit)	Case emergy flow ($\times 10^{14}$ seJ)	M-Low emergy flow ($\times 10^{14}$ seJ)	M-High emergy flow ($\times 10^{14}$ seJ)
LOCAL RENEWABLE FLOWS (R)								
1 Sun	J	9.7×10^{13}	1.4×10^{14}	6.6×10^{13}	1.0^a	1.0	1.4	0.7
2 Evapotranspiration	g	3.0×10^{10}	4.3×10^{10}	2.1×10^{10}	1.5×10^{5b}	44.0	62.6	29.8
3 Wind	J	3.9×10^{11}	5.5×10^{11}	2.6×10^{11}	2.5×10^{3c}	9.7	13.8	6.5
4 Geo-thermal heat	J	9.0×10^{10}	1.3×10^{11}	6.1×10^{10}	1.2×10^{4b}	10.8	15.4	7.3
SUM (excluding sun and wind)						54.8	78.0	37.1
PURCHASED MATERIALS (M)								
Cultivation Phase								
Miscellaneous materials								
5 Diesel, fields	J	4.0×10^{10}	2.7×10^{10}	1.5×10^{10}	1.8×10^{5d}	73.0	48.9	27.4
6 Lubricant and grease	J	1.9×10^{9}	1.1×10^{9}	5.7×10^{8}	1.8×10^{5d}	3.5	2.1	1.0
7 LPG	J	9.1×10^{8}	6.6×10^{9}	2.3×10^{9}	1.7×10^{5d}	1.6	11.3	3.8
8 Fleece and propagation tray	g	8.8×10^{3}	1.9×10^{5}	9.8×10^{4}	8.9×10^{9e}	0.8	16.7	8.7
9 Electricity	J	1.6×10^{10}	3.8×10^{9}	3.8×10^{9}	2.9×10^{5f}	45.4	10.9	10.9
10 Seedlings	pcs	0.0	3.2×10^{5}	1.8×10^{5}	9.6×10^{9g}	0.0	31.0	17.0
11 Seed	g	2.1×10^{4}	2.9×10^{4}	1.3×10^{4}	1.5×10^{9h}	0.3	0.4	0.2
12 Potato seeds	g	1.1×10^{6}	3.1×10^{6}	1.3×10^{6}	2.9×10^{9i}	30.1	89.4	37.3
SUM						154.6	210.7	106.3

		Unit	Case (Unit)	M-Low (Unit)	M-High (Unit)	UEV (seJ/unit)	Case emergy flow ($\times 10^{14}$ seJ)	M-Low emergy flow ($\times 10^{14}$ seJ)	M-High emergy flow ($\times 10^{14}$ seJ)
Irrigation									
13	Diesel	J	1.5×10^{10}	6.6×10^{8}	4.6×10^{8}	1.8×10^{5d}	27.8	1.2	0.8
14	Electricity	J	0.0	1.3×10^{10}	8.9×10^{9}	2.9×10^{5f}	0.0	36.9	25.9
15	Ground water	g	3.6×10^{9}	3.0×10^{9}	2.1×10^{9}	1.1×10^{6j}	41.5	34.5	24.2
16	Tap water	g	1.4×10^{9}	0.0	0.0	2.3×10^{6j}	30.8	0.0	0.0
SUM							100.0	72.6	50.9
Soil fertility enhancement									
17	Woodchips	J	3.7×10^{11}	0.0	0.0	1.1×10^{4k}	38.8	0.0	0.0
18	Lime	g	2.0×10^{4}	0.0	0.0	1.7×10^{9k}	0.3	0.0	0.0
19	Nitrogen (N)	g	0.0	18	2.8×10^{4}	4.1×10^{10l}	0.0	0.0	105.7
20	Phosphorus (P2O5)	g	0.0	3.0×10^{4}	7.4×10^{4}	3.7×10^{10l}	0.0	27.1	59.3
21	Potash (K2O)	g	6.6×10^{4}	2.5×10^{5}	2.9×10^{5}	2.9×10^{9k}	2.3	8.4	9.8
SUM							41.5	35.6	174.8
Farm Assets									
22	Tractors	g	1.4×10^{5}	8.2×10^{4}	3.8×10^{4}	8.2×10^{9m}	11.7	16.6	5.8
23	Other machinery	g	1.5×10^{5}	2.5×10^{5}	1.3×10^{5}	5.3×10^{9m}	8.0	11.4	5.4
24	Irrigation pipe	g	8.9×10^{3}	8.9×10^{3}	8.9×10^{3}	8.9×10^{9e}	0.8	0.8	0.8
25	Wood for buildings	J	9.9×10^{9}	9.9×10^{9}	9.9×10^{9}	1.1×10^{4k}	1.0	1.0	1.0
26	Glass for buildings	g	7.6×10^{4}	0.0	0.0	3.6×10^{9e}	2.8	0.0	0.0
27	Plastic for buildings	g	1.9×10^{4}	0.0	0.0	8.9×10^{9e}	5.8	0.0	0.0
28	Steel for buildings	g	2.5×10^{4}	0.0	0.0	3.7×10^{9m}	0.9	0.0	0.0
SUM							30.9	29.8	12.7

	Unit	Case (Unit)	M-Low (Unit)	M-High (Unit)	UEV (seJ/unit)	Case emergy flow ($\times 10^{14}$ seJ)	M-Low emergy flow ($\times 10^{14}$ seJ)	M-High emergy flow ($\times 10^{14}$ seJ)
Distribution Phase[q]								
29 Diesel	J	1.6	2.8×10^{10}		1.8×10^5[d]	28.6	50.3	
30 Gasoline	J	0.0	1.6×10^{10}		1.9×10^5[d]	0.0	29.6	
31 Electricity	J	0.0	3.6×10^9		2.9×10^5[f]	0.0	10.5	
32 Natural Gas	J	0.0	4.6×10^9		6.8×10^4[d]	0.0	3.2	
33 Machinery (van)	km	5.7×10^3	0.0		2.5×10^{10}[o]	1.4	0.0	
34 Machinery (truck)	tkm	0.0	1.5×10^4		4.1×10^9[p]	0.0	0.6	
SUM						30.0	94.2	
35 LABOR AND SERVICE (L&S)[q]	£	8.7×10^4	1.5×10^5		4.0×10^{12}[f]	3451.1	5,854.9	
SUM Purchased materials (M)						357.0	442.9	439.0
SUM Feedback from economy (M + L&S)						3808.2	6297.8	6293.9
TOTAL EMERGY USED (U) with L&S						3863.0	6375.8	6330.9
TOTAL EMERGY USED (U) without L&S						411.9	520.9	476.1

[a] *By definition,* [b] *Odum (2000) [46],* [c] *Odum (2000) [47],* [d] *Brown et al (2011) [48],* [e] *Buranakarn (1998) [49],* [f] *NEAD database [50],* [g] *This study. Based on input of 20 cm^3 peat and 1/774 l diesel per seedling [31],* [h] *Coppola (2009) [51],* [i] *This study, based on total M-High emergy use (less emergy for distribution phase and seed potato), allocated based on the yields and share of cultivated area used for main crop potato,* [j] *Buenfil (1998) [52],* [k] *Odum (1996) [23],* [l] *Brandt-Williams (2002) [53],* [m] *Kamp (2011) [54],* [n] *Bargigli (2003) [55],* [o] *This study. Weight of vehicle: 1500 kg, lifetime 500.000 km and same transformity as for tractors,* [p] *This study. Transformity per tkm calculated based on Pulselli (2008) [56].* [q] *M-Low and M-High have identical distribution system and need the same L&S calculated based on the consumer prices.*

TABLE 5: Emergy indices for the case system, M-Low and M-High with and without labor and service.

	With labor and service			Without labor and service		
	Case	M-Low	M-High	Case	M-Low	M-High
Emergy Yield Ratio (U/F)	1.01	1.01	1.01	1.15	1.18	1.08
Renewability (R/U)	1%	1%	1%	13%	15%	8%
Solar transformity (seJ/J)	5.20×10^6	8.58×10^6	8.52×10^6	5.54×10^5	7.01×10^5	6.40×10^5

When disregarding L&S, the case system uses less emergy to produce the total amount of vegetables sold compared to both model systems (Figure 4). This is especially due to a reduced emergy need for purchased seedlings and seed potato as compared to M-Low and a reduced emergy use for soil enhancements as compared to M-High. In addition, the case distribution system only use one third of the emergy used by the model supply chain. However, the case has a substantial higher consumption of on-farm fuel use and needs more emergy for water for irrigation (Figure 4).

The case-study farm uses significantly more diesel in the cultivation phase (Table 4). This may partially be explained by the tractors being less efficient than those assumed for the model systems. Another factor is that the diesel use per area is more or less independent of the yields, which means that high yielding crops tend to use less fuel per unit output. This is clearly reflected in the comparison of M-Low and M-High, but does not explain why M-Low uses less diesel than the case system.

On-farm electricity use (Figure 4) consists of electricity use for on-farm production of seedlings and offices (only for the case-study) as well as for irrigation (only model farms). Disregarding electricity for irrigation, the case uses significantly more electricity than the model systems. However, the electricity consumption of 4350 kWh is still relatively small as it corresponds to the average UK household (4391 kWh, [57]). The emergy needed for electricity in the case system (4.5×10^{15} seJ) is partly compensated by the emergy needed for purchased seedlings in M-Low (3.1×10^{15} seJ) and M-High (1.7×10^{15} seJ) (Table 4). The fact that 30% of the seed potatoes are farm-saved in the case system results in a considerable emergy saving as compared to both model systems. M-Low is particularly bad in this respect since it needs a larger area (Table 2) and thus more seed potatoes to produce the required amount of potatoes (Table 4).

M-High has the lowest emergy use for irrigation with M-Low using 50% more and the case using twice as much. The latter is in the first place a consequence of that the case uses more water (3.6×10^9 g groundwater and 1.4×10^9 g tap water) (Table 4). As the annual variation in precipitation is not considered in the model systems, the higher use of water for irrigation in the case system may reflect that the studied period, 2009–2010, was relatively dry, and for instance in 2008 the water use was 70% less. In addition, tap water, which accounts for 28% of total water used, has an UEV value twice as high as ground water due to the extra work that is needed for pumping and treatment (Table 4).

Emergy used for soil enhancement is the biggest input to M-High (Table 4). With 1.7×10^{16} seJ it is more than four times higher than the other systems. This reflects that fertilizer is a valuable resource, and that reducing the import of fertilizer is a key element in reducing emergy use in agricultural systems.

The model supply chain needs a total of 805 L diesel (calculated from Table 3) for HGV-transport. In addition, 932 L of gasoline is used for the 58% of the shopping trips done by car and 12.6 L of diesel for the 8% of the trips done by bus. The total use of liquid fuels in the model supply chain (4.5×10^{10} J) is thus bigger than on-farm use of diesel in the cultivation phase in all three systems (Table 4). The total emergy use for the distribution system is three times higher for the model systems than for the case system.

M-Low has the largest contribution of local renewable flows as these are calculated directly from the size of the farm (Table 4). Further, the emergy indices (Table 5) reveal that, disregarding L&S, M-High has the smallest share of local renewable inputs (8%). The renewable resources contribute only with 0.08 seJ per seJ invested from society (EYR = 1.08). M-low is in this respect a bit better than the case. The case on the other hand provides the vegetables with the highest resource efficiency (lowest UEV or transformity) and is as such overall more efficient than both model systems (Table 5). This is especially true when also considering L&S in which case the transformity of the case is 39% lower than for M-High.

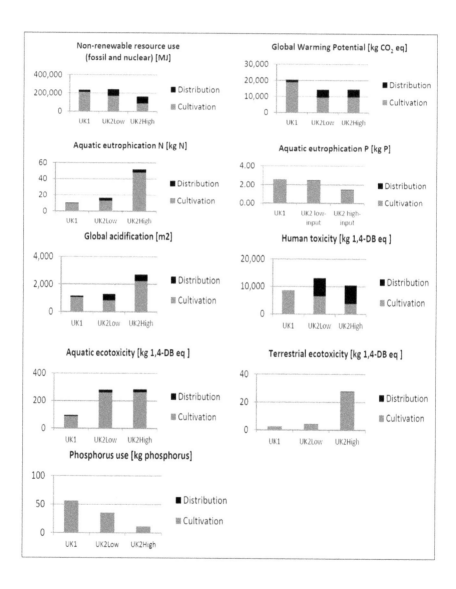

FIGURE 2: LCA results for the case, M-Low and M-High for the nine impact categories considered. Impacts per functional unit are divided into distribution phase and cultivation phase.

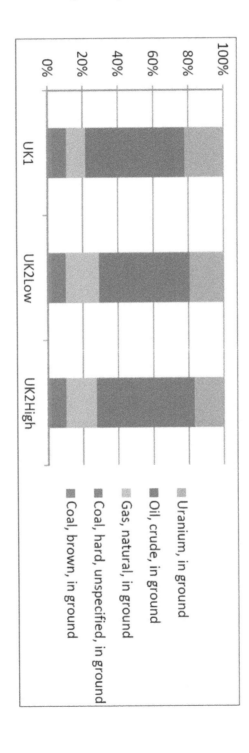

FIGURE 3: Contribution of raw materials to the overall result for the Life Cycle Impact category non-renewable resource use (fossil and nuclear) per functional unit for the case, M-Low and M-High.

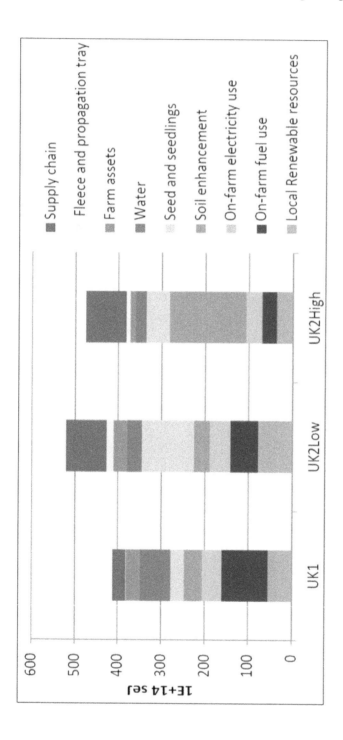

FIGURE 4: Emergy profiles without L&S for the case, M-Low and M-High.

1.5.2.2 BENCHMARKING BASED ON LCA

The distribution phase has an important contribution to the environmental impacts of the model systems and in particular for the impact categories non-renewable resource use, Global Warming Potential (GWP) and human toxicity (Figure 2). The use of non-renewable resources in the case system is similar to M-Low, while the impact of M-High is around 30% lower (Figure 2). The GWP of the case is about 40% higher than both model systems. The difference in GWP between M-Low and the case was related to differences in management processes. The on-farm production of seedlings and composting of woodchips, respectively, may not be as efficient as centralized production of seedlings and use of only green manure and rock phosphate for nutrient supply. The impact category Phosphorus use was calculated to be higher in the case-study as compared to the model systems due to the use of vermiculite, but it is necessary to bear in mind that the levels of phosphorus use were relatively low for all three analyzed systems. The case system and M-Low have significantly lower aquatic eutrophication N potential, terrestrial ecotoxicity and aquatic ecotoxicity than M-High. This is because these impact categories are more dependent on the applied fertilization and irrigation levels rather than on capital goods and on-farm diesel and electricity. Aquatic ecotoxicity and human toxicity effects of the case were also lower than both model systems. Aquatic ecotoxicity levels were found to be similar for model systems while human toxicity of M-Low was shown to be slightly higher than M-High.

In addition, for model systems, environmental impacts for all impact categories are clearly dominated by agricultural cultivation (Figure 2). As for the case-study farm, more than 50% of the non-renewable resource use in the model systems is from use of crude oil (Figure 4). Nearly 20% of non-renewable resource use in model systems comes from natural gas, while for the case it is only 10%. Uranium ore has a relatively high contribution, nearly 20% of the result for all systems, as the electricity mix in the UK includes nuclear energy. Hard coal contributes to around 9% of the resource use and the remaining raw materials play a minor role (5% or less).

Assessment of environmental impacts exclusively from the distribution phase reveals that the local distribution system provides significantly lower environmental impacts per functional unit for all of the impact categories considered (Table 6). The relative advantage of the case system compared to the model system reached from 69% for the non-renewable resource use up to 98% in the case of human toxicity potential.

TABLE 6: Environmental impacts per functional unit exclusively for the distribution phase (the same for M-Low and M-High).

Impact Category	Unit	Case Distribution	Model Distribution	Relative Advantage of Case System (%)
Non-renewable, fossil and nuclear	MJ eq	23,783	75,923	69
GWP 100	kg CO$_2$ eq	1629	4890	67
Acidification, GLO	m^2	92	469	80
Eutrophication aq. N, GLO	kg N	0.73	3.74	81
Eutrophication aq. P, GLO	kg P	0.00	0.05	91
Human toxicity	HTP	122	6646	98
Terrestrial ecotoxicity	TEP	2	21	92
Aquatic ecotoxicity	AEP	513	1445	64
Phosphorus	kg	0	0	71

1.6 DISCUSSION

1.6.1 LCA VERSUS EMERGY ASSESSMENT—HANDLING OF CO-PRODUCTS

The assessment of sustainability of the organic low-input vegetable supply system using emergy accounting and LCA has shown that the two methods lead to the same conclusion regarding the supply chain but differ to some extent in the assessment of the production systems. The sometimes contradictory results of the emergy and LCA results are to a large part due to differences in how co-products, e.g., manure, are accounted for. In emergy accounting, the focus is on the provision of resources, and a key principle in emergy algebra is that all emergy used in a process should be assigned to all co-products as long as they are considered in separate analyses [23]. As manure cannot be produced without producing meat and milk, the entire input to livestock production should be assigned to each of the three products. We have used this approach despite its disadvantages when comparing systems with or without inputs of manure [58]. As a proxy for the UEV of manure, we have combined the UEVs of mineral N, P and K. In the LCA approach, all environmental impacts from animal production were assigned to the main products of animal production being meat and milk. As a result, only emissions associated with transportation, storage and application are considered and the principle of no import of manure in the case system is only partially reflected in the LCA results. Further, this approach has lead to the counter-intuitive result that M-Low has higher phosphorus use than M-High (Figure 2) even though the latter system imports twice as much phosphorus as the former (Table A2). The assumption that manure is a waste may not reflect the actual situation for many organic growers who experience that the supply of N is often a limiting factor for maintaining productivity [59].

1.6.2 POTENTIALS FOR REDUCING RESOURCE USE IN THE CASE SYSTEM

Even though the case has a strong focus on minimizing use of purchased resources (M), their contribution is still more than six times larger than the contribution from local renewable resources (R) (Table 4). Disregarding L&S, then the largest potential for improving percentage of renewability is to reduce the amount of used fuels (Figure 2). However, to substitute fossil fuelled machinery with more labor intensive practices such as draft animals or manual labor, would under current socio-economic conditions increase overall resource consumption due to the high emergy flow associated with labor. In addition, draft animals would require that a considerable amount of land should be used for feed production.

Ground and tap water used for irrigation constitutes 17% of the total emergy use. Due to the differences in UEV between tap and ground water, the emergy use could be substantially reduced by using only ground water. Producing the woodchips, which accounts for 10% of the total emergy used, within the geographical boundaries of the farm would improve renewability. Currently they are residuals supplied from a local gardener who prunes and trims local gardens. In a larger perspective, there are, thus, few environmental benefits from becoming self-sufficient with wood chips in the case system. According to the LCA analysis, the composting process accounts for 30% of greenhouse gas emissions, and using less wood chip compost would reduce the overall global warming potential.

Electricity, which is primarily used for heating and lighting in the production of seedlings, constitutes 11% of the emergy used. It is no doubt convenient to use electricity for heating, but substituting the electricity with a firewood based system would largely reduce the emergy. Alternatively, it may be worthwhile to consider harvesting excess heat from the composting process to heat the green house.

As for the distribution phase, the case has the potential for decimating fossil fuel consumption by replacing the current customers with some of the many households located within few kilometers of the farm. This could dramatically reduce the 70 km round trip each week. The current way of organizing the distribution, however, is extremely efficient when compared to the alternative where customers would go by car each week and pick up the produce. The latter solution would require up to 1,000,000 car-km per year based on the case farmer's calculation. With a fuel efficiency of 15 km/L this translates to 66,666 liter of fuel. This is almost 40 times the fuel consumption for the model system (1737 L).

1.6.3 OUTLOOK FOR EMERGY USE FOR L&S

In a foreseeable future with increasing constraints on the non-renewable resources [6,60], which currently are powering the society with very high EYR-values [23], it is desirable or even necessary that agricultural systems become net-emergy providers, i.e., that more emergy is returned to society from local renewable resources than the society has invested in the production [61]. This requirement means that the contribution from R has to be bigger than F. Bearing in mind that R cannot be increased as the local renewable flows are flow limited, then achieving this can alone be achieved by reducing the emergy currently invested from society, F (3808.2×10^{14} seJ) to less than R (54.8×10^{14} seJ), i.e., by a factor of 70. Such an improvement seems out of reach without transforming the food supply system. Some improvements can be made on the farm as indicated, but the largest change will need to be in the society which determines the emergy use per unit labor. It is important to note that for the standard practices represented by the model system much larger reduction would be required.

Emergy used for L&S accounts for 89% of total emergy flow and constitutes by far the biggest potential for improvements. The L&S-component reflects the emergy used to support people directly employed on the farm and people employed in the bigger economy to manufacture and provide the purchased inputs. Due to the high average material living standard in UK with emergy use per capita being 8.99×10^{16} seJ/year [50], labor is highly resource intensive.

Emergy used for L&S can be reduced by reducing the revenue, but this is highly undesirable. Nevertheless, the employees already have a relatively low salary, which they accept because of the benefits enjoyed, e.g., free access to vegetables, cheap accommodation on the farm perimeter and in their opinion a meaningful job close to nature. Thus, the case system attracts people with a Spartan lifestyle with few expenses and thus below average emergy use.

In future, it is almost certain that the nation-wide emergy use per capita will be reduced. A likely future scenario for the UK is that the indigenous extraction of non-renewable resources continues to decline (down 23% from 2.4×10^{24} seJ in 2000 to 1.8×10^{24} seJ in 2008 [50]). This is a result of the oil extraction plunging from 2.6 to 1.5 million barrels per day (mbd) from 2000 to 2008 (in 2010 further down to 1.1 mbd) [62]. In the same period the extraction of natural gas dropped from 97.5 to 62.7 million tonnes oil equivalent (and to 40.7 by 2010) [62]. This decline has been compensated by increasing imports of fuels from 57×10^{22} seJ in 2000 to 95×10^{22} seJ by 2008 [50]. The UK has been able to maintain a high level of emergy use per capita by gradually substituting the decline in oil and gas production with imported fuels and services. Such a substitution may continue for some years, but in a longer time perspective a decline in global production of oil, gas and other non-renewable resource is inevitable. Coupled with an increased competition from a growing global population increasing in affluence, it is likely that the import of such resources will eventually decline for the UK as well as other industrialized nations [63].

Such a future scenario imply that the resource consumption per capita will be reduced and thereby that the emergy needed for supporting labor is reduced. However, it may also bring along transformations that are more substantial in the organization of the national economy and all its subsystems, not least the mass food supply system, which at present uses 9% of UK's petroleum products. In this perspective, the capacity of a system to adapt to changes is a crucial part of its sustainability, and this characteristic is not directly reflected in the quantitative indicators of emergy assessment and LCA.

1.6.4 LOCAL BASED BOX-SCHEME VERSUS NATIONAL-WIDE SUPERMARKET DISTRIBUTION—RESILIENCE

The supermarket based distribution system has during the previous decades been redesigned according to principles of Just-In-Time delivery (JIT). These principles aim at reducing the storage need and storage capacity at every link in a production chain, such that a minimum of capital investment is idle or in excess at any time. Less idle capital means fewer costs and fewer environmental impacts. While JIT may decrease environmental impacts per unit of produce for the particular system, as long as everything is running smoothly, it may compromise the system's resilience as it becomes more vulnerable to disturbance and systemic risks. Systemic risks include disruption in infrastructure supplying money, energy, fuel, power, communications and IT or transport as well as pandemics and climate change [64,65,66]. As can be imagined any disturbance caused by such events may quickly spread throughout the tightly connected network [65,67]. A loss of IT and communication would make it impossible for a national-wide JIT supply system to coordinate supplies [64]. A loss of money would make it impossible to conduct transactions with customers. A loss of fuel for transportation would results in large bulks of produce being stranded. A loss of power in the RDC would stop the entire chain and retailers would run out of products in a few days.

When benchmarking the case-study against the national wide supermarket system, the former may be more resilient than the latter as it is in a better position to handle infrastructure failures. Due to the higher degree of autonomy and fewer actors involved, the case system would be able to work around many events, which could bring the supermarket-based system to a halt. For instance, a loss of money supply could be handled by delaying payments until the system recovers. Loss of fuel would be difficult to overcome, but produce could still be collected on bike or by public transport.

However, from the consumer's point of view, the risks have a different nature. Consumers in the case system are vulnerable to a poor or failed harvest (e.g., caused by flooding, unusual weather conditions or pests). The supermarket supply system would be unaffected by a failed harvest at a single farm because of the large number of producers feeding into the system. However, the crop diversity of the case-study minimizes the risk of a complete harvest failure.

1.6.5 LIMITATIONS OF STUDY—VALIDITY OF MODEL SYSTEMS

The vegetables produced from the case-study farm have determined the design of the cultivation phase of the two model systems. It is very likely that other model systems would be developed if the systems were optimized for producing any mix of vegetables of certain food energy content but such analyses are outside the scope of this study. However, as the model systems are scale independent the results can be considered as representing the production from larger areas where different farms are producing different crops. Although in many cases the yields will vary from those presented, the high and low yield scenario should capture the range of possible outcomes.

For both emergy assessment and LCA the on-farm use of fuel and irrigation resulted in higher impacts for the case-study than for the model systems. This is to a large part due that the model systems are based on standard data for field operations and that annual variation in rainfall is not reflected. This implies that any additional driving of tractors and machinery, that may occur for various reasons in a real farming system are not included. This may result in an underestimation of the actual resource consumption, which real UK organic farms would have needed in the same years under the same conditions.

1.7 CONCLUSIONS

The results of the emergy analysis showed that the case study is more resource efficient than the modeled standard practices, and with the identified potential for further reducing the emergy use, the case-study farm can become substantially better. This is especially true when also considering emergy used to support labor and service. The results of the LCA for the cultivation phase were less conclusive as the case had neither consistently more nor consistently less environmental impacts compared to the model systems. However, for the distribution phase, both the emergy assessment and LCA evaluated the case to perform substantially better than model systems. In addition, we have argued that the case may be in a better position to cope with likely future scenario of reduced access to domestic and imported fossil fuels and other non-renewable resources.

The real value of the case study is that it points out that there are alternative ways of organizing the production and distribution of organic vegetables, which are more resource efficient and potentially more resilient. The case-study shows that it is possible to efficiently manage a highly diverse organic vegetable production system independently of external input of nutrients through animal manure, whilst remaining economically competitive. The success of the case system is to a large part due to management based on a clear vision of bringing down external inputs. This vision is generic but the specific practices of the case-study may not always be the most appropriate for a farm to improve its resource efficiency and resilience. For systems in other societal contexts, e.g., farms with livestock and crop production or farms in remote locations, other strategies will be needed.

REFERENCES

1. Pelletier, N.; Audsley, E.; Brodt, S.; Garnett, T.; Henriksson, P.; Kendall, A.; Kramer, K.J.; Murphy, D.; Nemecek, T.; Troell, M. Energy Intensity of Agriculture and Food Systems. Annu. Rev. Environ. Resour. 2011, 36, 223–246, doi:10.1146/annurev-environ-081710-161014.
2. Heller, M.C.; Keoleian, G.A. Assessing the Sustainability of the US Food System: A Life Cycle Perspective. Agric. Syst. 2003, 76, 1007–1041, doi:10.1016/S0308-521X(02)00027-6.
3. Barnosky, A.D.; Matzke, N.; Tomiya, S.; Wogan, G.O.U.; Swartz, B.; Quental, T.B.; Marshall, C.; McGuire, J.L.; Lindsey, E.L.; Maguire, K.C.; et al. Has the Earth's Sixth Mass Extinction Already Arrived? Nature 2011, 471, 51–57, doi:10.1038/nature09678.
4. Rockström, J.; Steffen, W.; Noone, K.; Persson, A.; Chapin, S.; Lambin, E.; Lenton, T.; Scheffer, M.; Folke, C.; Schellnhuber, H.J.; et al. Planetary Boundaries: Exploring the Safe Operating Space for Humanity. Ecol. Soc. 2009, 14, 1–33.
5. Sorrell, S.; Speirs, J.; Bentley, R.; Brandt, A.; Miller, R. Global Oil Depletion: A Review of the Evidence. Energ. Pol. 2010, 38, 5290–5295, doi:10.1016/j.enpol.2010.04.046.
6. Cordell, D.; Drangert, J.; White, S. The Story of Phosphorus: Global Food Security and Food for Thought. Global Environ. Change 2009, 19, 292–305, doi:10.1016/j.gloenvcha.2008.10.009.
7. Smil, V. Nitrogen and Food Production: Proteins for Human Diets. Ambio 2002, 31, 126–131.
8. IAASTD. International Assessment of Agricultural Science and Technology for Development, IAASTD; Island Press: Washington, DC, USA, 2009.
9. Defra. Food Industry Sustainability Strategy; Department for Envronment Food and Rural Affairs: London, UK, 2006.
10. Smith, A.; Watkiss, P.; Tweddle, G.; McKinnon, A.; Browne, M.; Hunt, A.; Treleven, C.; Nash, C.; Cross, S. The Validity of Food Miles as an Indicator of Sustainable Development: Final Report; AEA Technology Environment: London, UK, 2005.
11. Coley, D.; Howard, M.; Winter, M. Local Food, Food Miles and Carbon Emissions: A Comparison of Farm Shop and Mass Distribution Approaches. Food Policy 2009, 34, 150–155, doi:10.1016/j.foodpol.2008.11.001.
12. Mundler, P.; Rumpus, L. The Energy Efficiency of Local Food Systems: A Comparison between Different Modes of Distribution. Food Policy 2012, 37, 609–615, doi:10.1016/j.foodpol.2012.07.006.
13. Cardinale, B.J.; Duffy, J.E.; Gonzalez, A.; Hooper, D.U.; Perrings, C.; Venail, P.; Narwani, A.; Mace, G.M.; Tilman, D.; Wardle, D.A.; et al. Biodiversity Loss and its Impact on Humanity. Nature 2012, 486, 59–67, doi:10.1038/nature11148.
14. Østergård, H.; Finckh, M.R.; Fontaine, L.; Goldringer, I.; Hoad, S.P.; Kristensen, K.; Lammerts van Bueren, E.T.; Mascher, F.; Munk, L.; Wolfe, M.S. Time for a Shift in Crop Production: Embracing Complexity through Diversity at all Levels. J. Sci. Food. Agr. 2009, 89, 1439–1445, doi:10.1002/jsfa.3615.

15. Bjorklund, J.; Westberg, L.; Geber, U.; Milestad, R.; Ahnstrom, J. Local Selling as a Driving Force for Increased on-Farm Biodiversity. J. Sustain. Agr. 2009, 33, 885–902, doi:10.1080/10440040903303694.

16. Nemecek, T.; Dubois, D.; Huguenin-Elie, O.; Gaillard, G. Life Cycle Assessment of Swiss Farming Systems: I. Integrated and Organic Farming. Agric. Syst. 2011, 104, 217–232, doi:10.1016/j.agsy.2010.10.002.

17. Audsley, E.; Branderm, M.; Chatterton, J.; Murphy-Bokern, D.; Webster, C.; Williams, A. How Low Can We Go? An Assessment of Greenhouse Gas Emissions from the UK Food System and the Scope to Reduce Them by 2050; FCRN-WWF: London, UK, 2009.

18. Cavalett, O.; Ortega, E. Emergy, Nutrients Balance, and Economic Assessment of Soybean Production and Industrialization in Brazil. J. Clean. Prod. 2009, 17, 762–771, doi:10.1016/j.jclepro.2008.11.022.

19. Bastianoni, S.; Marchettini, N.; Panzieri, M.; Tiezzi, E. Sustainability Assessment of a Farm in the Chianti Area (Italy). J. Clean. Prod. 2001, 9, 365–373, doi:10.1016/S0959-6526(00)00079-2.

20. Rugani, B.; Benetto, E. Improvements to Emergy Evaluations by using Life Cycle Assessment. Environ. Sci. Technol. 2012, 46, 4701–4712, doi:10.1021/es203440n.

21. Raugei, M.; Rugani, B.; Benetto, E.; Ingwersen, W.W. Integrating Emergy into LCA: Potential Added Value and Lingering Obstacles. Ecol. Model. 2014, 271, 4–9, doi:10.1016/j.ecolmodel.2012.11.025.

22. Stockfree Organic Services. Available online: http://www.stockfreeorganic.net/ (accessed on 19 June 2013).

23. Odum, H.T. Environmental Accounting: Emergy and Environmental Decision Making; John Wiley & Sons, Inc.: New York, NY, USA, 1996.

24. ISO. ISO 14040: Environmental Management—Life Cycle Assessment—Principles and Framework; 2006.

25. Frischknecht, R.; Jungbluth, N.; Althaus, H.; Bauer, C.; Doka, G.; Dones, R.; Hischier, R.; Hellweg, S.; Humbert, S.; Margni, M.; et al. Implementation of Life Cycle Impact Assessment Methods; Swiss Center for Life Cycle Inventories: Dübendorf, Switzerland, 2007; p. 151.

26. IPCC. Climate Change 2001: The Scientific Basis. In Third Assessment Report of the Intergovernmental Panel on Climate Change (IPCC); Houghton, J.T., Ding, Y., Griggs, D.J., Noguer, M., van der Linden, P.J., Xiaosu, D., Eds.; Cambridge University Press: Cambridge, UK, 2001.

27. Guinée, J.; van Oers, L.; de Koning, A.; Tamis, W. Life Cycle Approaches for Conservation Agriculture. CML Report 171; Leiden University: Leiden, The Netherlands, 2006; p. 156.

28. Nemecek, T.; Freiermuth, R.; Alig, M.; Gaillard, G. The Advantages of Generic LCA Tools for Agriculture: Examples SALCAcrop and SALCAfarm. In Proceedings of the 7th Int. Conference on LCA in the Agri-Food Sector, Bari, Italy, 22–24 September 2010; pp. 433–438.

29. Pré Consultants. SimaPro V 7.3.3, Pré Consultants, Amersfoort, The Netherlands, 2007.

30. Frischknecht, R.; Jungbluth, N.; Althaus, H.; Doka, G.; Dones, G.; Heck, T.; Hellweg, S.; Hischier, R.; Nemecek, T.; Rebitzer, G.; et al. The Ecoinvent Database:

Overview and Methodological Framework. Int. J. LCA 2005, 10, 3–9, doi:10.1065/lca2004.10.181.1.

31. Stoessel, F.; Juraske, R.; Pfister, S.; Hellweg, S. Life Cycle Inventory and Carbon and Water Foodprint of Fruits and Vegetables: Application to a Swiss Retailer. Environ. Sci. Technol. 2012, 46, 3253–3262, doi:10.1021/es2030577.

32. Patel, M.; Bastioli, C.; Marini, L.; Würdinger, E. Life-cycle Assessment of Bio-based Polymers and Natural Fiber Composites. Biopolymers online 2003, doi:10.1002/3527600035.bpola014.

33. Wihersaari, M. Evaluation of Greenhouse Gas Emission Risks from Storage of Wood Residue. Biomass Bioenerg. 2005, 28, 444–453, doi:10.1016/j.biombioe.2004.11.011.

34. ISO. ISO 14044: Environmental Management—Life Cycle Assessment—Requirements and Guidelines; 2006.

35. Lampkin, N.; Measures, M.; Padel, S. 2011/12 Organic Farm Management Handbook, 9th ed. ed.; The Organic Research Center: Hamstead Marshall, UK, 2011.

36. Gerrard, C.L.; Smith, L.; Pearce, B.; Padel, S.; Hitchings, R.; Measures, M. Public goods and farming. In Farming for Food and Water Security; Lichtfouse, E., Ed.; Springer: London, UK, 2012; pp. 1–22.

37. Smith, L.; Padel, S.; Pearce, B.; Lampkin, N.; Gerrard, C.L.; Woodward, L.; Fowler, S.; Measures, M. Assessing the public goods provided by organic agriculture: Lessons learned from practice. In Organic is Life-Knowledge for Tomorrow, Proceedings of the Third Scientific Conference of ISOFAR, Namyangju, Korea, 28 September–1 October 2011; Volume 2, pp. 59–63.

38. Nemecek, T.; Kägi, T. Life Cycle Inventories of Swiss and European Agricultural Production Systems; Swiss Centre for Life Cycle Inventories: Zürich and Dübendorf, Switzerland, 2007.

39. Williams, A.; Audsley, E.; Sandars, D. Determining the Environmental Burdens and Resource Use in the Production of Agricultural and Horticultural Commodities; Department for Envronment Food and Rural Affairs: London, UK, 2006.

40. Milà i Canals, L. LCA Methodology and Modelling Considerations for Vegetable Production and Consumption; Center for Environmental Strategy (CES), University of Surrey: Guildford, UK, 2007.

41. Tassou, S.A.; De-Lille, G.; Ge, Y.T. Greenhouse Gas Impacts of Food Retailing; Defra: London, UK, 2009.

42. Terry, L.A.; Mena, C.; Williams, A.; Jenney, N.; Whitehead, P. Fruit and Vegetable Resource Maps: Mapping Fruit and Vegetable Waste through the Retail and Wholesale Supply Chain; WRAP: Cranfield, UK, 2011.

43. Mysupermarket. Available online: http://www.mysupermarket.co.uk/ (accessed on 3 June 2013).

44. UK National Statistics. Price Indices and Inflation: UK National Statistics Publication Hub. Available online: http://www.statistics.gov.uk/hub/economy/prices-output-and-productivity/price-indices-and-inflation/index.html (accessed on 30 May 2013).

45. World Food Programme. What is hunger? Available online: http://www.wfp.org/hunger/what-is (accessed on 10 September 2012).

46. Odum, H.T. Folio #2 Emergy of Global Processes. In Handbook of Emergy Evaluation: A Compendium of Data for Emergy Computation; Center for Environmental Policy, University of Florida: Gainesville, FL, USA, 2000.

47. Odum, H.; Mark, B.; Brandt-Williams, S. Folio #1 Introduction and Global Budget. In Handbook of Emergy Evaluation: A Compendium of Data for Emergy Computation; Center for Environmental Policy, University of Florida: Gainesville, FL, USA, 2000.

48. Brown, M.T.; Protano, G.; Ulgiati, S. Assessing Geobiosphere Work of Generating Global Reserves of Coal, Crude Oil, and Natural Gas. Ecol. Model. 2011, 222, 879–887, doi:10.1016/j.ecolmodel.2010.11.006.

49. Buranakarn, V. Evaluation of Recycling and Reuse of Building Materials using the Emergy Analysis Method. Ph.D. Thesis, University of Florida, Gainesville, FL, USA, 1998.

50. National Environmental Accounting Database. Available online: http://cep.ees.ufl.edu/nead/ (accessed on 3 June 2013).

51. Coppola, F.; Bastianoni, S..; Østergård, H. Sustainability of Bioethanol Production from Wheat with Recycled Residues as Evaluated by Emergy Assessment. Biomass Bioenerg. 2009, 33, 1626–1642, doi:10.1016/j.biombioe.2009.08.003.

52. Buenfil, A. Emergy Evaluation of Water. Ph.D. Thesis, University of Florida, Ganiesville, FL, USA, 1998.

53. Brandt-Williams, S. Folio #4 Emergy of Florida Agriculture. In Handbook of Emergy Evaluation: A Compendium of Data for Emergy Computation; Center for Environmental Policy, University of Florida: Gainesville, FL, USA, 2002.

54. Kamp, A.; Østergård, H.; Gylling, M. Sustainability Assessment of Growing and Using Willow for CHP Production. In Proceedings from the 19th European Biomass Conference and Exhibition, Berlin, Germany, 6–10 June 2011; pp. 2645–2656.

55. Bargigli, S.; Ulgiati, S. Emergy and Life-Cycle Assessment of Steel Production in Europe. In Emergy Synthesis 2: Theory and Application of the Emergy Methodology, Proceedings from the Second Biennial Emergy Analysis Research Conference, Gainesville, FL, USA, 20–22 September 2001; pp. 141–156.

56. Pulselli, R.M.; Simoncini, E.; Ridolfi, R.; Bastianoni, S. Specific Emergy of Cement and Concrete: An Energy-Based Appraisal of Building Materials and their Transport. Ecol. Ind. 2008, 8, 647–656, doi:10.1016/j.ecolind.2007.10.001.

57. Regional and Local Authority Electricity Consumption Statistics: 2005 to 2011. Available online: https://www.gov.uk/government/statistical-data-sets/regional-and-local-authority-electricity-consumption-statistics-2005-to-2011 (accessed on 3 June 2013).

58. Kamp, A.; Østergård, H. How to Manage Co-Product Inputs in Emergy Accounting Exemplified by Willow Production for Bioenergy. Ecol. Model. 2013, 253, 70–78, doi:10.1016/j.ecolmodel.2012.12.027.

59. Berry, P.M.; Sylvester-Bradley, R.; Philipps, L.; Hatch, D.J.; Cuttle, S.P.; Rayns, F.W.; Gosling, P. Is the Productivity of Organic Farms Restricted by the Supply of Available Nitrogen? Soil Use Manag. 2002, 18, 248–255, doi:10.1079/SUM2002129.

60. Brown, M.T.; Ulgiati, S. Understanding the Global Economic Crisis: A Biophysical Perspective. Ecol. Model. 2011, 223, 4–13, doi:10.1016/j.ecolmodel.2011.05.019.

61. Odum, H.T.; Odum, E.C. The Prosperous Way Down. Energy 2006, 31, 21–32, doi:10.1016/j.energy.2004.05.012.

62. Statistical Review of World Energy 2012. Available online: http://www.bp.com./bp_internet/globalbp/globalbp_uk_english/reports_and_publications/statistical_energy_review_2011/STAGING/local_assets/pdf/statistical_review_of_world_energy_full_report_2012.pdf (accessed on 3 June 2013).

63. UK Ministry of Defense. Regional Survey: South Asia Out to 2040. Strategic Trends Programme. 2012. Available online: https://www.Gov.uk/government/uploads/system/uploads/attachment_data/file/49954/20121129_dcdc_gst_regions_sasia.pdf (accessed on 3 June 2013).

64. Peck, H. Resilience in the Food Chain: A Study of Business Continuity Management in the Food and Drink Industry: Final Report to the Department for Environment, Food and Rural Affairs; Defra: Shrivenham, UK, 2006; p. 171.

65. Goldin, I.; Vogel, T. Global Governance and Systemic Risk in the 21st Century: Lessons from the Financial Crisis. Global Pol. 2010, 1, 4–15, doi:10.1111/j.1758-5899.2009.00011.x.

66. Fantazzini, D.; Höök, M.; Angelantoni, A. Global Oil Risks in the Early 21st Century. Energ. Pol. 2011, 39, 7865–7873, doi:10.1016/j.enpol.2011.09.035.

67. Korhonen, J.; Seager, T.P. Beyond Eco-Efficiency: A Resilience Perspective. Bus. Strat. Environ. 2008, 17, 411–419, doi:10.1002/bse.635.

68. Souci, S.; Fachmann, W.; Kraut, H. Food Composition and Nutrition Tables, 6th ed. ed.; CRC Press: Stuttgart, Germany, 2000.

69. Regional Climates: Southern England. Available online: http://www.metoffice.gov.uk/climate/uk/so/ (accessed on 28 February 2014).

70. Dunn, S.M.; Mackay, R. Spatial Variation in Evapotranspiration and the Influence of Land use on Catchment Hydrology. J. Hydrol. 1995, 171, 49–73, doi:10.1016/0022-1694(95)02733-6.

71. British Geological Survey. Available online: http://www.bgs.ac.uk/research/energy/geothermal/ (accessed on 28 February 2014).

72. Shawbury Garden Center. Gardening Supplies. Available online: http://www.shawburygardencentre.co.uk/ (accessed on 28 February 2014).

73. Brown, M.; Bardi, E. Folio #3 Emergy of Ecosystems. Handbook of Emergy Evaluation: A Compendium of Data for Emergy Computation; Center for Environmental Policy, University of Florida: Gainesville, FL, USA, 2001.

74. Engineeringtoolbox. Fuels-Higher Calorific Values. Available online: http://www.engineeringtoolbox.com/fuels-higher-calorific-values-d_169.html (accessed on 28 February 2014).

75. Francescato, V.; Antonini, E.; Luca, B. Wood Fuels Handbook; Italian Agriforestry Energy Association: Legnaro, Italy, 2008; p. 79.

76. Skov-og Naturstyrelsen. Træ som brændsel. Available online: http://skov-info.dk/haefte/18/kap07.htm (accessed on 28 February 2014). (In Danish).

77. The Physics Factbook. Density of Glass. Available online: http://hypertextbook.com/facts/2004/ShayeStorm.shtml (accessed on 28 February 2014).

78. OANDA. Historical Exchange Rates. Available online: http://www.oanda.com/currency/historical-rates/ (accessed on 28 February 2014).

There are several supplemental files and appendices that are not available in this version of the article. To view this additional information, please use the citation information cited on the first page of this chapter.

CHAPTER 2

SOIL ORGANISMS IN ORGANIC AND CONVENTIONAL CROPPING SYSTEMS

WAGNER BETTIOL, RAQUEL GHINI,
JOSÉ ABRAHÃO HADDAD GALVÃO,
MARCOS ANTÔNIO VIEIRA LIGO,
AND JEFERSON LUIZ DE CARVALHO MINEIRO

2.1 INTRODUCTION

Contamination of the water-soil-plant system with pesticides and fertilizers, in addition to breaking up the soil structure due to inadequate use of machinery and implements, is one of the main problems caused by intensive agriculture. The implementation of integrated cropping systems and the reduction of the external energy requirements have been suggested to minimize these problems. The organic cropping system is defined as a production system that is sustainable in time and space, by means of management and protection of the natural resources, without the use of chemicals that are aggressive to humans and to the environment, retaining fertility increases, soil life and biological diversity. Thus, the use of highly soluble fertilizers, pesticides and growth regulators must be excluded in

This chapter was originally published under the Creative Commons Attribution License. Bettiol W Ghini R, Haddad Galvão JA, Vieira Ligo MA and de Carvalho Mineiro JL. Soil Organisms in Organic and Conventional Cropping Systems. Scientia Agricola 59,3 (2002). http://dx.doi.org/10.1590/S0103-90162002000300023.

this system (Paschoal, 1995). Not only does the system have to satisfy the need for reducing the environmental negative-impact problems caused by intensive agriculture, it must also be economically competitive. In comparing the organic and the conventional cropping systems, an important step is to establish which social, economic and ecological factors influence the production systems the most. Besides, a knowledge of those factors allows for a better understanding of how the production systems are structured and how they work.

With respect to the biological activity, in studies to compare the conventional, integrated and organic cropping systems, Bokhorst (1989) found that the number of worms in a soil planted with sugar beets was five times higher in the organic system than in other systems, and that the percentage of wheat and potato roots infected with arbuscular mycorrhizae was twice as high in the organic as compared to the conventional and integrated systems. Gliessman et al. (1990, 1996), working with similar objectives, compared conventional and organic strawberry cropping systems in areas where farmers became organic producers, and verified an increase in the number of plants infected with mycorrhizae. Swezey et al. (1994) found higher microbial biomass in the soil and in arbuscular mycorrhizae in the organic system than in the conventional, in an area being changed from conventional into an organic apple growing area. All these studies emphasize the biological elasticity in the organic systems as a fundamental characteristic, influencing the occurrence of pests and diseases.

With regard to soil organisms, Brussaard et al. (1988, 1990) verified that the total biomass of soil organisms was higher for the integrated than for the conventional cropping system, with figures averaging 907 kg C ha^{-1} and 690 kg C ha^{-1}, respectively. Of these biomasses, bacteria accounted for over 90%, fungi represented approximately 5% and protozoa were less than 2% of the total biomass. El Titi & Ipach (1989) studied the effect of a cropping system with low input rate index as well as the conventional system on the soil fauna components and observed there were smaller populations of nematodes pathogenic to plants, higher worm biomass, and larger populations of collembolans and Mesostigmata mites in the system with low input index. Collembola is a microarthropod related to the soil's capacity to suppress *Rhizoctonia solani* (Lartey et al., 1994). Rickerl et al. (1989) found that populations of this organism were 29% larger in soils under minimum tillage as compared to soils under conventional tillage. Ladd et al. (1994) verified that the C biomass of microbial populations was greater in soils under crop rotation than in soils under continuous monoculture; greater in soils where plant residues were incorporated or remained on the soil surface than where they were removed; and smaller in a nitrogen-fertilized soil than in non-fertilized ones. This information is important because these are characteristics that contribute to soil biological equilibrium, nutrient mineralization and suppressive capacity toward plant pathogens, among others, making the system less dependent on external input.

The objective of this work was to evaluate the influence of the organic and the conventional cropping systems, for tomato and corn, on the community of soil organisms.

2.2 MATERIALS AND METHODS

The experiment was carried out in Jaguariúna, SP, Brasil, latitude 22° 41' S, longitude 47° W Gr., and an altitude of 570 m, on a dystrophic Ultisol, with the following chemical properties of the 0–0.2 m topsoil layer, before liming: pH ($CaCl_2$) 4.4; OM 0.6%; P (resin) 1 mg cm^{-3}; K 0.5; Ca 7; Mg 7; H + Al 28; CEC 43 and S 15 mmol dm^{-3} of soil; and V 35%. The studies were conducted from January 1993 to September 1995.

The experiment was set up as randomized blocks with six replicates, and plots measuring 25 x 17 m. Tomato planting pits were spaced 0.5 m apart with 1.20 m between rows. Each plot was split in two halves, the first 12.5 x 17 m half being planted with the variety Débora and the other planted with the variety Santa Clara. Therefore, each of the twelve rows contained 17 planting pits for each variety. The edging between plots was 10 m wide and was planted with sorghum. Two tomato plants were transplanted per pit. The tomato crop was conducted using the stake system, with one or two stems/plant. The number of stems was determined based on the successful establishment of the seedlings. Furrow irrigation and plant pruning were performed as often as necessary.

The entire area received 4.2 t ha^{-1} lime and 2 kg per meter, 110 and 12 days before planting, respectively. Fertilization in the organic system employed 2.5 L of organic compost (pH=6.4; C=29.6%; N=1.6%; P_2O_5=1.8; K_2O=0.17% and U=25.3%) plus 130 g of single superphosphate/pit; additionally, 2.5 L of organic compost, 60 g of single superphosphate, and 60 g of dolomitic lime/pit were applied as sidedressing; plants were sprayed twice a week with biofertilizer (Bettiol et al., 1997), at concentrations of 5 or 10%. In the conventional system, fertilization consisted of 200 g 4-14-8 (NPK)/pit and, after planting, a sidedressing application of 30 g N, 33 g K and 10.5 g P/pit; 52 days after planting and beyond, plants were sprayed once a week with foliar fertilizer [5-8-0,5 (NCaB)] at a rate of 3 mL L^{-1}.

In the conventional system, 0.15g/pit of active ingredient of the insecticide carbofuram were applied before planting. According to the procedures utilized by conventional local growers, a blend of insecticides, fungicides and miticides was sprayed twice a week, after planting. Active ingredients of fungicides sprayed during the crop cycle were metalaxyl, mancozeb, chlorothalonil, copper oxychloride, kasugamycine, cuprous oxide, methyl thyophanate, iprodione, benomyl, cymoxamil, maneb and monohydrate zinc sulphate, at the rates recommended by the manufacturers. Insecticides used were deltamethrin, permethrin, methomyl, methamidophos, acephate, avermectin and cartap, also at the recommended rates.

Extracts of black pepper, *Eucalyptus*, garlic and fern; Bordeaux mixture, and biofertilizer were applied twice a week (Bettiol et al., 1997; Abreu Junior, 1998) to control diseases and pests in the organic system. These applications were performed according to the program adopted by organic producers in the region.

Weed control was carried out by mechanical weeding and with the herbicide glyphosate (directed spray) on post-planting in the conventional system, and with mechanical weeding in the organic system.

After harvesting the tomato the area was planted with 'BR 201' corn; sowing occurred 178 days after planting the tomatoes. The organic system plots received an application of 4 m^3 of organic compost and single superphosphate at the rate of 20 g per meter; in addition, the biofertilizer was sprayed at 10% as sidedressing. In the conventional system fertilization consisted of 500 kg ha^{-1} of the 4-14-8 NPK rate applied pre-planting and 15 g m^{-1} urea as sidedressing. Weed control used the herbicide paraquat (directed spray) in the conventional system, and mechanical weeding was used in the organic.

After harvesting the corn, 'Débora' tomatoes were again cultivated, as previously described. Transplantation was made 401 days after the initial tomato planting.

2.2.1 SOIL MICROORGANISMS

A sample composed of 20 sub-samples of soil taken at the planting row from the 0–7 cm depth layer was obtained for each plot. Samples were placed in plastic bags and immediately transported to the laboratory. Assessments were performed within 24 hours after collecting the samples.

Populations of fungi, bacteria and actinomycetes: The populations of fungi, bacteria and actinomycetes were quantified through the serial-dilution method, followed by plating in culture medium. Martin's culture medium (Tuite, 1969) added of 100 mg mL^{-1} streptomycine was used for fungi; for bacteria, the agar nutrient medium added of nistatin (42 mg L^{-1}) was used; for the actinomycetes, the alkalized agar-water medium was utilized. Aliquots (0.1 mL) from three dilutions, for each soil sample, were transferred to the culture media in three replications. Assessments were performed by counting the number of colonies per Petri dish and expressed as colony-forming units/g of dry soil (CFU g^{-1} dry soil).

Total respiratory activity: Total microbial respiration was evaluated according the method described by Grisi (1978). Soil samples (200 g) were incubated for 10, 20, and 30 days within tightly sealed containers holding 10 mL of a 0.5 mol L^{-1} (10 mL) KOH solution. At 10-day intervals, the solution was substituted and titrated with 0.1 mol L^{-1} of HCl. Incubation was conducted in the dark, at 25°C. This parameter was expressed as g CO_2 (g dry soil^{-1}) (day^{-1}). Since the more substantial changes happened in the first days, only readings up to the tenth day were used to determine mean values. For the statistical analysis, data were transformed into square root (x + 0.5) and subjected to analysis of variance and Duncan's mean comparison test.

Soil microarthropods: Collecting was made with a Uhland-type, stainless steel auger 5 cm in diameter and 10 cm in height, totaling four samples per plot. Samples were placed in plastic bags and taken to the laboratory. Collecting was between 8:00 and 10:30 h, 82 days before and 325 days after the first tomato seeding, for a total of 16 evaluations. Extraction was according to Tullgren's modified method, which uses heat and desiccation to force the animals to leave the soil. Samples remained in the extractor for 72 hours. An alcohol:glycerin (1:1) aqueous solution was used for specimen preservation. After extraction, the animals were counted and separated into groups with the use of a stereoscopic microscope. Mites and other smaller animals were fixed on permanent slides for identification. Data were expressed as number of individuals per 785 cm^3 soil. Shannon's diversity index (Shannon & Weaver, 1949) was calculated for a better understanding of the variations in the soil microarthropod populations.

Organic matter decomposition rate estimate: The decomposition rate was estimated via loss of organic content from leaf litter confined in nylon bags, 20 x 20 cm, with a 1 mm mesh, where 10 g of elephant grass dried at 60°C for three days. The field-collected samples, were collected every 20 days and transported to the laboratory, dried at 105°C for 24 hours and ashed at 600°C for 4 hours. The loss of organic matter estimate was calculated using the equation described by Santos & Whitford (1981), which corrects for the adhesion of soil particles to the organic matter.

Evaluation of earthworms in the soil: The first evaluation was carried out 81 days before the first planting, i.e., before plowing and liming. A hand excavator was used to collect samples; two samples were collected from each plot, up to a depth of 20 cm, with 20 cm diameter. Shortly after planting the tomatoes, and 90 days later, samples were taken at about 40 cm depth, with a diameter of 10 cm. Three samples were collected from the compost: one from the pile surface; another at a layer up to 35 cm, and the third at a depth of 90 cm. The worm populations were determined 370, 407, and 471 days after the first tomato planting.

2.3 RESULTS AND DISCUSSION

The populations of fungi, bacteria and actinomycetes were similar for the two cropping systems over the entire period of study, with populations of fungi varying from 10^4 to 10^5, whereas populations of bacteria and actinomycetes varied from 10^5 to 10^7 CFU g^{-1} dry soil (Figure 1).

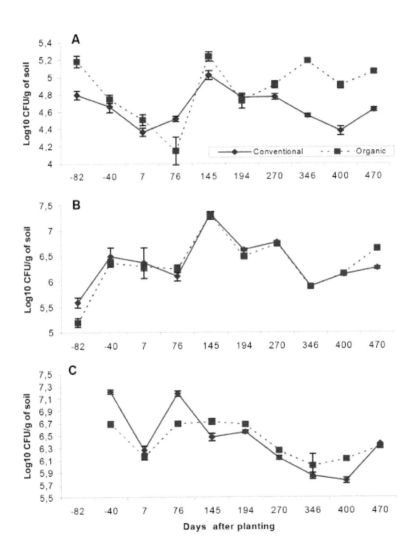

FIGURE 1: Dynamic population of fungi, bacteria and actinomycetes in soil from organic (- - -) and conventional ( ) cropping systems for tomato and corn. CFU: Colony Forming Units. A=Fungi; B=Actinomycetes; C=Bacteria. The data represent the mean of six replicates. The bars indicate the standart deviation.

Similar results were obtained by Castro et al. (1993), when several types of soybean management were compared, and by Cattelan & Vidor (1990) on soils cultivated with different crop rotation systems. Grigorova & Norris (1990) justified not adopting this method for evaluating soil microorganisms, because only a small fraction of microbial biomass could be cultivated on a selective medium. However, Cattelan & Vidor (1990) demonstrated the effectiveness of the method in studies with different cropping systems. In spite of a similar behavior in regard to microbial populations, starting 145 days after planting the tomatoes, the bacteria populations (Figure 1 C) were higher in the organic system as compared to the conventional. This could be due to soil plant cover, like Cattelan & Vidor (1990) who found a smaller bacterial population on naked as compared to cultivated soil.

Soil total respiratory activity continued higher in the organic system during the crop cycles, showing in some evaluations twice as much as the evolution observed in the conventional system (Figure 2). Differences were found during the intermediate period, that is, between 142 and 400 days after planting. There were no statistical differences between treatments at the initial periods or at the end. The higher respiratory rate in the organic system could be due to the addition of an exogenous source of organic matter to the soil and the consequent stimulation of heterotrophic microorganisms (Lambais, 1997).

Observed organic matter decomposition rates ranged from 15 to 45% of organic carbon loss in a 20-day period. Rodrigues et al. (1997) observed, in corn cultivated during the summer, values reaching 70% of carbon loss in a period of 30 days. There was no difference among results from the organic and the conventional systems (Figure 3). However, regardless of the system, there was an influence of time on the organic matter decomposition rate was, although no interaction between time and the treatments was found. This suggests that variations found during the study period could be related to the humidity and temperature fluctuations that occur in the field, thus providing no evidence that the adopted management forms influenced decomposition rate.

The CO_2 release method used in this study to evaluate respiratory activity favors the microorganism population, since soil manipulation can eliminate the majority of the microarthropod community. Several authors have, in microcosmos studies, demonstrated the role microarthropods in soil organic matter decomposition process. A low fungivore density (Collembola) has a stimulating effect on microbial respiration, whereas high densities inhibited microorganism respiration Barsdate et al, 1974; Hanlon & Anderson, 1979).

Mites and insects, belonging to various families, were the two main groups of arthropods found in the soil in 1993 and 1994 (Tables 1 and 2). In general, rates and numbers of individuals from these groups were higher in the organic cropping system, reflecting on Shannon's diversity indices, which were higher in the organic system on all sampling dates (Figure 4), but not on the soil organic matter decomposition (Figure 3).

The largest populations of insects were from the Order *Collembola*, and the number of individuals found in the organic system was three times as high as that in the conventional system, during the first nine months (Table 1). During the following six months, the number of collembolans remained 20% higher in the organic cropping system than in the conventional (Table 2). These data agree with El Titi & Ipach (1989), who verified larger populations of collembolans for the low-input system than for the conventional. Collembolans contribute to the soil's ability of suppressing plant pathogens such as *Rhizoctonia solani, Fusarium oxysporum f.* sp. *vasinfectum*, and *Pythium* (Wiggins & Curl, 1979; Curl et al., 1985a, b; Rickerl et al., 1989; Lartey et al., 1994), because these organisms are, for the most part, mycophagous, modifying the community of fungi. Because in this work the practices in the organic system stimulated the community of collembolans, it can be inferred that these organisms are responsible, at least in part, for the suppression ability in soils enriched with organic matter. Still, in regard to insects, the number of individuals was low for the rest of the orders (Tables 1 and 2).

During the first nine months of evaluation (Table 1), for both cropping systems, the largest mite population was of the superfamily *Oribatuloidea*, followed by the family *Galumnidae* and by the superfamily *Passalozetoidea*, all in the suborder *Oribatida* and with similar behavior between cropping systems. In the suborder *Gamasida* the most abundant population was *Laelapidae* and in *Actinedida* the most abundant was *Pygmephoridae*, both more numerous in the organic system. Populations in the suborders *Acaridida* and *Ixodida* were very small.

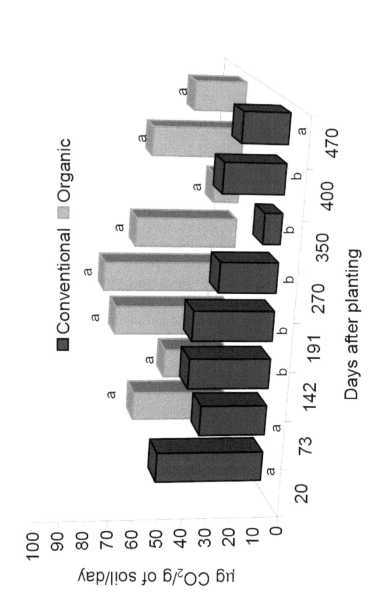

FIGURE 2: CO_2 evolution from soil microorganisms of organic and conventional systems for tomato and corn crops. Results were obtained though soil incubation at 25°C for 10 days. For each planting time, data followed the same letter did not differ (Duncan 5%).

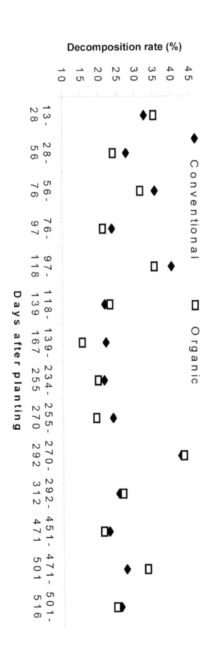

Figure 3: Organic matter decomposition rate soil of organic and conventional cropping systems.

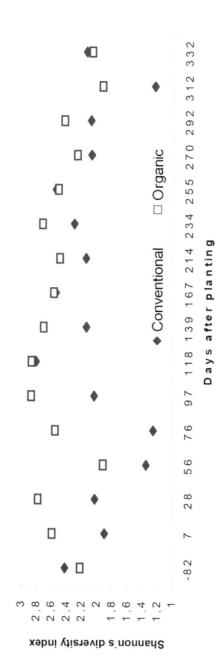

FIGURE 4: Shannon's diversity index for soil microarthropods of the organic and conventional cropping systems.

In the six subsequent months (Table 2), when only the families of mites were quantified, the largest population was of Scheloribatidae followed by Galumnidae, with similar behavior between the systems. The expressive number of individuals in the families *Galumnidae* and *Scheloribatidae* for both cropping systems is due to the characteristic these families exhibit toward occupying space in agroecosystems. In the orders *Actinedida* and *Gamasida*, families *Cunaxidae* and *Laelapidae* were the largest, respectively. In general, mite population densities in the classes *Gamasida* and *Actinedida* were higher in the organic system. The fact that the *Gamasida* showed high numbers is possibly due to a large *Collembola* population, because these organisms are a source of food for this class of mites. El Titi & Ipach (1989) verified the existence of larger populations of collembolans and Gamasida mites in the low-input system than in the conventional.

Due to the more abundance of microarthropods in the organic system, it was believed that the organic matter decomposition rate would be higher in this system, because these organisms contribute for organic matter degradation and stimulate microbial activity in the soil (Nosek, 1981). Accordingly, when the presence of *Oribatida* and *Collembola* in litterbags incorporated into the organic and the conventional systems was evaluated, a larger number of individuals in the litterbags was found for the organic system (Melo & Ligo, 1999), indicating that this system contributes for an increase in biological diversity. Since the presence of these organisms in larger numbers was not accompanied by a higher decomposition of organic matter, one can say that the differences in arthropod density found in the soil between the organic and the conventional systems did not reflect on the organic matter decomposition rate, as evaluated by the litterbag method. The community of microarthropods in the soil might have, among other factors, influenced microbial activity, since the organic system showed a higher microbial activity potential than the conventional system. The influence of the soil fauna on the organic matter decomposition rate of forest soils is well documented, but this is not true for agricultural ecosystems (Crossley et al., 1989). In agroecosystems the effect of the fauna on the organic matter decomposition rate seems not to be very significant and consequently, there are many points that need to be clarified when it comes to the role of fauna in agricultural soils. Occasionally, and similarly among the crop systems evaluated, individuals belonging in the groups *Aranae, Chilopoda, Diploploda, Diplura, Pauropoda, Protura* and *Symphila* were collected. In addition to these, individuals of the insect orders *Dermaptera, Hemiptera, Homoptera, Isoptera* and *Thysanoptera* were found in limited numbers.

The higher biological diversity in the organic system is important because it contributes to keeping the biological equilibrium, essential in an agroecosystem. This equilibrium may bring about greater stability for the system and consequently fewer problems with diseases and pests.

With respect to the worm community, after a one-year period of cropping, the soil in the organic system showed at least a ten-fold higher number of specimens per 3140 mL soil sample than the conventional system. After 370, 407 and 471 days from planting a total of 18, 24 and 101 specimens were found in the organic cropping system, and 1, 2 and 12 specimens were found in the conventional system, respectively. These data agree with Bokhorst (1989), who found that the number of worm individuals per square meter, in a soil planted with sugar beets, was five times higher in the organic system as compared to the conventional. Also, El Titi & Ipach (1989) observed the existence of greater worm biomass in a low-input system than in the conventional. The higher number of species in the organic system is possibly due to the availability of organic substrates for them to breed on and the absence of pesticides. On the other hand, the presence of pesticides explains the small number of species in the conventional system, since the worms are sensitive to the products used in the conventional system (Lee, 1985). These organisms are important because they not only improve the physical properties (Lee, 1985), but also contribute to the soil's ability to suppress pathogens, such as *R. solani*, among others (Stephens et al., 1993). No worm specimens were found in the recently plowed soil (81 days before planting) and in the evaluations carried out at planting time and 90 days after planting the seeding, as well as in the organic compost.

TABLE 1: Number of soil microarthropods in the tomato organic (O) and conventional (C) cropping systems.

Class/Order	Superfamily/Family	-84 C	-84 O	7 C	7 O	28 C	28 O	49 C	49 O	70 C	70 O	91 C	91 O	112 C	112 O	134 C	134 O	162 C	162 O
I-Arachnida																			
I.1 Acari																			
Acaridida	Anoetidae	0	0	0	0	1	0	1	0	0	0	1	0	0	0	0	0	0	0
	Acaridae	0	0	0	1	0	0	0	1	0	1	0	1	5	0	0	0	0	3
Actinedida	Cunaxidae	1	0	2	1	4	2	1	0	1	0	9	1	4	1	18	1	16	0
	Eupodidae	0	0	0	1	25	0	2	0	0	0	1	0	2	0	0	5	5	3
	Nanorchestidae	0	4	18	0	19	0	4	1	4	0	6	7	1	1	2	2	2	1
	Rhagidiidae	0	0	0	0	0	0	0	0	5	0	1	0	0	0	0	0	0	3
	Pygmephoridae	0	1	2	1	37	0	2	0	13	0	30	1	8	0	0	0	0	1
	Scutacaridae	0	0	0	0	1	0	1	0	0	0	0	0	1	0	0	0	0	0
	Tarsonemidae	0	0	0	2	0	0	0	0	0	0	1	2	1	1	5	0	0	0
Gamasida	Ascidae	0	0	2	0	1	1	4	0	0	0	11	1	7	0	1	0	1	0
	Digamasellidae	0	0	1	0	0	0	0	0	0	0	0	0	0	0	0	0	0	0
	Laelapidae	1	6	1	9	8	4	7	1	15	3	13	2	27	1	5	4	2	0
	Macrochelidae	1	2	1	0	1	0	1	0	1	0	0	0	0	0	0	0	0	0
	Ologamasidae	6	8	1	1	6	2	3	0	6	0	2	1	5	0	4	0	4	3
	Pachylaelapidae	12	20	2	0	2	0	2	0	0	0	3	0	1	0	0	0	0	0
	Parasitidae	0	1	0	1	1	1	0	2	5	0	2	9	5	2	0	0	0	0
	Parhoslaspididae	1	0	1	0	1	0	0	0	0	0	0	0	0	0	0	0	0	0
	Phytoseiidae	0	0	0	0	6	0	0	0	0	0	0	1	1	0	0	0	3	1
	Rhodacaridae	0	0	0	0	0	0	0	0	0	0	1	0	0	0	0	0	0	0

TABLE 1: *Cont.*

Class/Order	Superfamily/Family	Days after the first tomato planting								
		-84	7	28	49	70	91	112	134	162
	Uropodidae	1	6	13	1	7	0	3	15	2
	Immature	0	0	5	6	8	20	9	3	8
	Male	4	5	1	0	0	1	0	0	0
Ixodida	Ixodidae	0	0	0	0	0	0	0	0	0
Oribatida	Brachichythonidadae	0	0	0	0	0	0	0	0	0
	Euphthiracaridae	1	3	2	1	0	1	3	1	0
	Galumnidae	15	33	26	22	16	10	19	14	9
	Haplochthoniidae	0	0	1	1	0	0	0	0	0
	Microzetidae	0	0	0	0	1	1	2	1	0
	Oppiidae	0	4	11	13	4	5	3	0	0
	Oribatuloidea	21	46	47	106	78	26	31	28	17
	Oribatellidae	0	1	1	1	0	1	0	0	2
	Passalozetoidea	6	13	6	1	4	37	24	4	0
	Suctobelbidae	0	0	2	1	1	7	0	1	0
	Scheloribatidae	21	48	47	106	78	26	31	28	17
	Thrypochthoniidae	0	0	0	0	0	16	14	0	1
	Immature	0	1	2	0	0	0	1	0	0
	Oribatida	2	1	2	2	2	3	0	7	0
II. Insecta										
Coleoptera	Carabidae	0	0	0	1	1	1	0	2	0

TABLE 1: *Cont.*

Class/Order	Superfamily/Family	\-84	7	28	49	70	91	112	134	162
	Cicindelidae	0	1	0	0	0	1	0	0	0
	Hydroscaphidae	0	0	1	0	0	0	0	0	0
	Nitidulidae	0	1	0	0	0	0	1	0	0
	Scarabaedidae	0	0	0	1	0	0	1	0	0
	Scolytidae	0	0	0	0	0	1	0	0	0
	Staphylinidae	0	0	2	0	0	3	5	2	4
	Larva	4	1	6	1	1	10	11	17	4
Collembola	Entomobryidae	7	10	43	2	2	19	36	35	52
	Isotomidae	4	71	93	3	0	38	16	30	32
	Poduridae	1	39	34	0	1	5	60	44	13
Diptera	Adult	0	5	10	0	0	9	12	8	20
	Larva	1	0	5	1	1	0	0	1	1
Homoptera	Homoptera	0	1	0	0	0	1	0	2	3
Hymenoptera	Formicidae	0	0	4	0	0	3	9	1	3
	Psocoptera	1	1	0	0	0	2	2	5	5

Days after the first tomato planting

Data expressed in number of individuals per 785 mL soil and represent the mean of six replicates.

TABLE 2: Number of soil microarthropods in the corn organic (O) and conventional (C) cropping systems.

Class/order	Family	197		218		240		262		284		304		325	
		C	O	C	O	C	O	C	O	C	O	C	O	C	O
1-Arachnida															
I.1 Acari															
Acaridida	Anoetidae	0	0	0	0	0	0	1	0	0	0	0	0	0	0
	Acaridae	0	0	0	0	1	0	0	0	0	0	0	0	0	0
Actinedida	Cunaxidae	4	9	8	7	17	17	16	14	7	17	7	18	6	8
	Erythraeidae	0	0	1	1	2	0	0	0	0	1	0	0	0	0
	Eupodidae	0	2	0	0	0	11	1	11	3	4	1	4	9	10
	Nanorchestidae	0	2	0	5	5	10	1	3	1	3	1	2	2	2
	Rhagidiidae	0	1	0	0	2	2	0	0	0	0	1	2	1	0
	Pygmephoridae	2	0	0	2	1	2	3	1	7	3	0	4	0	0
	Tarsonemidae	0	0	0	0	2	10	0	4	0	1	0	0	1	0
Gamasida	Ascidae	0	1	0	0	4	5	5	11	4	4	5	12	1	2
	Amoroseiidae	0	0	1	0	0	0	0	0	0	0	0	0	0	0
	Digamasellidae	0	0	0	0	0	0	0	1	0	0	0	0	0	0
	Laelapidae	3	1	2	13	17	48	44	41	5	20	1	18	7	12
	Macrochelidae	0	0	0	0	0	0	0	0	0	1	0	1	1	2
	Ologamasidae	1	4	1	1	2	5	9	12	6	5	5	4	1	2
	Pachylaelapidae	0	2	0	0	0	1	1	1	0	0	0	0	4	7
	Parasitidae	0	0	1	17	4	10	1	7	2	0	4	7	0	0
	Parhoslaspididae	0	1	0	0	3	0	0	0	0	1	0	1	0	4
	Phytoseiidae	0	0	0	0	1	0	4	1	0	5	4	4	2	0

Days after the first tomato planting

TABLE 2: *Cont.*

Class/order	Family	Days after the first tomato planting												
		197		218		240		262		284		304		325
	Rhodacaridae	0	0	0	0	0	0	4	0	0	1	0	7	0	1
	Uropodidae	15	0	37	2	24	1	8	1	27	0	17	0	1	1
	Immature	0	2	21	1	16	19	24	7	4	9	5	7	3	0
Oribatida	Male	1	4	0	5	6	2	4	3	0	0	1	0	0	0
	Brachichythonidadae	0	0	0	0	0	0	0	0	0	0	0	0	1	0
	Euphthiracaridae	0	0	0	0	5	1	1	2	0	3	5	4	3	5
	Galumnidae	123	11	19	94	45	51	60	56	31	64	20	108	18	106
	Haplochthoniidae	1	0	0	0	0	0	0	0	0	0	0	0	0	0
	Microzetidae	0	0	1	1	1	0	1	5	0	0	0	0	0	0
	Oppiidae	0	0	0	0	7	2	0	5	1	1	1	0	0	0
	Oribatellidae	1	1	3	4	8	3	15	6	5	7	1	10	6	1
	Phthriracaridae	0	0	0	1	0	0	0	0	0	0	0	1	2	0
	Suctobelbidae	0	0	2	0	0	0	0	0	0	0	0	0	0	0
	Scheloribatidae	53	48	144	132	210	163	365	272	175	140	90	122	68	88
	Thrypochthoniidae	0	0	1	0	0	0	0	0	0	0	0	0	0	0
	Immature	20	17	35	33	139	82	99	125	54	29	10	26	5	11
	Oribatida	0	4	3	4	15	4	1	2	1	0	0	4	2	4
II.Insecta															
Coleoptera	Carabidae	1	2	2	0	3	5	1	1	2	1	0	7	1	6
	Cicindelidae	0	1	3	2	0	0	0	0	0	0	0	0	0	0

TABLE 2: *Cont.*

Class/order	Family	Days after the first tomato planting						
		197	218	240	262	284	304	325
	Nitidulidae	0	1	0	0	0	0	1
	Scarabaedidae	1	0	0	0	0	1	0
	Scolytidae	0	2	1	0	0	1	0
	Staphylinidae	1	1	7	1	0	1	0
	Larva	1	4	6	5	6	3	1
Collembola	Entomobryidae	0	8	20	28	16	3	1
	Isotomidae	15	104	223	266	326	516	92
	Poduridae	0	8	234	128	54	33	174
Diptera	Adult	31	68	20	16	7	8	7
	Larva	0	0	9	5	3	2	3
Hymenoptera	Formicidae	4	8	16	17	26	13	12
	Wasp	1	0	3	1	3	3	2
Psocoptera		15	12	22	7	0	1	1

Data expressed in number of individuals per 785 mL soil and represent the mean of six replicates.

REFERENCES

1. Abreu JR., H. Práticas alternativas de controle de pragas e doenças na agricultura. Campinas: EMOPI, 1998. 112p.
2. Bettiol, W.; Tratch, R.; Galvao, J.A.H. Controle de doenças de plantas com biofertilizantes. Jaguariúna: EMBRAPA, CNPMA, 1997. 22p.   (Circular Técnica, 2).
3. Barsdate, R.J.; Pretki, R.T.; Fenchel, T. Phosphorus cycle of model ecosystems: significance for decomposer food chains and effect of bacterial grazers. OIKOS, v.25, p.239-251, 1974.
4. Bokhorst, J.G. The organic farm at Nageli. In: Zadoks, J.C. Development of farming systems. Pudoc: Wageningen, 1989. p.57-65.
5. Brussaard, L.; Van Veen, J.A.; Kooistra, M.J.; Lebbink, E. The Dutch programme on soil ecology of arable farming systems: I. Objectives, approach and some preliminary results. Ecological Bulletins, v.39, p.35-40, 1988.
6. Brussaard, L.; Bouwman, L.A.; Geurs, M.; Hassink, J.; Zwart, K.B. Biomass, composition and temporal dynamics of soil organisms of a silt loam soil under conventional and integrated management. Netherlands Journal of Agricultural Science, v.38, p.283-302, 1990.
7. Castro, O.M. de; Prado, H. do; Severo, A.C.R.; Cardoso, E.J.B.N. Avaliação da atividade de microrganismos do solo em diferentes sistemas de manejo de soja. Scientia Agricola, v.50, p.212-219, 1993.
8. Cattelan, A.J.; Vidor, C. Sistemas de culturas e a população microbiana do solo. Revista Brasileira de Ciência do Solo, v.14, p.125-132, 1990.
9. Crossley JR., D.A.; Coleman, D.C.; Hendrix, P.F. The importance of the fauna in agricultural soils: research approaches and perspectives. Agriculture, Ecosystems and Environment, v.27, p.47-55, 1989.
10. Curl, E.A.; Harper, J.D.; Peterson, C.M.; Gudauskas, R.T. Relationships of mycophagous collembola and Rhizoctonia solani populations in biocontrol. Phytopathology, v.75, p.1360 (Abstract). 1985a.
11. Curl, E.A.; Gudauskas, R.T.; Harper, J.D.; Peterson, C.M. Effects of soil insects on populatons and germination of fungal propagules. In: Parker, C.A.; Rovira, A.D.; Moore, K.J.; Wong, P.T.W. Ecology and management of soilborne plant pathogens. St. Paul: APS, 1985b. p.20-23.
12. El Titi, A.; Ipach, U. Soil fauna in sustainable agriculture: results of na integrated farming system at Lautenbach, F.R.G. Agriculture, Ecosystems and Environment, v.27, p.561-572, 1989.
13. Gliessman, S.R.; Swezey, S.L.; Allison, J.; Cochran, J.; Farrell, J.; Kluson, R.; Rosado-May, F.; Werner, M. Strawberry production systems during conversion to organic management. California Agriculture, v.44, p.4-7, 1990.
14. Gliessman, S.R.; Werner, M.R.; Swezey, S.L.; Caswell, E.; Cochran, J.; Rosado-May, F. Conversion to organic strawberry management changes ecological processes. California Agriculture, v.50, p.24-31, 1996.
15. Grigorova, R.; Norris, J.R. Methods in microbiology. Techniques in microbial ecology. London: Academic Press, 1990, 627p.

16. Grisi, B.M. Método químico de medição da respiração edáfica: alguns aspectos téc-
    nicos. Ciência e Cultura, v.30, p.82-88, 1978.
17. Hanlon, R.D.G.; Anderson, J.M. The effects of Collembola grazing on microbial
    activity in decomposing leaf litter. Oecologia, v.30, p.93-99, 1979.
18. Ladd, J.N.; Amato, M.; Li-Kai, Z.; Zchultz, J.E. Differential effects of rotation, plant
    residue and nitrogen fertilizers on microbial biomass and organic matter in na Aus-
    tralian Alfisol. Soil Biology & Biochemistry, v.26, p.821-831, 1994.
19. Lambais, M.R. Atividades microbiológicas envolvidas na mineralização da matéria
    orgânica: potenciais indicadores da qualidade de solos agrícolas. In: Martos, H.L.;
    Maia, N.B. Indicadores ambientais. Sorocaba: PUC, 1997. p.167-174.
20. Lrtey, R.T.; Curl, E.A.; Peterson, C.M. Interacitons of mycophagous collembola and
    biological control fungi in the suppression of Rhizoctonia solani. Soil Biology &
    Biochemistry, v.26, p.81-88, 1994.
21. Lee, K.E. Earthworms: Their ecology and relationships with soils and land use. Syd-
    ney: Academic Press, 1985. 411p.
22. Melo, L.A.S.; Ligo, M.A.V. Amostragem de solo e uso de "litterbags" na avaliação
    populacional de microartrópodos edáficos. Scientia Agricola, v.56, p.523-528, 1999.
23. Nosek, J. Ecological niche of Collembola in biogeocoenoses. Pedobiologia, v.21,
    p.166-171, 1981.
24. Paschoal, A.B. Modelos sustentáveis de agricultura. Agricultura Sustentável, v.2,
    p.11-16, 1995.
25. Rickerl, D.H.; Curl, E.A.; Touchton, J.T. Tillage and rotation effects on Collembola
    populations and Rhizoctonia infestation. Soil & Tillage Research, v.15, p.41-49,
    1989.
26. Rodrigues, G.S.; Ligo, M.A.V.; Mineiro, J.L.de C. Organic matter decomposition
    and microartropod community structure in corn fields under low input and intensive
    management in Guaíra, SP. Scientia Agricola, v.54, p.69-77, 1997.
27. Santos, P.F.; Whitford, W.G. The effects of microarthopods on litter decompositions
    in a Chihuahuan desert ecosystem. Ecology, v.62, p.654-663, 1981.
28. Shannon, C.E.; Weaver, W. The mathematical theory of communications. Urbana:
    The University of Illinois Press, 1949. 117p.
29. Stephens, P.M.; Davoren, C.W.; Doube, B.M.; Ryder, M.H.; Benger, A.M.; Neate,
    S.M. Reduced severety of Rhizoctonia solani disease on wheat seedlings associated
    with the presence of the earthworm Aporrectodea trapezoides (LUMBRICIDAE).
    Soil Biology & Biochemistry, v.25, p.1447-1484, 1993.
30. Swezey, S.L.; Rider, J.; Werner, M.R.; Buchanan, M.; Allison, J.; Gliessman, S.R.
    Granny Smith conversions to organic show early success. California Agriculture,
    v.48, p.36-44, 1994.
31. Tuite, J. Plant pathological methods fungi and bacteria. Minneapolis: Burgess, 1969.
    239p.
32. Wiggins, E.A.; Curl, E.A. Interactions of Collembola and microflora of cotton rhi-
    zosphere. Phytopathology, v.69, p.244-249, 1979.

CHAPTER 3

LOCALIZING THE NITROGEN IMPRINT OF THE PARIS FOOD SUPPLY: THE POTENTIAL OF ORGANIC FARMING AND CHANGES IN HUMAN DIET

G. BILLEN, J. GARNIER, V. THIEU, M. SILVESTRE, S. BARLES, AND P. CHATZIMPIROS

3.1 INTRODUCTION

Food supply is a major factor in shaping cities (Steele, 2010) and determining their relationships with surrounding (close or distant) rural territories. The agricultural development in the city hinterland as well as the construction of large transport infrastructures has been largely dictated by the requirements of urban food markets (Keene, 2011; Charruadas, 2011; Billen et al., 2011). Because cities consume most of the final products of agriculture and dictate its specialisation and location, urbanisation is a major driver of the human perturbation of the nitrogen cycle (Svirejeva-Hopkins and Reis, 2011). As shown by several authors, the anthropogenic introduction into the biosphere of reactive nitrogen, which subsequently

This chapter was originally published under the Creative Commons Attribution License. Billen G, Garnier J, Thieu V, Silvestre M, Barles S, and Chatzimpiro P. Localizing the Nitrogen Imprint of the Paris Food Supply: The Potential of Organic Farming and Changes in Human Diet. Biogeosciences 9 (2012); 607–616. doi:10.5194/bg-9-607-2012.

cascades through a number of environmental compartments until it is finally denitrified back to the atmosphere, causing multiple harmful effects, is nowadays among the most severe environmental threats (Galloway et al., 2002; Sutton et al., 2011). Any attempt to eliminate or reduce the source of the nitrogen cascade should take into account the urban food supply issue.

There is presently a lively debate on the possibility of local sourcing of the urban food supply. Food-Miles, i.e. the total transport distance covered by foodstuffs from their production to their consumption sites, have been proposed as an indicator of sustainability of the human food system (Paxton, 1994; Smith et al., 2005). However, in an analysis of the US food supply chain, Weber and Matthews (2008) showed that foodstuff transportation is only a minor term in its total carbon imprint and that changes in agricultural practices or in the composition of human diet would have a much more pronounced effect on greenhouse gas emission than reduction of the food transport distance. They conclude that the issue of localisation of the food supply is not a question of climate impact optimisation but is conditioned by the political will to support local agricultural communities and to restore the link between cities and their rural hinterland. The latter concern is revealed by the recent and quite rapid development in the Western world of citizen organisations aiming to reconnect food production and urban consumption, as well as directly controlling the quality of products and the environmental impact of their production, through direct contracts between farmers and groups of concerned consumers (Community Supported Agriculture in North America and the UK, www.localharvest. org; AMAP in France, www.reseau-amap.org; Gruppi di Acquisto Solidali in Italy, www.retegas.org; Teikei in Japan, etc.) (www.urgenci.net) (Watts et al., 2005).

As water is also a food, drinking water provision is another important aspect of the food supply. Because of their intensive use of fertilisers and pesticides, modern agricultural practices often lead to severe degradation of the quality of water produced by rural territories. A conflict therefore exists in the allocation of land areas for either drinking water or food production. Some large cities, such as New York in the US, made the choice of reserving certain nearby lands for clean drinking water production, excluding agricultural activities (Swaney et al., 2012). Others, such as Athens in Greece (Stergiouli et al., 2012) and Barcelona in Spain (Tello et al.,

2011), extend their water supply areas over a very long distance, annexing the water resources of other watersheds. In a few instances, such as in Munich (http://www.partagedeseaux.info/article226.html), a deliberate policy was implemented to reconcile water and food production in the surrounding hinterland, with organic agriculture being promoted through the establishment of a strong public urban market demand.

The case of the huge Paris agglomeration offers an interesting example to study the potentiality of changes in the relationships between a large city and its food-supplying area, in particular in terms of its contribution to the nitrogen cascade. Previous studies (Billen et al., 2009, 2011) have traced back the geographical areas supplying food (measured as protein N content) to the Paris agglomeration during the last two centuries, a period when the city grew from 700 000 inhabitants in 1786 to 11 500 000 at the present time: its per capita N consumption rose from 5.4 to 7.7 kgNcap^{-1} yr^{-1}, and the share of animal products in the diet increased from 39 to 65% of total protein consumption. Part of the consumption increase is due to higher intake, while another part is the result of higher waste generation all along the food chain (currently estimated to about 30% of the total in Europe, Gustavsson et al., 2011). Surprisingly, these major changes in the city's food demand were not accompanied by a considerable extension of the mean food supply distance, but rather by deep reorganisation, specialisation and opening of agriculture in the hinterland. On the other hand, the Seine watershed still represents the only source of drinking water for Paris, with two-thirds coming from surface water and one-third from groundwater resources, both threatened by increasing nitrate contamination (Billen et al., 2007; Ledoux et al., 2007). Previous studies have also shown that only radical changes in agricultural practices would be able to durably reduce nitrate pollution of ground and surface water in the Seine watershed (Thieu et al., 2010a, b; Lancelot et al., 2011).

This paper pursues the analysis of the environmental imprint of Paris's food consumption on the biogeochemical functioning of its rural hinterland and attempts to bring out the relations between urban demand for foodstuffs, nitrogen cycling in agricultural systems, and nitrate contamination of water resources. We then explore the potentialities of radically changing both the agricultural practices and the urban diet patterns for reducing the environmental imprint of urban food consumption.

3.2 APPROACH AND METHODS

The approach applied herein takes inspiration from the concepts and methods of territorial ecology as defined by Barles (2010). Starting from the urban demand, we first delimited the territory supplying food to the Paris agglomeration, then analysed the fluxes of material involved in its production and supply. Based on the results of this analysis, alternative scenarios able to fulfil the same functions while meeting additional constraints are imagined.

To define the historical and current food supply areas of the Paris agglomeration and to analyse the functioning of their agricultural system, a combination of agricultural and transport statistics were used. In particular for the present period, we made use of the national agricultural statistics, available at the département level (Agreste, Ministry of Agriculture, www.agreste.agriculture.gouv.fr/). The French départements are territorial and administrative divisions of about 7000 km^2, except for Paris which is a much smaller département in itself. For estimating the origin of agricultural goods consumed in Paris, the SitraM database on commodity transport between French départements (French Ministry of Environment, www.statistiques.developpement-durable.gouv.fr/) and the FAO statistics on production and international trade of food products (http://faostat.fao.org/) were compiled (Billen et al., 2011). All crop and animal product quantities, originally provided in tons, were converted to their nitrogen weight equivalent using the N content figures (1.8%N for cereals, 3.4%N for meat, 0.5% for milk and 0.1–0.4% for most fruits and vegetables) presented in Billen et al. (2009). The relationship between crop production (in terms of N content) and total N fertilisation of arable land (manure, synthetic fertilisers, symbiotic N fixation and atmospheric deposition) was established at the d épartement scale. Data on the use of synthetic fertilisers by département were provided by Unifa (2008). Symbiotic N fixation was estimated from the yield of each legume crop, considering that underground and unharvested parts together represent 40% of the yield. The yield-fertilisation relationship was then used to calculate agricultural production in the different scenarios explored (see below) and to assess the environmental losses of nitrogen by the agro-systems, considering that all N used in food production that is not in the crop product is lost to the environment.

3.3 THE CURRENT NITROGEN IMPRINT OF PARIS FOOD CONSUMPTION

Until the middle of the 20th century, Paris was still mainly supplied by an area roughly corresponding to a 150-km circle around the city, including most of the Seine watershed. Although a certain degree of agricultural specialisation was already present within this area, the agricultural system was fundamentally based on a close relation between cereal cultivation and livestock farming, the latter producing the organic fertilisers required by the former. The increase of grain yield during the 19th century mostly resulted from an improvement in animal feeding and manure management, owing to the replacement of the triennial fallow by a N_2-fixing legume fodder crop, which provided a considerable increase in livestock density (Billen et al., 2009, 2011).

The 20th century agrarian revolution, following the generalisation of the use of synthetic N fertilisers, led to pronounced spatial segregation between the regions of the central Paris basin—specialised in cereal cultivation at the expense of livestock farming—and the more peripheral regions, which specialised in cattle rearing and imported a large share of the feed from other regions, countries or continents. Today, Paris's main food-supplying area is composed of three areas with quite different biogeochemical functioning (Billen et al., 2011):

1. The directly surrounding area corresponding to the Seine watershed is now mainly devoted to intensive cereal farming and provides most of the vegetal proteins consumed in Paris. It exports 82% of its production.
2. The area made up of Brittany, Normandy and Nord- Pas-de-Calais produces the largest fraction of the animal proteins consumed in Paris, but depends on massive imports of feed from other regions.
3. Among the imports of vegetal proteins required for feeding the livestock of the latter territory, soybean and oilcake meals from

Brazil and Argentina are the most significant: the size of the agricultural surfaces involved is similar to that of the two other territories.

These three areas together contribute 70% of the needs of Paris consumption. The remaining 30% is supplied half by other French départements and half by foreign countries.

From these data, the environmental imprint of the Paris food supply can be approximately represented in terms of (i) the land area required and (ii) per capita nitrogen fluxes involved, exactly as is currently done for wastewater domestic effluents (Fig. 1). To satisfy the needs in vegetal products of one Paris inhabitant, only 0.05 ha of the territory of the Seine watershed is required, but 0.33 ha of territory in polycultureanimal farming areas such as those of Brittany, Normandy and Nord-Pas-de-Calais are required to satisfy the needs in meat and milk of the same person (although some surplus crop production in this area is available to supply other regions); in addition, an area of about 0.12 ha in South America is required to supplement the feed of the livestock in the latter region. This indicates that by far the largest territorial area required to feed Paris is for producing animal products. This is not surprising as for the whole of Europe, 83% of crop production is destined to feeding livestock (Sutton et al., 2011).

To estimate the losses occurring through leaching and volatilisation from agricultural soils in the two French supply areas, we calculated the difference between total fertilisation and crop production of agricultural land (both expressed in N content) at the département scale. The relationship between crop yield and total fertilisation follows an asymptotic curve (Fig. 2a), with the surplus (i.e. the difference between fertilisation and yield) increasing regularly with increasing fertilisation (Fig. 2b). The surplus is generally lower in the cereal crop regions of the centre of the Paris basin than in the intensive livestock farming areas of the West of France. If the surplus is assumed to be entirely diluted in the infiltrating water depth (as a mean $300 mmyr^{-1}$ for the areas described herein), the corresponding theoretical sub-root nitrate concentration in water produced by agland surfaces can be calculated. In most d épartements, the value is far above the drinking water standard of 11 $mgNl^{-1}$ (50 $mgNO_3 l^{-1}$). In these regions, agland occupies about half the total area.

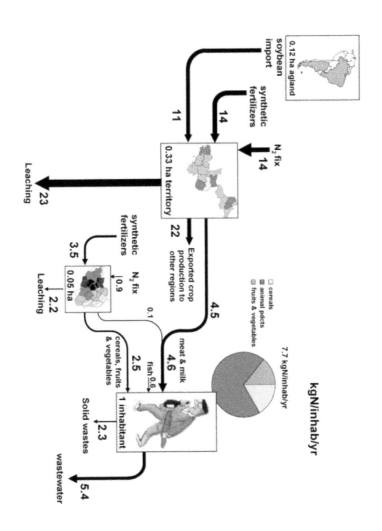

FIGURE 1: Schematic representation of the nitrogen imprint of the food supply of one Paris inhabitant. Nitrogen fluxes involved are expressed in kgNinhab⁻¹ yr⁻¹. The figures are derived from the description established in Billen et al. (2011).

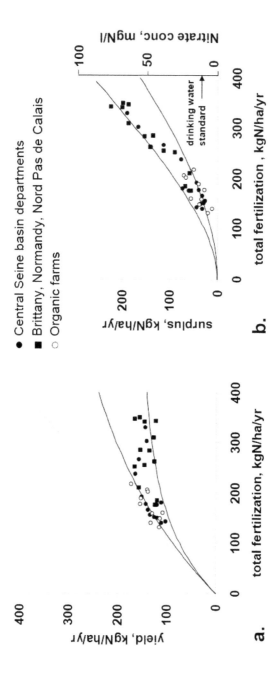

FIGURE 2: (a) Relationship between crop yield and total fertilisation (both expressed in kgNha⁻¹ yr⁻¹) of agricultural land of the different French d'epartements mainly involved in the Paris food supply. The following mathematical function has been fitted to give the upper and lower bounds of the data points: yield = Ymax · (1−exp (−Fert/Ymax)), with Ymax = 350–150 kgNha−1 yr⁻¹. (b) Agricultural surplus (total fertilisation minus crop yield) plotted against fertilisation. The corresponding theoretical sub-root water concentration, calculated assuming a mean runoff depth of 300mmyr⁻¹, is also indicated (right scale). The data available for a number of organic farms in the Seine watershed (unpublished results) are also shown for comparison (open symbols).

For the two French agricultural areas supplying food to Paris, the estimated environmental losses of nitrogen are estimated at 25 kgNcapita^{-1} yr^{-1}, i.e. five times the load discharged as urban wastewater by one inhabitant. The largest part (92 %) of these losses occurs in the livestock farming supply territory. Admittedly, because some surplus vegetal production of this territory is exported to other regions, a fraction of the losses occurring there should be ascribed to the population of these regions.

3.4 A LOCAL AND ORGANIC FARMING SCENARIO FOR THE SEINE WATERSHED

Thieu et al. (2010b) have shown that organic agriculture, if generalised to the entire agriculture area of the Seine watershed, has the potential to restore nitrogen contamination to a level below the ecological water resources standards. In order to explore the potential of reducing the nitrogen imprint of the Paris food supply, we constructed a hypothetical scenario based on the following constraints:

The organic and local scenario (OrgLoc) first requires that most of the food supply be produced within the limits of the Seine watershed, Paris's traditional food-producing hinterland. This necessarily implies restoring livestock farming within the area. The scenario assumes, however, that this livestock be reared with only local feed: no import of proteins from outside the Seine watershed limits would be allowed. To calculate the feed consumed by this livestock, we used a nitrogen conversion ratio of 17 %, the current average value calculated for the Brittany-Normandy-Nord-Pas-de-Calais regions.

The scenario also assumes organic agricultural practices, implying no use of synthetic fertilisers, all the fertilisation of arable land being ensured through symbiotic nitrogen fixation and manure application. Our calculations assume that the currently observed relationship between yield and total fertilisation shown in Fig. 2 also holds for organic practises. This assumption implies that at a similar fertilisation rate, whether the fertilisation is organic or mineral, the yields are identical. In other words, the often observed lower yields of organic vs. conventional farming are the results of lower fertilisation rather any other difference. The upper relationship illustrated in Fig. 2

Yield = Ymax (1−exp(−fertilization/Ymax))

(with Ymax = 350 $kgNha^{-1} yr^{-1}$) is used in the calculation, once the total fertilisation (manure, legume N_2 fixation and atmospheric deposition) is known, to calculate both the yield of arable land and the nitrogen surplus defined as the difference between fertilisation and yield. In the construction of the OrgLoc scenario, the first step is to assess the size of the livestock required to sustain the needs of the local population. The current area of permanent grassland in the Seine watershed is maintained unchanged. The size of the nitrogen-fixing arable land (temporary grassland involved in crop succession and legume fodder crops) is adjusted so as to meet the feed requirements of livestock with (i) the production of the permanent grassland, (ii) 70% of the production of nitrogen-fixing crops (30% is considered to be used as green manure, thus not available for livestock feeding) and (iii) an adjustable fraction of the production of non-legume crops. The total fertilisation of arable land can thus be calculated as well as the yield and the nitrogen surplus of arable land. The theoretical sub-root nitrate concentration below arable land is evaluated by dividing the surplus by the infiltrated water depth. Finally, the calculated production of non-fixing cropland which has not been allocated to human or animal feeding is considered available for export. Table 1 and Fig. 3 summarise the calculations and compare the results with the current situation.

These calculations show that it is quite possible to conceive a scenario of organic farming locally meeting the quantitative food requirements of the current population of the Paris agglomeration and of the other cities of the Seine watershed, totalling 16.9 million inhabitants, provided that the livestock density is increased from the current value of 18 LU km^{-2} to 50 LU km^{-2}. The basin should still be able to export about 1630 $kgNkm^{-2} yr^{-1}$. This should be compared to the current cereal export of 4820 $kgNkm^{-2}$ yr^{-1}, an amount reduced to a net protein export of 800 $kgNkm^{-2} yr^{-1}$ if the import of feed and animal products, as well as the vegetal proteins required for the production of the latter, are deducted (Fig. 3a). In this OrgLoc scenario, however, the nitrogen surplus, although lower than in the current situation, would still represent a sub-root water nitrate concentration double the drinking water standard.

TABLE 1: Nitrogen budget of the agricultural system of the Seine watershed for the current situation (2006) and the organic and local (OrgLoc) and organic, local and demitarian (OrgLocDem) scenarios. The 2006 situation is primarily based on agricultural statistics (Agreste, 2007), with a few items calculated as explained in the text and summarised in the "remarks" column. The OrgLoc and OrgLocDem scenarios are calculated starting from the current situation and reallocating land use and livestock numbers in order to meet the following constraints: no use of synthetic fertilisers (organic fertilisation only), no import of feed (self-sufficiency in animal nutrition), no import of animal products (self-sufficiency in supplying animal products for local human consumption). Cereal production is calculated from the yield/fertilisation relationship observed for the Seine basin départements in the current situation. Cereal production in excess of livestock and human requirements is exported. Nitrate concentration in infiltrating sub-root water is estimated from the difference between fertilisation and yield in arable land, taking into account an infiltration rate of 300 mmyr⁻¹.

2006				Org-Loc	OrgLocDem
Main characteristics of the territory					
total area	km²	92,381	Agreste (2007) (all dép.mts)	92,381	92,381 unchanged
agricultural area	1000 ha	5119	Agreste (2007)	5,119	5,119 unchanged
population density	hab km⁻²	183	INSEE (2007)	183	183 unchanged
per capita protein consumption	kgNcap⁻¹ yr⁻¹	7	FAOstat (2012)	7	7 unchanged
% animal protein in diet	%	65	FAOstat (2012)	65	40 unchanged for OrgLoc, reduced to 40% for OrgLocDem
Livestock farming					
livestock density	LU km⁻²	18	Agreste (2007)	49	30 adjusted to meet population requirements
meat and milk production	ktonNyr⁻¹	15	Agreste (2007)	77	48 assuming yield of 17% (mean value for France)
manure production	ktonNyr⁻¹	138	85 kgNLU⁻¹ yr⁻¹	387	238 85 kgNLU⁻¹ yr⁻¹
feed consumption	ktonNyr⁻¹	153	production + manure	464	286 production + manure

TABLE 1: *Cont.*

2006 Agriculture				Org-Loc	OrgLocDem	
permanent meadows						
area	1000 ha	800	Agreste (2007)	800	800	same as current situation
yield	$kgNha^{-1}\,yr^{-1}$	118	Agreste (2007)	118	118	same as current situation
production	$ktonNyr^{-1}$	95	Agreste (2007)	95	95	area×yield
use as feed	%	100		100	100	
N_2 fixing crops + temporary meadows						
area	1000 ha	334	Agreste (2007)	2000	1150	adjusted to meet livestock requirements
yield	$kgNha^{-1}\,yr^{-1}$	171	production/area	150	150	conservative estimate
production	$ktonNyr^{-1}$	57	Agreste (2007)	300	173	area×yield
N_2 fixation	$ktonNyr^{-1}$	80	1.4×production (aerial + root contrib.)	420	242	1.4×production (aerial + root contribution)
use as feed	%	25	Agreste (2007)	70	70	conservative estimate, remaining part used as green manure
non fixing crops						
area	1000 ha	3985	Agreste (2007)	2319	3169	agricultural area − (N_2 fixing crops + temporary and permanent meadows)
average yield	$kgNha^{-1}\,yr^{-1}$	131	production/area	144	100	calculated from total fertilization, using relationship of Fig. 2
production	$ktonNyr^{-1}$	523	Agreste (2007)	334	317	yield×area

TABLE 1: *Cont.*

2006		2006	Org-Loc	OrgLocDem	
use as feed	%	7	45	20	Agreste (2007) — adjusted to meet livestock needs, remaining used for human needs
total arable land arable area	1000 ha	4319	4319	4319	N_2 fixing and non fixing crops
total N_2 fixation	$kgNha^{-1}yr^{-1}$	19	97	56	=N_2 fixation/arable area
synthetic fertilizers	$kgNha^{-1}yr^{-1}$	161	0	0	Unifa (2009) — no synthetic fertilizers
manure	$kgNha^{-1}yr^{-1}$	32	87	54	manure production/arable area
atmospheric deposition	$kgNha^{-1}yr^{-1}$	12	12	12	Ntot deposition, EMEP
total fertilization	$kgNha^{-1}yr^{-1}$	224	196	122	sum
total crop exportation	$kgNha^{-1}yr^{-1}$	124	126	101	N_2 fixing used as feed and non fixing crops
System performance					
coverage of livestock needs					
local feed production	$ktonNyr^{-1}$	145	455	279	sum
local coverage of feed requirements	%	95	100	100	N_2 fixing used as feed and non fixing crops
import of feed	$ktonNyr^{-1}$	8	0	0	no import of feed

TABLE 1: *Cont.*

2006				Org-Loc	OrgLocDem	
fcoverage population needs						
human needs for animal products	$ktonNyr^{-1}$	=population×animal pdcts consumptn	77	77	47	=population×animal pdcts consumption
local coverage in animal pdcts	%	=local prod of animal pdcts/human needs	20	100	100	=local prod of animal pdcts/human needs
human needs for vegetal products	$ktonNyr^{-1}$	=population×vegetal pdcts consumptn	41	41	71	=population×vegetal pdcts consumption
local coverage in vegetal pdcts	%	=local crop prod not used as feed/human needs	1176	446	357	=local crop prod not used as feed/human needs
import(+)/export(−) animal pdcts	$ktonNyr^{-1}$	=local animal pdcts prod−local consumption	62	0	0	=local animal pdcts prod−local consumption
import(+)/export(−) vegetal pdcts	$ktonNyr^{-1}$	=local vegetal food prod−local consumption	−445	−143	−182	=local vegetal food prod−local consumption
hydrosphere contamination						
N surplus on arable land	$kgNha^{-1} yr^{-1}$	=total fertilization−total crop prodctn	99	70	20	=total fertilization−total cop production
infiltration rate	$mmyr^{-1}$	average for Seine watershed	300	300	300	average for Seine watershed
sub-root NO_3 conc	$mgNl^{-1}$	=Nsurplus/infiltration rate	33	23	7	=Nsurplus/infiltration rate

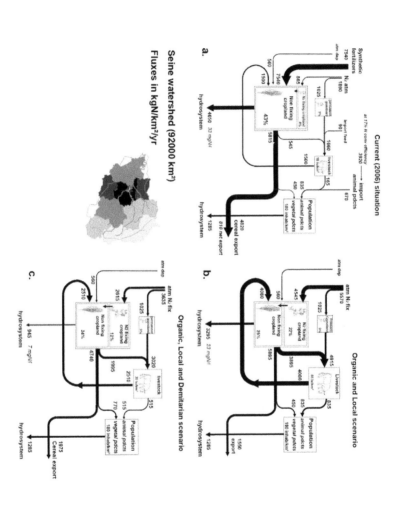

FIGURE 3: Nitrogen fluxes in the agricultural system of the Seine watershed (fluxes expressed in kgN per km2 of watershed area and per year). (a) Current situation (2006). (b) Organic and local scenario (OrgLoc). (c) Organic, local and demitarian scenario (OrgLocDem).

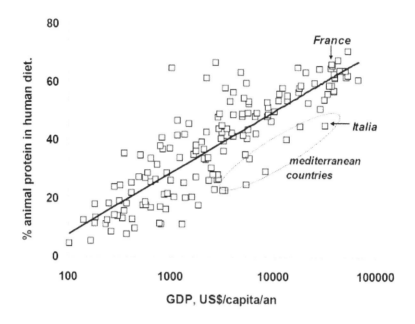

FIGURE 4: Share of animal products in human total protein consumption of the world's countries in 2003, plotted against their GDP (sources: FAOstat, 2012).

3.5 CHANGING THE HUMAN DIET

The present human consumption of animal products in France accounts for 65% of the total protein intake. This rate is among the highest in the world (Fig. 4). The trend shown in Fig. 4 suggests that the share of animal product consumption in the human diet increases with the gross domestic product, as if meat and milk consumption were a sign of economic wealth. Yet the adverse public health effects of a high animal-protein diet are well documented both in terms of cardiovascular diseases (Lloyd-Williams et al., 2008) and colorectal cancer risks (Norat et al., 2005). A reduction of animal protein consumption in industrialised countries is also advocated for environmental reasons by NinE (2009) (The Barsac Declaration for a "demitarian" diet of 35–40% animal proteins), and as a condition for world food security and equity in the Agrimonde scenario (Paillard et al., 2010).

We therefore decided to vary the share of animal products in the human diet and calculate the resulting effect on the organic-localised scenario described above. Figure 5 shows the effect of reducing the proportion of animal protein in the total human diet of the Seine basin inhabitants from the present value of 65% to a "demitarian" diet of 35–40% (NinE, 2009). Obviously, the area's self-sufficiency in terms of meat and milk production can be achieved in this scenario with lower livestock density, decreasing the total fertilisation rate of arable land and the nitrogen surplus, without decreasing the area's cereal export capacity. With a demitarian diet, the Seine watershed should be able to feed its population, still export about 1950 $kgNkm^{-2} yr^{-1}$ as cereal, and produce sub-root water meeting the best quality standards (Table 1 and Fig. 3).

The per capita imprint of the individual Parisian according to this organic-localised-demitarian (OrgLocDem) scenario is represented in Fig. 6. Compared with the current imprint as represented in Fig. 1, the area required to feed the individual Parisian in this scenario is similar to that required today, but can be mainly restricted to the surrounding traditional hinterland of Paris. The total reactive nitrogen injected into the process of food production is reduced to 20 $kgNcapita^{-1} yr^{-1}$ compared to the 30 $kgNcapita^{-1} yr^{-1}$ required in the current situation. The losses of reactive nitrogen by soil leaching are much lower, accounting for 40% of the losses at the final consumption stage, while these losses are 300% of this value in the current situation.

3.6 DISCUSSION AND CONCLUSION

The impassioned controversy about localisation of food supply has been very well analysed by Cowell and Parkinson (2003). The defenders of the so-called Food-Miles movement put forward the benefits to be expected from localising agriculture in terms of both food security and environmental impact. They claim that regional self-sufficiency of food production and consumption is more likely to increase food security than a globalised food system. They also stress the negative environmental impacts of transporting foodstuffs over long distances. Finally they wish to reconnect people with food, neighbouring producers and seasonality, as well as to reduce the potential of environmental degradation and human exploitation by avoiding "out of sight, out of mind" effects produced by long-distance trade.

The opponents criticise the Food-Miles logic for its negation of productivity differentials between geographical locations, claiming that feeding a rapidly growing world population in a sustainable manner requires long-distance trade to ensure that food is produced most efficiently in the most suitable locations (Desrochers and Shimizu, 2008). Ballingal and Winchester (2008) go further by stating that local preference in food choices in Europe would lead to "starving the poor" by depriving Southern countries of important commercial income.

One particular aspect of the debate, the question of greenhouse gas (GHG) emissions related to the transportation of foodstuffs, is probably the simplest to assess. All lines of evidence show that the benefits of localising food production are minor in that respect. For instance, Heyes and Smith (2008) showed that daily familial shopping trips between home and supermarket use as much energy as the overseas transport of the same amount of food, while Weber and Matthews (2008) demonstrated that transport is responsible for only a minor share of GHG emissions of the whole food production chain in the US and concluded that changes in agricultural practices or in the consumer diet composition have a much greater impact than localising production.

However, the debate has never included the question of drinking water supply even though drinking water production often competes with food production in the same land areas, particularly in the densely populated regions of Europe. This paper addresses the question of the compatibility of agriculture and water production in rural areas. For the Seine watershed, both the traditional feeding hinterland of Paris and its sole drinking water supply, we show that it is possible to conceive an agricultural system able to reconcile the dual function of rural areas, namely feeding the city and producing high-quality water.

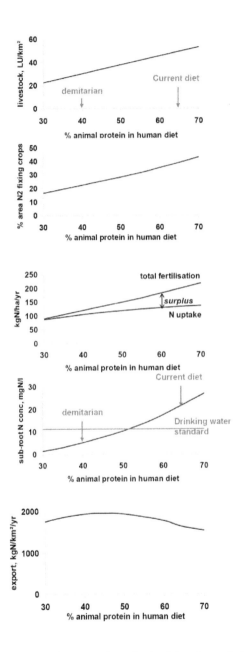

FIGURE 5: Effect of reducing the proportion of animal-based protein in the human diet on the N fluxes through the agricultural system in an organic, local scenario of the Seine watershed.

The Paris basin area is emblematic because it is currently one of the most intensive cereals-producing areas in the world; some people even consider as its vocation to export grain to the rest of the world. This view should be tempered by the fact that most of these exports are currently intended for other European countries where a large part is used for animal feed. Nevertheless, what is striking in the OrgLoc and OrgLocDem scenarios is that they allow the Seine basin area to continue exporting a large share of its cereals production and in fact have a larger net export of proteins than in the current situation.

This purely biogeochemical approach is obviously restrictive and does not take into account the reluctance of either farmers, invested in their current technical, economic and social network, to shift their practices to organic farming, or consumers to change their eating habits so deeply em-

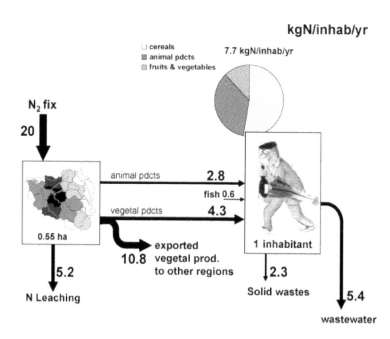

FIGURE 6: Schematic representation of the nitrogen imprint of the food supply of one Paris inhabitant in the organic-localdemitarian scenario. The nitrogen fluxes involved are expressed in kgNinhab^{-1} yr^{-1}.

bedded in both culture and commercial advertising influences. We simply delineate here the physical possibilities of a particular area.

In this respect also we are aware that the conclusions drawn herein can be quite easily caricatured. We know for instance that a number of foodstuffs consumed in Paris are imported from far away because they cannot be grown in the North of France. Our analysis of the nitrogen flows involved in the import of such products shows that these are minor (Billen et al., 2011), while they can account for a significant share of the monetary fluxes involved in feeding the city. This has always been the case, as Abad (2001) showed that already at the end of the 18th century, import to Paris of exotic and luxury agricultural goods (such as citrus fruits, tea, coffee, spices, etc.) already accounted for a significant value in the food "Great Market". Our approach is focused on the major biogeochemical fluxes associated with food production and consumption, because these are the most significant in terms of environmental impact, including water quality. There is no reason in this respect to advocate the complete cessation of exotic product imports. The localisation option explored in this paper concerns the general organisation of the agro-food system feeding Paris and the general balance of its inhabitants' diet, not the details of their occasional consumption of luxury products. In that sense, the suggestion that a shift to localising the food supply in Europe would "starve the poor" and "increase global inequality", as suggested by Ballingall and Winchester (2008), is irrelevant.

In a recent report (Westhoek et al., 2011), a scenario of 50% reduction of animal protein in the human diet in the EU has been explored at the global scale using the IMAGE model of the Netherlands Environmental Assessment Agency. The results are a strong decline in the import of protein-rich feed into Europe and an increase in cereal, meat and milk exports from Europe, so that the scenario would mainly have effects on land use and environmental quality outside the EU. Our purely biogeochemical approach differs from this one not only because we are not taking into account any economic mechanisms, but also because we first constrained the local agricultural system to self-sufficiency for feed and for meat and milk products. This additional constraint explains the strong local response of the system to decreasing the animal protein consumption observed in our scenario, in particular in terms of water quality. It therefore appears that localising as far as possible the major fluxes of food and feed supply at the regional scale is a pre-requisite to get full environmental benefits from a change to lower animal products in the human diet.

REFERENCES

1. Abad, R.: Le Grand March é: l'approvisionnement de Paris sous l'Ancien R égime, Fayard, Paris, 2002 (in French).
2. Agreste: Ministry of Agriculture, available at: www.agreste.agriculture.gouv.fr/, 2006.
3. Ballingall, J. and Winchester, N.: Food miles: Starving the poor?, University of Otago Economics Discussion Papers No. 0812, available at: www.business.otago.ac.nz/econ/research/discussionpapers/DP 0812.pdf, 2008.
4. Barles, S.: Ecologie territoriale, in: Dictionnaire de l'urbanisme et de l'am énagement, edited by: Merlin, P. and Choay, C., 3rd edn., PUF, Paris. 843 pp., 2010.
5. Billen, G., Garnier, J., Nemery, J., Sebilo, M., Sferratore, A., Benoit, P., Barles, S., Benoit, M.: A long term view of nutrient transfers through the Seine river continuum, Sci. Total Environ., 275, 80–97, 2007.
6. Billen, G., Barles, S., Garnier, J., Rouillard, J., and Benoit, P.: The Food-Print of Paris: Long term Reconstruction of the Nitrogen Flows imported to the City from its Rural Hinterland, Reg. Environ. Change, 9, 13–24, 2009.
7. Billen, G., Barles, S., Chatzimpiros, P., and Garnier, J.: Grain, meat and vegetables to feed Paris: where did and do they come from? Localising Paris food supply areas from the eighteenth to the twenty-first century, Reg. Environ. Change, online first: doi:10.1007/s10113-011-0244-7, 2011.
8. Charruadas, P.: The cradle of the city: the environmental imprint of Brussels and its hinterland in the High Middle Ages, Reg. Environ. Change, online first: doi:10.1007/s10113-011-0212-2, 2011.
9. Cowell, S. J. and Parkinson, S.: Localisation of UK food production: an analysis using land area and energy as indicators, Agr. Ecosyst. Environ., 94, 221–236, 2003.
10. Desrochers, P. and Shimizu, H.: Yes We Have No Bananas: A Critique of the Food Mile Perspective, Mercatus Policy Series, Policy Primer No. 8, available at: http://www.mercatus.org/uploadedFiles/Mercatus/Publications/Yes%20We%20Have%20No%20Bananas %20A%20Critique%20of%20the%20Food%20Mile%20Perspective.pdf, last access: January 2012, 2008.
11. FAOstat: Statistics on production and international trade of food products, available at: http://faostat.fao.org/, 2012.
12. Galloway, J. N. and Cowling, E. B.: Reactive nitrogen and the world: 200 years of change, Ambio, 31, 64–71, 2002.
13. Gustavsson, J., Cederberg, C., Sonesson, U., van Otterdijk, R., and Meybeck, A.: Global food losses and food wastes. Extent, causes and prevention, FAO, Rome, 2011.

14. Heid, P.: Organic agriculture protects drinking water around Munich, Germany, Ecology and Farming, 14, 24–34, 1997.

15. Heyes, J. A. and Smith, A.: Could Food Miles become a Non-Tariff Barrier?, SHS, Acta Hortic., 768, 431–36, 2008.

16. Keene, D.: Medieval London and its supply hinterlands, Reg. Environ. Change, online first: doi:10.1007/s10113-011-0243-8, 2011.

17. Lancelot, C., Thieu, V., Polard, A., Garnier, J., Billen, G., Hecq, W., and Gypens, N.: Ecological and economic effectiveness of nutrient reduction policies on coastal Phaeocystis colony blooms in the Southern North Sea: an integrated modeling approach, Sci. Total Environ., 409, 2179–2191, 2011.

18. Ledoux, E., Gomez, E., Monget, J. M., Viavatenne, C., Viennot, P., Ducharne, A., Benoit, M., Mignolet, C., Schott, C., and Mary, B.: Agriculture and groundwater contamination in the Seine basin. The STICS-MODCOU modelling chain, Sci. Total Environ., 409, 33–47, 2007.

19. Lloyd-Williams, F., Mwatsama, M., and Birt, C.: Estimating the cardiovascular mortality burden attributable to the Common Agricultural Policy on dietary saturated fats, B. World Health Organ., 86, 535–545, 2008.

20. NinE: The Barsac Declaration: Environmental Sustainability and the Demitarian Diet, available at: www.nine-esf.org/, 2009.

21. Norat, T., Bingham, S., and Ferrari, P.: Meat, fish and colorectal cancer risks: the European Prospective Investigation into Cancer and Nutrition, J. Natl. Cancer I., 97, 906–916, 2005.

22. Paillard, S., Treyer, S., and Dorin, B.: Agrimonde: Scénarios et d éfis pour nourrir le monde en 2050, Quae, Paris, 2010 (in French).

23. Paxton, A.: The Food Miles Report: the dangers of long distance food transport, Safe Alliance, London, 1994.

24. SitraM: Data base on commodity transport between French départements (French Ministry of Environment), www.statistiques.developpement-durable.gouv.fr/, 2006.

25. Smith, A., Watkiss, P., Tweddle, G., McKinnon, A., Browne, M., Hunt, A., Treleven, C., Nash, C., and Crossal, S.: The Validity of Food Miles as an Indicator of Sustainable Development, Final Report produced for DEFRA, ED50254, 103 pp., 2005.

26. Steele, C.: Hungry City: How food shapes our lives, Vintage books, London, 2008.

27. Stergiouli, M. L. and Hadjibiros, K.: The growing water imprint of Athens, the capital of Greece, The increasing flux of water resources from its hinterland throughout history, Reg. Environ. Change, online first: doi:10.1007/s10113-011-0260-7, 2012.

28. Sutton, M. A., Howarth, C. M., Erisman, J. W., Billen, G., Bleeker, A., Grennfelt, P., van Grinsven, H., and Grizzetti, B.: The European Nitrogen Assessment: Sources, Effect and Policy perspectives, Cambridge University Press, 612 pp., 2011.

29. Svirejeva-Hopkins, A. and Reis, S.: Nitrogen flows and fate in urban landscapes, in: The European Nitrogen Assessment: Sources, Effect and Policy perspectives, edited by: Sutton, M. A., Howard, C. M., Erisman, J. W., Billen, G., Bleecker, A., Grennfelt, P., van Grinsven, H., and Grizzetti, B., Chapter 12, 249–270, Cambridge University Press, 2011.

30. Swaney, D., Santoro, R. L., Howarth, R. W., Hong, B., and Donaghy, K. P.: Historical changes in the food and water supply systems of the New York metropolitan area, Reg. Environ. Change, online first: doi:10.1007/s10113-011-0266-1, 2012.

31. Tello, E. and Ostos, J. R.: Water consumption in Barcelona and its regional environmental imprint: a long-term history (1717– 2008), Reg. Environ. Change, online first: doi:10.1007/s10113- 011-0223-z, 2011.

32. Thieu, V., Billen, G., and Garnier, J.: Nutrient transfer in three contrasting NW European watersheds: The Seine, Somme, and Scheldt Rivers. A comparative application of the Seneque/Riverstrahler model, Water Res., 43, 1740–1748, 2009.

33. Thieu, V., Garnier, J., and Billen, G.: Assessing the effect of nutrient mitigation measures in the watersheds of the Southern Bight of the North Sea, Sci. Total Environ., 408, 1245–1255, 2010.

34. Thieu, V., Billen, G., Garnier, J., and Benoˆıt, M.: Nitrogen cycling in a hypothetical scenario of generalised organic agriculture in the Seine, Somme and Scheldt watersheds, Reg. Environ. Change, 11, 359–370, 2011.

35. Unifa: Livraisons d'engrais min éraux en France métropolitaine par département et par région. Union des Industries de la fertilisation, Paris, 2008 (in French).

36. Watts, D. C. H., Ilbery, B., and Maye, D.: Making reconnections in agro-food geography: Alternative systems of food provision, Prog. Hum. Geog., 29, 22–40, 2005.

37. Weber, C. and Matthews, H.: Food-Miles and the Relative Climate Impacts of Food Choices in the United States, Environ. Sci. Technol., 42, 3508–3513, 2008.

38. Westhoek, H., Rood, T., van de Berg, M., Janse, J., Nijdam, D., Reudink, M., and Stehfest, E.: The Protein Puzzle: the consumption and production of meat, dairy and fish in the European Union, The Hague: PBL Netherlands Environmental Assessment Agency, 2011.

CHAPTER 4

COMPARATIVE GROWTH ANALYSIS OF *CALLISTEPHUS CHINENSIS* L. USING VERMICOMPOST AND CHEMICAL FERTILIZER

DULAL CHANDRA DAS

4.1 INTRODUCTION

Modern day agriculture throughout the globe is fully dependent up on the chemical fertilizers and pesticides. Immense and uninterrupted uses of chemical fertilizers and pesticides have led to diverse pernicious effects on soil and environment. Inorganic agriculture created the planetary environmental menace like loss of biodiversity, desertification, climate and transboundary pollution. Increased use of inorganic fertilizers in crop production deteriorated soil health, caused health hazard and created imbalance to environment by polluting air, water, soil etc. The continuous use of chemical fertilizers badly affected the texture and structure, reduced organic matter content and decreased microbial activities of soil. The sol degradation threatens the sustainability of cropping system (1-8). At present to control this alarming situation and to overcome the dangerous affects of modernized agriculture a new farming system "Organic farming" has

This chapter was originally published under the Creative Commons Attribution License. Das DC. Comparative Growth Analysis of Callistephus chinensis L. Using Vermicompost and Chemical Fertilizer. International Journal of Bioassays 2,2 (2013); 398-402.

been developed. This system is ecofriendly, ecologically sustainable and economically viable and is based on the concept of production, cultivation and use of biofertilizers, organic manure, green manure and biopesticides.

Organic matter has a property of binding mineral particles like calcium, magnesium and potassium in the form of colloids of humus and clay, facilitating stable aggregates of soil particles for desire porosity to sustain plant growth. Soil microbial biomass and enzyme activity are important indicators of soil improvement as a result of addition of organic matter (9).

Use of manures is generally seen as a key practice for maintaining soil fertility and agricultural sustainability in the wheat maize rotation and rice based cropping system (2, 4, 10-14).

Combined application of inorganic fertilizers with manure may not result in an increase of soil organic carbon because organic matter accumulation depends on the net incorporation of organic matter in the soil, which in turn depends on the cropping and management system employed (15-17).

In recent years vermicompost an organic amendment has been selectively and effectively used in soil conditioning and in varying degrees to influence the soil properties. It promote humification, increased microbial activity and enzyme production, which in turn, increases the aggregate stability of soil particles, resulting in better aeration (9, 18-20).

Vermicompost is prepared using earthworms. Earthworms are "The intestine of earth" and are considered as agents to restore the soil fertility. Earth worm is physically an aerator, crusher and mixer, chemically a degrader and biologically a stimulator for decomposition of various organic wastes.

Vermicompost is usually a finely divided peat-like material with excellent structure, porosity, aeration, damage and moisture holding capacity (21). It usually contains higher levels of most of the mineral elements, which are in available forms than the parent material (22,23). Vermicompost improves the physical, chemical and biological properties of soil (24). There is a good evidence that vermicompost promotes growth of plants (25-27) and it has been found it has a favorable influence on all yield parameters of crops like wheat, paddy and sugarcane (28,29). The objective of the present study was to investigate the comparative growth performance showed by vermicompost and chemical fertilizer and to observe the position of vermicompost in respect of functional status of both the chemical fertilizer and vermicompost.

4.2 MATERIALS AND METHODS

4.2.1 PLANT MATERIAL

The species *Callistephus chinensis* L. (Asteraceae) was selected as experimental tool. The purified seeds of the species were collected from the National Seed Corporation in Paschim Medinipur District of the State West Bengal. The seeds were sown and grown in the Experimental Garden. The plant is herbaceous, annual, seasonal flowering winter crop and has high commercial potentiality in India and abroad.

4.2.2 LOCATIONS AND EXPERIMENTAL DESIGN

Field experiments were established at the experimental Garden of Raja Narendralal Khan Womens College, Midnapore, West Bengal, India (21°47¢ North latitude and 87°42¢ East longitude) during Jan to Mar of five consecutive years 2007-2010 (30). The present investigation was carried out under the agro ecological sub region of Eastern Plateau, hot dry and sub humid ecosystem with red and lateritic soils.

Five experimental sets like T1, T2, T3, T4 and T5 with three replicas of each set were done. The container of each experimental set was polythene bag of 250 mm in diameter and 250 mm in height. Each bag was filled up with 4 kg. of soil. The T1 set was as control filled up with soil only. The T2 set was filled up with 4 kg soil and 1.5g chemical fertilizer, the T3 set with 4 kg soil and 7.5 g vermicompost, the T4 set with 4 kg soil and 3g chemical fertilizer and T5 set with 4 kg soil and 15g vermicompost were prepared. The chemical fertilizer was DAP and its NPK value was 10:26:26. The vermicompost was procured from the Agriculture Department of the Indian Institute of Technology, Kharagpur, India and it had NPK value of 2:1:1.

4.2.3 CULTIVATION PRACTICES

The replicas of each experimental sets were placed adjacently into five groups. The sets were amended as per schedule from the very beginning except the control. Initially the seedlings of *Callistephus chinensis* were about 50 to 55 mm height and planted in each set. The amendments were applied for 6 times with 12 days of interval starting from the planting of the species. The irrigation was done as and when required.

4.2.4 DATA COLLECTION

The soil samples were collected just after planting and after harvesting of the crop. The samples were subjected to physic-chemical analysis e.g. organic carbon (OC), pH, and electrical conductivity (EC). The OC content in the soil (31) and the pH were measured (32). In course of time at regular 12 days of interval the observations on plant height, number of leaves developed, size and longevity of the leaves, number of axillary bad developed, number of flowers appeared, diameter of each flower and the life span of the flowers were recorded for the comparative growth analysis.

4.2.5 STATISTICAL ANALYSIS:

The data were plotted in Table 1 to 9 and for comparative growth analysis the statistical package for social science, version 17.0 (SPSS software) was used.

4.3 RESULTS

In the present investigation ten parameters were used to study the comparative growth efficiency of vermicompost and chemical fertilizer. The experimental results have been presented in 9 Tables. The growth efficiency of the vermicompost and chemical fertilizer was compared and discussed below.

The chemical analyses of the soil were done within 15 to 20 days after planting the plant material into the soil and just after harvesting of the crop. The status of the OC, pH and EC were measured.

4.3.1 ORGANIC CARBON

The OC content in T3 and T5 sets increased nearly in double. Whereas is T2 and T4 increased in little bit (Table1). The T5 (vermicompost with full dose) was more efficient than T3 (vermicompost with half dose) whereas T4 (chemical fertilizer with full dose) and T2 (chemical fertilizer with half dose) remained equally active. The results reflected the organic manure was more efficient to incorporate the more percentage of OC into the soil than the chemical fertilizer (33,34).

4.3.2 PH

In Table-2, the pH values of different soil samples have been recorded. In all the cases pH values decreased but significant by decreased in T2 and T4 sets. The normal pH of the soil ranges from 6.5-7.5. In this respect organic manure vermicompost are better than the chemical fertilizer for the sustenance of better soil health status.

4.3.3 ELECTRICAL CONDUCTIVITY

The EC values in Table-3 significantly decreased in T5 and T3 than the T2 and T4 sets. The high value of the EC is deleterious for the sound health status of the soil and is never recommended for the normal growth and development of the plants. In this respect, vermicompost treated soil in T3 and T5 showed better performance for the sustenance of better health status of the soil.

4.3.4 HEIGHT

The results of growth analysis in height have shown in Table 4, the growth efficiency in terms of days in Table 4a and comparative growth analysis in height have documented. The T5 set with full dose and T3 set with half dose of vermicompost manifested significant growth in height than others till the end of 72 days (30). T4 set had stopped its growth after 60 days but expressed maximum growth efficiency during 24-36 days. T2 set expressed maximum growth efficiency during 36-48 days and declined thereafter. The T1 without any amendments represented very poor results. In respect of the growth efficiency the experimental sets have compared and arranged accordingly: T5>T3>T4>T2>T1.

4.3.5 DEVELOPMENT OF LEAVES

Maximum numbers of leaves (twenty eight) were developed in T5 set and have shown in Table 5, 5a. Cessation of leaf development after 48 days in T2 and after 36 days in T4 set were observed but the growth activity was continued in T3 and T5 sets even up to 72 days. In terms of leaves development the sets were compared and arranged as follows: T5>T3>T4>T2>T1.

4.3.6 LEAF SIZE AND LONGEVITY

Maximum growth in leaf area were observed in T5 considering the first, second, third and fourth leaves. In T5 at the end of 48 days first leaf increased in surface area of 1898 square mm, after 60 days second leaf covered 1704 square mm, and after 72 days third and fourth leaves covered 4656 and 3230 square mm respectively (Table 6). In T2 and T4 the increment in surface area of leaves usually stopped after 36 days of appearance. Regarding longevity the leaves of T3 and T5 survived much longer periods than the others. The intensity of the green pigments in the leaves usually express the vitality of the plants. The leaves of T3 and T5 were deep green rather than others. The comparative statements of this parameter ranking the sets as follows: T5>T3>T4>T2>T1.

4.3.7 AXILLARY BUDS

Maximum number (fourteen) of axillary buds were developed in T5 which was followed by T3 (twelve) during 72 days (Table 7). Initially during 24 DAP T5 and T3 had only single axillary bud and showed slow growth where as T4 and T2 showed rapid growth and had developed 2 and 4 buds respectively. The ranking order of the sets to be clear from the performance record as: T5>T3>T4>T2>T1.

4.3.8 DEVELOPMENT OF FLOWERS

Maximum numbers of flowers (seven) were developed in T5 which was followed by T3 (six) (Table 8). In all the sets flowers appeared first during 48 DAP but T2 and T4 was much more efficient initially than T3 and T5. The diameter of the flowers in T3 and T5 were much more (100 mm) than the others (30).

4.3.9 LONGEVITY OF THE FLOWERS:

In all the sets flowers appeared during 48 DAP but the flowers became almost dried within 60 DAP in T1 and T2 sets and already dried within 72 DAP in T4 set (Table 9). The flower buds and flowers were very soft, became flaccid and were not subjected to long lasting whereas the flowers of T3 and T5 sets were hardy, stiff and were freshly long lasted even up to 72 DAP (30).

4.4 DISCUSSION

The results of the present work reflected that the chemical fertilizer had taken away large amount of nutrients from the soil with high initial yields which had resulted in the gradual depletion of some nutrients (35). As a result the yields were significantly lowered during the later stages, where as the vermicompost had showed initially the low release of soil OC but gradually became increased in the later stages and persisted up to the longer periods and yield performance became more. The soil OC dynamics is affected by cropping practice and other environmental conditions (33, 34), the present investigation had also proved the regular use of manure resulted in larger increase in soil OC and had changed the environment of the soil. Without application of manure typically caused a decline of soil OC content (1, 36) and hampered the net yield in T2 and T4 sets. The rate of decomposition of manure was slow due to which the nutrients were released slowly and lately and the output was also expressed so lately. The input of nutrients in T5 was more than the T3, for that reason the output in T5 was more.

The full dose of chemical in T4 was more dangerous than the half dose in T2 set due to which the plant had stopped its growth during the period of 60-72 days. The chemical fertilizer decreased the exchangeable Ca and Mg as well as the pH in the soil which resulted in the decline of yield (37, 38) and had been proved in T2 and T4 sets.

The main conclusions from the present work are the application of vermicompost had positive effects of the soil OC, sustained the positive health status of the soil, had better yield performance and was much more efficient in comparison to the chemical fertilizer.

REFERENCES

1. Duxbury JM, Abrol IP, Gupta RK, Bronson KF, Analysis of longterm soil fertility experiments with rice-wheat rotations in South Asia, In: Long-term Soil Fertility Experiments with Rice- Wheat Rotations in South Asia, Rice-Wheat Consortium Paper Series No. 6. (Ed. IP Abrol, KF Bronson, JM Duxbury, RK Gupta), Rice-Wheat Consortium for the Indo-Gangetic Plains, New Delhi, 2000, pp. 7-22.
2. Yadav RL, Dwivedi BS, Pandey PS, Rice-Wheat cropping system: assessment of sustainability under green manuring and chemical fertilizer inputs, Field Crop Res, 2000, 65, 15-30.
3. Timsina J, Connor DJ, Productivity and management of ricewheat cropping systems: Issues and challenges, Field Crop Res, 2001, 69, 93-132.
4. Bhandari AL, Ladha JK, Pathak H, Padre AT, Dawe D, Gupta RK, Yield and soil nutrient changes in a long-term rice-wheat rotation in India, Soil Sci Soc Am J, 2002, 66, 162-70.
5. Ladha J K, Dawe D, Pathak H, Padre AT, Yadav RL, Singh B, Singh Y, Singh Y, Singh P, Kundu AL, Sakal R, Ram N, Regmi AP, Gami SK, Bhandari AL, Amin R,

Yadav CR, Bhattarai EM, Das S, Aggarwal HP, Gupta RK, Hobbs PR, How extensive are yield declines in long-term rice-wheat experiments in Asia? Field Crop Res, 2003, 81, 159-80.

6. De Costa WAJM, Sangakkara UK, Agronomic regeneration of soil fertility in tropical Asian small holder uplands for sustainable food production, J Agri Sci, Camb, 2006, 144, 111-33.

7. Ghosh PK, Manna MC, Dayal D, Wanjari RH, Carbon sequestration potential and sustainable yield index for groundnut and fallow-based cropping systems. J Agri Sci, Camb, 2006, 144, 249-59.

8. Taylor BR, Younie D, Matheson S, Coutts M, Mayer C, Watson CA, Walker RL, Output and sustainability of organic ley/arable crop rotations at two sites in northern Scotland, J Agri Sci, Camb, 2006, 144, 435-47.

9. Perucci P, Effect of the addition of municipal solid-waste compost on microbial biomass and enzyme activities in soil, Biol Fertil Soils, 1990, 10, 221.

10. Regmi AP, Ladha JK, Pathak H, Pasuquin E, Bueno C, Dawe D, Hobbs PR, Joshy D, Maskey SL, Pandey SP, Yield and soil fertility trends in a 20-year rice-rice-wheat experiment in Nepal, Soil Sci Soc Am J, 2002, 66, 857-67.

11. Sarkar S, Singh SR, Singh RP, The effect of organic and inorganic fertilizers on soil physical condition and the productivity of a rice-lentil cropping sequence in India, J Agri Sci, Camb, 2003, 140, 419-25.

12. Saleque MA, Abedin MJ, Bhuiyan NI, Zaman SK, Panaullah GM, Long-term effects of inorganic and organic fertilizer sources on yield and nutrient accumulation of lowland rice, Field Crop Res, 2004, 86, 53-65.

13. Yadvinder S, Bijay S, Ladha JK, Khind CS, Gupta RK, Meelu OP, Pasuquin E, Long-term effects of organic inputs on yield and soil fertility in the rice-wheat rotation, Soil Sci Soc Am J, 2004, 68, 845-53.

14. Jiang D, Hengdijk H, Dai TB, de Boer W, Jing Q, Cao WX, Longterm effects of manure and inorganic fertilizers on yield and soil fertility for a winter wheat-maize system in Jiangsu, China, Pedosphere, 2006, 16, 25-32.

15. Dick RP, A review: long-term effects of agricultural systems on soil biochemical and microbial parameters Agriculture Ecosystems & Environment, 1992, 40, 25-36.

16. Paustian K, Collins HP, Paul EA, Management controls on soil carbon. In: Soil Organic Matter in Temperate Agroecosystems. (Ed. EA Paul, K Paustian, ET Elliot, CV Cole) CRC Press. Boca Raton, FL, 1997, pp. 15-49.

17. Edmeades DC, The long-term effects of manures and fertilisers on soil productivity and quality: a review, Nutr Cycl Agroecosys, 2003, 66, 165-80.

18. Tisdale JM, Oades JM, Organic matter and water-stable aggregates in soil, J Soil Sci, 1982, 33, 141.

19. Dong A, Chester G, Simsiman V, Soil dispersibility, J Soil Sci, 1983, 136, 208.

20. Haynes RJ, Swift RS, Stability of soil aggregates in relation to organic constituents and soil water content, J Soil Sci, 1990, 41, 73.

21. Edwards CA, Production of earthworm protein for animal feed from potato waste. In: Upgrading waste for feed and food, (Ed. DA Ledward, AJ Taylor, RA Lawrie,) Butter worths, London, 1982.

22. Buchanan MA, Russell E, Block SD, Chemical characterization and nitrogen mineralization potentials of vermin composts derived from different organic wastes, In:

Earthworms in environmental and waste management. (Ed. CA Edwards, E Neu-hauser) The Netherlands SPB Acad Publ, Netherland, 1988, pp. 231-239.

23. Edwards CA, Bohlen PJ, Biology and Ecology of Earthworm, 3rd ed. Chapman and Hall, London, 1996.

24. Kale RD, Earthworm Cinderella of Organic Farming, Prism Book Pvt Ltd, Banga-lore, India, 1998.

25. Reddy MV, The effect of casts of Pheretima alexandri on the growth of Vinca rosea and Oryza sativa. In: Earthworms in environmental and waste management, (Ed. CA Edwards, EF Neuhauser) SPB Baker, The Netherlands, 1988, pp. 241-248.

26. Lalitha R, Fathima K, Ismail SA, Impact of biopesticides and microbial fertilizers on productivity and growth of Abelmoschus esculentus, Vasundhara- The Earth, 2000, 1-2, 4-9.

27. Rajkhowa DJ, Gogoi AK, Kandal R, Rajkhowa RM, Effect of vermicompost on Greengram nutrition, J Indian Soc Soil Sci, 2000, 48, 207-8.

28. Garg K, Bhardwaj N, Effect of vermin compost of Parthenium on two cultivars of wheat. Indian J Ecol, 27, 2000, 177-80.

29. Ansari AA, Urban Planning and Environment – Strategies and Challenges, Macmil-lan India Ltd., New Delhi, 2007.

30. Das D, Chakraborty K, Samanta A, Samanta SS, Jana K, Juxtaposition of Vegetative growth and reproductive parameters of Zinnia pauciflora L. using Vermicompost and Chemical fertilizer, Environment & Ecology, 2010, 28 (4A), 2478-81.

31. Walkley A, Black IA, An examination of the Degtjareff method of determining soil organic matter and a proposed modification of the chromic acid titration method, Soil Sci, 1934, 37, 29-38.

32. Lu RK, Analytical Methods of Soil Agricultural Chemistry, Agricultural Science and Technology Press, Beijing, China, 2000.

33. Olk DC, Van Kessel C, Bronson KF, Managing soil organic matter in rice and non rice soils: agronomic questions, In: Carbon and Nitrogen Dynamics in Flooded Soils. (Ed. GJD Kirk, DC Olk) International Rice Research Institute, Makati City, Philippines, 2000, pp. 27-47.

34. Powlson S, Olk DC, Long-term soil organic matter dynamics, In: Carbon and Nitro-gen Dynamics in flooded soils, (Ed. GJD Kirk, DC Olk) International Rice Research Institute, Makati City, Philippines. 2000, pp. 49-63.

35. Xu MG, Liang GQ, Zhang FD, Variation of Soil Fertility in China. Agricultural Sci-ence and Technology Press. Beijing, China, 2006.

36. Ram N, Effect of continuous fertilizer use on soil fertility and productivity of a Mol-lisol, In: Long-Term Soil Fertility Management through Integrated Plant Nutrient Supply. (A Swarup, DD Reddy , RN Prasad) Indian Institute of Soil Science, Bhopal, India, 1998, pp. 229-237.

37. Wang BR, Xu MG, Wen SL, Effect of long time fertilizers application on soil charac-teristics and crop growth in red soil upland, J Soil Water Conserv, 2005, 19, 97-100.

38. Zhang HM, Wang BR, Xu MG, Effects of inorganic fertilizer inputs on grain yields and soil properties in a long-term wheatcorn cropping system in south China. Comm Soil Sci Plant Anal, 2008, 39, 1583-99.

CHAPTER 5

COMPARISON OF THE FARMING SYSTEM AND CARBON SEQUESTRATION BETWEEN CONVENTIONAL AND ORGANIC RICE PRODUCTION IN WEST JAVA, INDONESIA

MASAKAZU KOMATSUZAKI AND M. FAIZ SYUAIB

5.1 INTRODUCTION

During the latter half of the 20th century, intensive agriculture increased crop yields and was successful in meeting the growing demand for food, but it also degraded the natural resources upon which agriculture depends—soil, water resources, and natural genetic diversity [1,2]. Today, conventional agriculture is built around two related goals: the maximization of production and the maximization of profit. In pursuit of these economic goals, a host of practices have been developed without regard for their unintended long-term consequences and without consideration of the ecological dynamics of agroecosystems. Millennium Ecosystem Assessment [3] revealed that the overuse and mismanagement of pesticides

This chapter was originally published under the Creative Commons Attribution License. Komatsuzaki M and Syuaib MF. Comparison of the Farming System and Carbon Sequestration between Conventional and Organic Rice Production in West Java, Indonesia. Sustainability 2 (2010). doi:10.3390/su2030833.

poisons the water and soil, while nitrogen and phosphorus inputs and live-stock wastes have become major pollutants of surface water, aquifers, and coastal wetlands and estuaries. These situations are also serious and cause severe ecological problems in the tropical biological environment, especially in Indonesia.

With a population of 230 million, Indonesia is the world's fourth most populous country, and its population is growing at a rate of 1.4% per year. Agriculture plays a substantial role in the Indonesian economy, involving about 45% of the population, and accounting for 19% of the gross domestic product and more than 60% of the value of non-oil exports. Over the last two decades, annual agricultural output has grown by 4% [4]. In Indonesia, modern farming technologies have kept production apace with population growth, but major problems with food distribution still plague many communities and regions.

Indonesia is also a richly bio-diverse tropical agro-ecosystem. About 95% of the land surface of Indonesia is still covered by vegetation, either as tropical rain forest, woodland, mangrove, agricultural fields or grass-land, which contain a variety of indigenous flora and fauna, many of which are unique species that do not occur in any other place in the world [5].

However, increasing synthetic chemical input to cropland to meet the increasing demand for food has led to decreasing biodiversity in agricultural areas. The soil management system, which is overly dependent on chemical fertilizers, has also led to decreasing soil organic matter and soil quality [6]. Currently, Indonesia has a total of 40,000 hectares under organic cultivation, about 0.12 percent of its total land area [7].

After the "Green Revolution" program was launched in the late 60s, the application of chemical fertilizer dramatically increased due to governmental encouragement to achieve food self-sufficiency. Since then, farmers have been using blended fertilizer (N, P_2O_5, and K_2O) with the recommended composition. Fertilizer consumption in the agricultural sector increased five-fold between 1975 and 1990 and increased slightly further afterwards. However, as a result of the Asian economic crisis in 1998, the government reduced the subsidies for fertilizers, resulting in increasing agricultural input costs. Since that time, farmers have been reducing the use of chemical fertilizers and have started to utilize more organic fertilizer and improve the methods for its application [5].

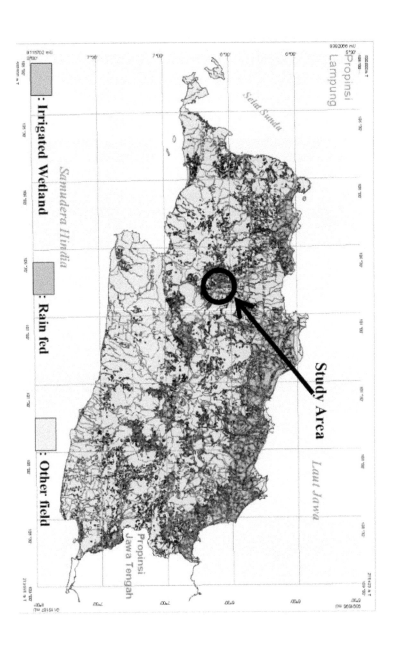

FIGURE 1: Geographical location of studied sites in west Java, Indonesia.

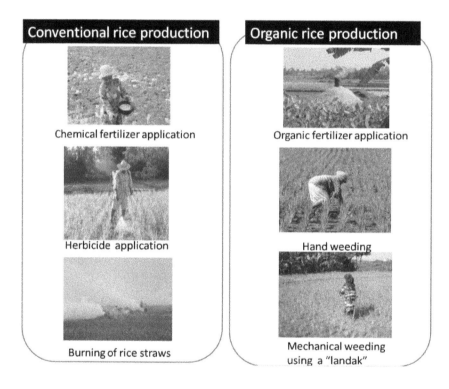

FIGURE 2: Comparison of farm management systems between conventional and organic rice production in west Java, Indonesia.

Public awareness of what "organic agriculture" means, as well as consumer demand for organic crops, are currently very low in Indonesia, where the benefits of organic farming are understood by only few who are concerned about food safety for their own health. However, through the efforts of NGOs and the government of Indonesia, people are developing interest in environmentally friendly organic farming [7]. Organic farming provides a lot of benefits to the farming system in Indonesia, because it can improve soil and food quality, and increase the soil organic carbon (SOC) storage in the soil. For global environmental conservation, this soil management strategy has great potential to contribute to carbon sequestration, because the carbon sink capacity of the world's agricultural and degraded soil is 50–66% of the historic carbon loss of 42–72 pentagrams (1 Pg = 10^{15} g), although actual carbon storage in cultivated soil may be smaller if climate change leads to increasing mineralization [8]. The importance of SOC in agricultural soil is, however, not controversial, because SOC helps to sustain soil fertility and conserve soil and water quality, and these compounds play a variety of roles in the nutrient, water, and biological cycles.

Organic farming also has great potential to improve soil carbon storage [9,10]. However, only a few studies have been conducted in Indonesia, and there is little data for comparing soil carbon storage between organic farming and conventional farming. In addition, the organic farming system and associated farm work had not been studied in Indonesia. Therefore, this research was designed to compare soil carbon sequestration levels in conventional and organic farming systems for rice production on the island of Java, Indonesia

5.2 MATERIALS AND METHODS

5.2.1 STUDY AREA

The study area was the city of Bogor, located in the Cisadane watershed, west Java, Indonesia. The Cisadane River flows through urban areas from Bogor to Jakarta, the capital of Indonesia, and is a major rice and vegetable production area. Soil type is Latosole. Organic and conventional rice farmer groups for the study were selected from the Situgede district of Bogor. Figure 1 illustrates the geographical location of the studied sites in west Java, Indonesia.

5.2.2 FIELD INVESTIGATIONS

Field investigations were conducted in March and September 2008 and September 2009, and details of farming practices and chemical or organic fertilizer application were obtained from interviews with conventional and organic rice farmers. For organic rice production, farmers use a self-produced organic fertilizer called "bochashi," which is composed of 10% rice bran, 20% rice chaff, and 70% cow manure (in volume). Nutrient component of this organic fertilizer are 28.29% carbon, 0.35% nitrogen, 0.17% phosphorus, 2.31% potassium, 1.87% calcium, and 0.42% magnesium (on a dry matter basis). Based on the interview data, the cost for farming and amount of labor required for farming were calculated. The prices of materials were also obtained through interviews with merchants. Figure 2 illustrates the difference between conventional and organic rice production systems. We also interviewed about the true state of organic rice production to workers.

5.2.3 SOIL AND CROP YIELD ANALYSIS

Organic fields were converted to organic from conventional farming after August 2003, and we examined the 13th rice crop grown after the switch to organic management. Organic fields did not receive the organic certification. Conventional fields continued to be fertilized with chemical input and no organic fertilizer. Both fields of organic and conventional rice production were located in the same area.

Topsoil samples (from a depth of 0–10 cm) were taken from three conventional and three organic rice fields in September 2008 with a 55.8 mm diameter soil core sampler. Soil core volume was registered and water content determined (gravimetrically) on sample aliquots, to calculate bulk density (BD). Soil samples were dried at 50°C, ground to pass a 2 mm sieve and stored until analysis. Organic carbon content was analyzed by the Walkley-Black Procedure [11] and soil organic carbon storage (SCS) was calculated as following.

$$SCS \ (Mg \ ha^{-1}) = BD \times SOC \times DP \times 100$$

where, BD: bulk density (g mL^{-1}); SOC: soil organic carbon content (%); DP: soil depth (m).

To obtain the soil carbon distribution between conventional and organic rice fields, soil samples were collected from different fields managed by the fields mentioned above (where topsoil samples were taken), using a post hole auger at depth of 0–2.5, 2.5–7.5, 7.5–15, and 15–30 cm at harvest of rice in September 2009. These soil samples were also analyzed for organic carbon content by the above mentioned methods.

Rice yields were estimated by air dried subsamples taken using a 1 m row section after manually harvesting whole fields in September 2008.

5.3 RESULTS AND DISCUSSION

The soil in the organic farming system showed higher soil carbon content than conventional soils after four years of continuous organic farming, however, there were no significant differences in soil bulk density between the two farming systems (Table 1). Soil carbon storage in organic farming was significantly increased compared with conventional farming. These differences were also obvious in soil carbon content profile, which was significantly higher in the organic field than the conventional field, especially in the top 10 cm soil layer (Figure 3). However, soil carbon contents are somewhat lower in Figure 3 compare with Table 1. This difference suggests that there was diversification of soil carbon content depending on the site specific location, however, organic fields showed higher carbon content than the conventional fields in both 2008 and 2009.

TABLE 1: Comparison of soil carbon sequestration between organic and conventional rice fields in the top 10 cm soil depth.

	Soil Bulk Density g mL^{-1}	Carbon Content %	Soil Carbon Storage Mg ha^{-1}
Organic	0.88	2.89	25.0
Conventional	0.80	2.22	17.6
Significance	NS	*	**

**, * and NS indicate significance at 1% and 5% level and not significant, respectively.

Using Table 1 data, we can estimate the soil carbon sequestration ability. Organic farming can increase 1.85 ton C per hectare per year compared with the conventional farming system. This value also agrees with the data that Shirato et al. [12] obtained from paddy fields in Thailand. The rate of increase in SOC stock resulting from changes in land-use and adoption of recommended farming practices, follows a sigmoid curve that attains the maximum 5–20 years after adoption of recommended farming practices [8]. In addition, the amount of organic carbon stored in paddy soils is greater than in dry field soils, due to different biochemical processes and mechanisms specifically caused by the presence of floodwater in paddy soils [13]. These results show organic rice farming has a lot of potential to improve soil carbon sequestration and it may also mitigate global warming in Indonesia.

The humid tropic condition covers a large area in west Java with very high annual rainfall (>200 cm). Tropical evergreen forests are the climax vegetation, and crop production is limited by low fertility and soil acidity due to leaching and rapid invasion by weeds. On suitable soils, notably on rice paddies, intensive food cropping systems exist [14]. In general, if crop residues are returned and supplemented with nutrient inputs, these systems maintain adequate soil organic matter and production levels in the humid tropic regions [15]. However, burning crop residue is common practice in the conventional rice production system in west Java. When stubble is removed or burned, only the root systems are recycled. The SOC contents of these managed paddy soils were very low, for example, Kyuma [16] described low SOC content in Indonesia (Ave. = 1.39%, n = 44) resulting from the widespread practice of burning rice straw at the end of the cropping cycle. Our data reveals that SOC contents in the organic rice paddy soils were higher than in the conventional managed soils, due to appropriate organic matter input such as bochashi.

Organic farming helped to reduce the cost of rice production (Figure 4). For example, conventional farmers had to pay 1,410,000 Rp (rupiah) for chemical fertilizers, while organic farmers only had to pay 30,000 Rp for the bochashi organic fertilizer. Therefore, organic farming could cut 90% of the total cost of rice production.

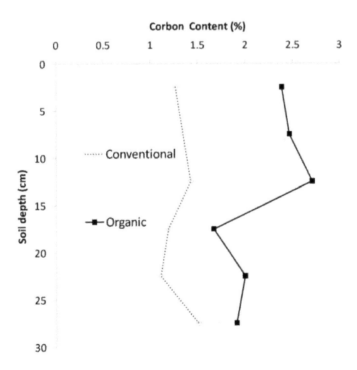

FIGURE 3: Comparison of soil carbon content profile between organic and conventional rice fields in the top 30 cm soil depth.

According to the latest data, farming with chemical fertilizers is on average twice as expensive as the use of organic products, while production levels are the same, if not a fraction higher in the organic sector [17]. This indicates that the economic crisis has helped to boost the growth of Indonesia's organic farming sector. The organic system required more labor to apply the organic fertilizer and weeding. The amount of organic fertilizer applied was 2 Mg ha^{-1} for each rice growing season, which was four-times greater than conventional farming, due to the lack of appropriate technology for application of the organic fertilizer. Weeding in Indonesia is mainly done by hand, and while there are also traditional weeding tools called "landak" (Figure 2), these tools still require a lot of manual labor. Total labor time for rice cultivation was 768 man hours ha^{-1} for the conventional system while it was almost twice as high, 1,406 man hours ha^{-1}, for the organic system.

Table 2 shows the gross profit and wages per working hour between the organic and conventional farming systems. The yield in the organic farming system was lower than in the conventional, while the price of organic rice was 18% higher than conventionally-grown rice, resulting in almost the same gross profits for organic and conventional farming.

TABLE 2: Comparison of gross profit and hourly wages between organic and conventional farming system.

Management	Yield[1] (without husk)	Price (Rp kg^{-1})	Gross Profit (Rp ha^{-1})	Hourly Wage[2] (Rp)
Organic	3.2 Mg	6,500	20,800,000	2,955
Conventional	4.1 Mg	5,500	22,550,000	5,872

[1] Yields were obtained by sampling in September, 2008.
[2] Hourly wages were calculated as the cost share of total benefits to the landowner, manager and worker.

Wages per working hour in the organic system, however, were significantly lower, only about half those in the conventional system. The increased man-hour required in the organic system results in a lower hourly

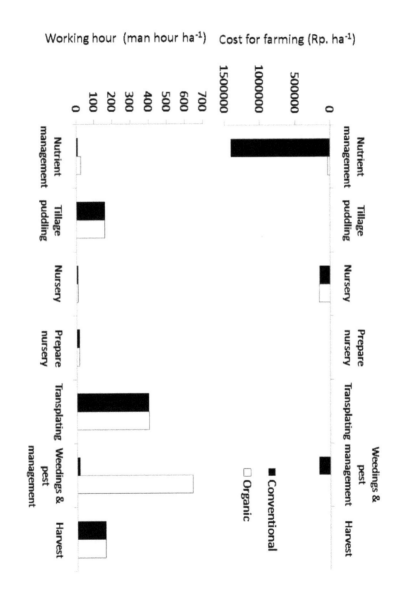

FIGURE 4: Costs for farming (Rp per hectare; upper panel) and labor (hrs per hectare; lower panel) needed for organic and conventional rice farming in west Java.

return and has been a major factor limiting the expansion of this farming system in west Java.

According to the soil analysis, organic farming showed significantly higher SOC storage, which may help not only to mitigate global warming, but also to establish a sustainable food system in Indonesia. Organic paddy rice farmer also has potential to improve soil quality, reduce the cost of chemicals that have recently been increasing with the price of fossils fuels, and increase farmers' incomes due to its higher price. However, organic farming requires intensive labor such as weeding and applying bocashi fertilizer to the fields.

In the study area, profits from rice production were shared among the land owners, farmers (managers), and workers, but workers received only about 20% of the yield base of rice production. This suggests that by converting conventional farming to organic farming, land owners and farmers can increase their profits, while workers receive relatively little added benefit, despite more work. Thus, while organic farming has great potential to improve environmental quality, it also creates social justice problems in Indonesia.

From an interview with a worker who was a member of women group workers in Shitogede, a strongly positive opinion regarding organic farming was received (Figure 5). The worker stated clearly "the quality of organic rice is excellent, organic rice can maintain the good flavors after cooking for long hours." Although we need a scientific evaluation of the rice quality of organic and conventional rice, her opinion suggests that the organic rice farming in Bogor Indonesia was strongly supported by the disinterest works that were provided by many female workers. These disinterest works also brought the benefit to improve the soil quality.

As local environmental quality becomes increasingly degraded by agricultural practices, the importance of protecting and restoring soil resources is being recognized by the world community [18-20]. Sustainable management of soil received strong support at the Rio Summit in 1992 as well as at Agenda 21 [21], the UN Framework Convention on Climate Change [22], in articles 3.3 and 3.4 of the Kyoto Protocol [23], and elsewhere. These conventions are indicative of the recognition by the world community of the strong links between soil degradation and desertification on the one hand, and loss of biodiversity, threats to food security, increases in poverty, and risks of accelerated greenhouse effects and climate change on the other. This situation suggests that a global support network

FIGURE 5: One of the female organic rice production workers (47 years old) replied "Organic rice is delicious compare with conventional rice using chemical fertilizer. I do not care so much about weeding because we really want to have organic rice with my family" (Situgede, Bogor, Indonesia).

system is needed to conserve the local environment such as Indonesia's organic farmlands.

To develop the renewable agriculture for enhancing sustainability, the combination approach of biophysical with socio-economic aspects of an agricultural system will be needed. Although this research data is limited to the discussion of organic farming in Indonesia, this research data highlights the cutting edge of benefits and problems of organic farming in Indonesia. Therefore, further collaboration researches between tropical countries and developed countries will be necessary to develop the sustainable agriculture for the global environment.

5.4 CONCLUSIONS

The use of bochashi organic fertilizer as an alternative in tropical paddy soils of west Java has resulted in increased soil organic carbon storage and gross benefits for farming. Therefore, organic farming for rice production may help not only to mitigate global warming by carbon sequestration, but also to establish a sustainable food system in Indonesia. Although, land owners and farmers can increase their profits by converting conventional farming to organic farming, the workers must work harder and they receive relatively little added benefit due to the sharecropping system. Therefore, political and social incentives will be required based on the common understanding that management of soil and agro-ecosystems will be essential to develop a society in this century where nature and humans coexist.

REFERENCES

1. Gliessman, S.R. Agroecology: The Ecology of Sustainable Food System; CRC press: Boca Raton, FL, USA, 2006.
2. Pimentel, D.; Harvey, C.; Resosudarmo, P.; Sinclair, K.; Kurz, D.; McNair, M.; Crist, S.; Shpritz, L.; Fitton, L.; Saffouri, R.; Blair. R. Environmental and economic costs of soil erosion and conservation benefits. Science 1995, 267; 1117–1122.
3. Millennium Ecosystem Assessment: Ecosystems and Human Well-Being: Synthesis; World Resource Institute: Washington, DC, USA, 2005.
4. Syuaib, M.F. Perspective of sustainable agriculture in Indonesia: Keep growing in harmony with environment. In Final Report of International Post Graduate GP Workshop on "From Environmental to Sustainable Science: Thinking the Shift and the Role of Asian Agricultural Science"; Ibaraki University: Ibaraki, Japan, 2009; pp. 93–99.
5. Syuaib, M.F. Farming system in Indonesia and its carbon balance feature. In Final Report of International Symposium "Food and Environmental Preservation in Asian Agriculture"; Ibaraki University: Ibaraki, Japan, 2006; pp. 26–30.
6. Komatsuzaki, M.; Ohta, H. Soil management practice for sustainable agroecosystem. Sustain. Sci. 2007, 2, 103–120.
7. Heieh, S.C. Organic farming for sustainable agriculture in Asia with special reference to Taiwan experience, 2005; Available online: http://www.agnet.org/library/eb/558/ (accessed on 31 January 2010).
8. Lal, R. Soil carbon sequestration impacts on global climate change and food security. Science 2004, 304, 1623–1627.
9. Marriott, E.E.; Wander, M.M. Total and labile soil organic matter in organic and conventional farming systems. Soil Sci. Soc. Am. J. 2006, 70, 950–959.
10. Pimentel, D.; Hepperly, P.; Hanson, J.; Douds, D.; Seidel, R. Environmental, energetic, and economic comparisons of organic and conventional farming systems. BioScience 2005, 55, 573–582.

11. Nelson, D.W.; Sommers, L.E. Total carbon, organic carbon, and organic matter. In Methods of Soil Analysis Part 2. Chemical and Biological Properties, 2nd ed.; Page, A.L., Miller, R.H., Keeney, D.R., Eds.; American Society of Agronomy: Madison, WI, USA, 1982; pp. 539–579.

12. Shirato, Y.; Paisancharoen, K.; Sangtong, P.; Nakviro, C.; Yokozawa, M.; Matsumoto, N. Testing the Rothamsted Carbon Model against data from long-term experiments on upland soils in Thailand. Eur. J. Soil Sci. 2005, 56, 179–188.

13. Katoh, T. Carbon accumulation in soils by soil management, mainly by organic matter application—Experimental results in Aichi prefecture. Jpn. J. Soil Sci. Plant Nutr. 2003, 73, 193–201.

14. Bronson, K.F.; Cassman, K.G.; Wassmann, R.; Olk, D.C.; van Noordwijk, M.; Garrity, D.P. Soil carbon dynamics in different cropping systems in principal ecoregions of Asia. In Management of Carbon Sequestration in Soil; Lal, R., Kimble, J.M., Follett, R.F., Stewart, B.A., Eds.; CRC Lewis Publishers: Boca Raton, FL, USA, 1998; pp. 35–57.

15. Hutchinson, J.J.; Campbell, C.A.; Desjardins, R.L. Some perspectives on carbon sequestration in agriculture. Agri. Forest Meteorol. 2007, 142, 288–302.

16. Kyuma, K. Paddy soils of Japan in comparison with those in Tropical Asia. In Proceedings of First International Symposium on Paddy Soil Fertility, Chiangmai, Thailand, 6–13 December 1988; pp. 5–19.

17. The Jakalta post economic crisis helps boost growth in Indonesia's organic fertilizer sector, 2009; Available online: http://www.thejakartapost.com/news/2009/03/02/economic-crisis-helps-boost-growth-indonesia039s-organic-fertilizer-sector.html (accessed on 31 January 2010).

18. Barford, C.C.; Wofsy, S.C.; Goulden, M.L.; Munger, J.W.; Pyle, E.H.; Urbanski, S.P.; Saleska, S.R.; Fitzjarrald, D.; Moore, K. Factors controlling long- and short-term sequestration of atmospheric CO2 in a mid-latitude forest. Science 2001, 294, 1688–1691.

19. Lal, R. Soil erosion impact on agronomic productivity and environment quality. Crit. Rev. Plant Sci. 1998, 17, 319–464.

20. Lal, R. World cropland soils as source or sink for atmospheric carbon. Adv. Agron. 2001, 71, 145–191.

21. UNCED Agenda 21: Programme of action for sustainable development, Rio declaration on environment and development, statement of principles. In Final Text of Agreement Negotiated by Governments at the United Nations Conference on Environment and Development (UNCED); United Nations Development Programme (UNDP): New York, NY, USA, 1992; pp. 3–14.

22. UNFCCC United Nations Framework Convention on Climate Change; United Nations: New York, NY, USA, 1992.

23. UNFCCC Kyoto Protocol to the United Nations Framework Convention on Climate Change; United Nations: New York, NY, USA, 1998; Available online: http://unfccc.int/resource/docs/ convkp/kpeng.html (accessed on 31 January 2010).

PART II

BIOFERTILIZERS

CHAPTER 6

RESIDUAL INFLUENCE OF ORGANIC MATERIALS, CROP RESIDUES, AND BIOFERTILIZERS ON PERFORMANCE OF SUCCEEDING MUNG BEAN IN AN ORGANIC RICE-BASED CROPPING SYSTEM

MOHAMMADREZA DAVARI, SHRI NIWAS SHARMA, AND MOHAMMAD MIRZAKHANI

6.1 INTRODUCTION

The rice (*Oryza sativa*)-wheat (*Triticum aestivum*) cropping system (RWCS) occupies about 28.8 million hectares (m ha) in five Asian countries, namely, India, Pakistan, Nepal, Bangladesh, and China (Prasad 2005). These five countries represent 43% of the world population on 20% of the world's arable land (Singh and Paroda 1994). In India, RWCS occupy 12 m ha and contributes about 31% of the total food grain production (Kumar and Yadav 2006). Similarly in China, RWCS occupies about 13 m ha (Jasdan and Hutchaon 1996) and contributes about 25% of the total cereal production in the country (Lianzheng and Yixian 1994). Thus,

This chapter was originally published under the Creative Commons Attribution License. Davari M, Sharma SN, and Mirzakhani M. Residual Influence of Organic Materials, Crop Residues, and Biofertilizers on Performance of Succeeding Mung Bean in an Organic Rice-Based Cropping System. International Journal of Recycling of Organic Waste in Agriculture *1,14 (2012). doi:10.1186/2251-7715-1-14.*

RWCS are of considerable significance in meeting Asia's food requirements. However, practice of following a cereal-cereal cropping system on
the same piece of land over years has led to soil fertility deterioration, and
questions are being raised on its sustainability (Duxbury and Gupta 2000;
Ladha et al. 2000; Prasad 2005). Efforts were, therefore, made to find
out alternate cropping systems. Sharma and Prasad (1999) recommended
that growing a short-duration mung bean after wheat and incorporating
of its residue in succeeding rice made rice-wheat cropping system more
productive, remunerative, and soil recuperative than traditional rice-wheat
cropping system.

Organic farming of Basmati rice-based cropping system is another
alternative system for sustainability of crop production and natural resources. Moreover, there is a great demand of organically grown food in
European and Middle East countries and offer two to two and a half times
higher prices for organic produce (Partap 2006). Organic farming often has
to deal with a scarcity of readily available nutrients in contrast to inorganic
farming which relies on soluble fertilizers. The aim of nutrient management in organic systems is to optimize the use of on-farm resources and
minimize losses (Kopke 1995). Maximum use of crop residues has been
suggested towards building soil fertility (Jasdan and Hutchaon 1996). Rice
and wheat straw have large potential for plant nutrients in organic farming
of rice-wheat system. The straw in the system accounts about 35% to 40%
N, 10% to 15% of P, and 80% to 90% of K removal by these crops (Sharma and Sharma 2004). Incorporation of straw, thus, results in recycling
of a sizable amount of plant nutrients. However, there is a great difficulty
in using the plant residue of cereals due to higher C/N ratio. Hence, there
is an urgent need to develop a suitable technology to use crop residue in
organic farming. We have to mix the plant residues of cereals with well-
decomposed farmyard manures or plant residue of legumes for narrowing
down of C/N ratio so as to overcome the adverse effect of immobilization
of native plant nutrients. Sharma and Prasad (1999) reported that incorporation of mung bean residue was found to be at par with Sesbania green
manure in rice-wheat system.

The responses of the succeeding crops in a cropping system are influenced greatly by the preceding crops and the inputs applied therein. There-

fore, recently greater emphasis is being laid on the cropping system as a whole rather than on the individual crops. In addition, organic manures and biofertilizers have carry-over effect on the succeeding crops. Jamaval (2006) reported that around 30% of the applied nitrogen as manure may become available to the immediate crop and rest to the subsequent crops. Maintenance of soil fertility is important for obtaining higher and sustainable yield due to large turnover of nutrient in the soil-plant system. Mung bean (*Vigna radiate* L.), commonly known as green gram, is an important conventional pulse crop in India. It has an edge over other pulses because of its high nutritive value, digestibility, and non-flatulent behavior. It is grown principally for its protein-rich edible seeds (Haq 1989). An important feature of the mung bean crop is its ability to establish a symbiotic partnership with specific bacteria, setting up the biological N_2 fixation in root nodules that supplies required nitrogen to the plant (Mandal et al. 2009). The present investigation was therefore undertaken to assess the residual influence of organic materials and biofertilizers applied to rice and wheat on yield, nutrient status, and economics of succeeding mung bean in an organic cropping system.

6.2 METHODS

6.2.1 EXPERIMENTAL DESIGN

The field experiment was conducted during spring 2007 and 2008 at the research farm of the Indian Agricultural Research Institute, New Delhi, India to study the residual effects of organic materials, crop residues, and biofertilizers applied to the cropping system on performance of succeeding mung bean crop (Figure 1). It is situated at 28.4°N latitude and 77.1°E longitude at an elevation of 228.6 m above the mean sea level (Arabian Sea) for two years (2006 to 2007 and 2007 to 2008). The soil was medium in organic C (5.1-mg kg^{-1} soil), low in available nitrogen (73.1-mg kg^{-1} soil), medium in available phosphorus (8.42-mg kg^{-1} soil), available potassium (108.87-mg kg^{-1} soil), and had a pH 8.16. The experiment was laid out in a randomized block design with three replications.

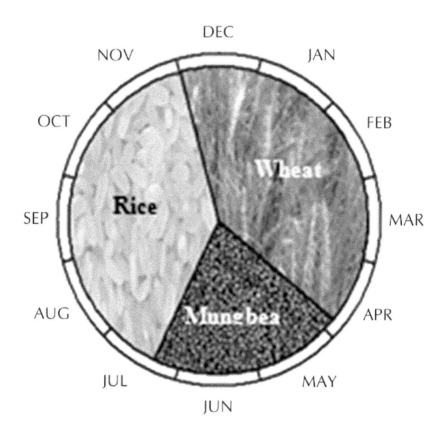

FIGURE 1: Timetable of the cropping system.

6.2.2 TREATMENT REGIMEN

Treatments consisted of six combinations of different residual organic materials and biofertilizers and control as follows:

1. Farmyard manure (FYM) applied on nitrogen basis at 60 kg ha^{-1} to each rice and wheat crops.
2. Vermicompost (VC) applied on nitrogen basis at 60 kg ha^{-1} to each rice and wheat crops. FYM and VC were applied to both rice and wheat, whereas mung bean was grown on residual fertility.
3. FYM + crop residue (CR) whereas rice and wheat residues were applied at 6 t ha^{-1} to wheat and mung bean, respectively, and mung bean residue applied at 3 t ha^{-1} to rice.
4. VC + CR.
5. FYM + CR + biofertilizers (B) including cellulolytic culture (CC) and phosphate-solubilizing bacteria (PSB) were used in all the crops, whereas BGA, Azotobacter, and Rhizobium applied in rice, wheat, and mung bean, respectively.
6. VC + CR + B.
7. Control (no fertilizer applied).

As biofertilizers, CC and PSB were used in all the crops, whereas BGA, *Azotobacter*, and *Rhizobium* applied in rice, wheat, and mung bean, respectively. Cellulolytic culture containing four fungi (*Aspergillus awamori, Trichoderme viride, Phanerochaete chrysosporium*, and *Aspergillus wolulens*) was inoculated at the time of residue incorporation, whereas *Azotobacter, Rhizobium*, and PSB (*Pseudomonas striata*) were used to inoculate the seeds as per the treatments. Farmyard manure used in the previous crops was well decomposed. It contained N (6,150 mg kg^{-1}), P (2,600 mg kg^{-1}), K (3,150 mg kg^{-1}), Mn (11.5 mg kg^{-1}), Zn (39.5 mg kg^{-1}), Cu (2.65 mg kg^{-1}), and Fe (21.5 mg kg^{-1}) and had a C/N ratio of 23.5. Similarly VC contained N (1,1950 mg kg^{-1}), P (6,283mg kg^{-1}), K (6,950 mg kg^{-1}), Mn (37.5 mg kg^{-1}), Zn (87 mg kg^{-1}), Cu (8.5 mg kg^{-1}), and Fe (57.5 mg kg^{-1}) and had a C/N ratio of 17.5. The chemical compositions of crop residues are given in Table 1.

TABLE 1: Chemical composition of crop residues applied in rice-wheat-mung bean cropping system

Composition (mg kg^{-1})	2006 to 2007			2007 to 2008		
	Rice	Wheat	Mung bean	Rice	Wheat	Mung bean
Total N	4,700	3,900	15,000	5,000	4,100	15,200
Total P	680	490	1,100	700	500	1,200
Total K	14,600	15,600	4,400	14,650	15,700	4,500
Organic C	408,000	400,000	401,000	410,000	403,000	403,000
Fe	434.23	349.80	849.56	437.41	372.97	876.21
Zn	100.52	29.89	69.13	105.09	34.67	72.04
Mn	58.23	73.69	79.65	60.69	78.52	88.62
Cu	40.02	16.85	22.23	40.67	17.43	23.04

6.2.3 ANALYTICAL TECHNIQUE

Grain and stover samples of mung bean were dried in hot air oven at 60°C for 6 h and ground in a Macro-Wiley Mill (Paul N. Gardner Company, Inc., FL, USA) to pass through a 40-mesh sieve. A representative sample of 0.5-g grain and straw was taken for the determination of nitrogen, phosphorus, and potassium. The nitrogen concentration in grain and straw samples was determined by modified Kjeldahl method (Jackson 1973); total phosphorus, by Vanadomolybdo phosphoric acid yellow color method and flame photometry method, as described by Prasad et al. (2006). The NPK concentration in grain and straw was expressed in percentage. Iron, zinc, manganese, and copper were determined in diacid digest of plant tissues using AAS as in the case of soil analysis. The N, P, and K uptake in grain or straw was worked out by multiplying their percent concentrations with the corresponding yield. The total uptake of N, P, and K was obtained by adding up their respective uptake in grain and straw. This was expressed in kilogram per hectare. Protein content in mung bean grains was obtained by multiplying the N concentration of grain with a factor 6.25 (Juliano 1985). The protein yield of mung bean was calculated by multiplying its protein concentration with grain yield.

The cost of mung bean cultivation was calculated on the basis of prevailing rates of inputs, and gross income was calculated on the basis of procurement price of mung bean grain and prevailing market price of mung bean stover. The income was obtained by subtracting the cost of cultivation from the gross income, i.e.,

$$\text{Net income} = \text{gross income} - \text{cost of cultivation} \tag{1}$$

The net profit of the rotation was calculated by adding the net profits of the rice and wheat together.

6.2.4 STATISTICAL ANALYSIS

The data relating to each character were analyzed by applying the technique of 'analysis of variance' for randomized block design as described by Cochran and Cox (1957). Critical difference at 5% level of significance was calculated for comparing the mean of difference presented in the summary table.

6.3 RESULTS AND DISCUSSION

6.3.1 YIELD ATTRIBUTES

During both years of study, application of FYM had no significant effect on the number of pods of mung bean, whereas all other combinations of organic manures and biofertilizers significantly ($p = 0.05$) increased the number of pods per plant over the control (Table 2). During the second year of experiment, the combination of VC + CR + B was significantly superior to FYM and VC alone. The increased pod formation in treatments, where organic manures and crop residues were applied, may be attributed due to better plant development through efficient utilization of soil resources by the plant, where primary growth elements were available in sufficient amount. During the first year, FYM, VC, FYM + CR, VC + CR, and FYM + CR + B had no significant effect on the number of grains per pod, whereas VC + CR + B significantly increased the number of grains per pod. During the second year, the results were similar to those observed in the first year except that VC + CR, FYM + CR + B also significantly increased the number of grains per pod over the control. Similar results were reported by Srinivas and Shaik (2002). During both years, FYM and VC applied to rice and wheat did not affect the test weight of mung bean, whereas all other combinations of organic manures and biofertilizers significantly increased the test weight of mung bean over the control. Effects of different combinations of organic manures and biofertilizers applied to rice, wheat, and mung bean on the grain yield of mung bean were greater in the second year as compared to the first year. Residual effect of organics was also noticed by Reddy and Reddy (2005) wherein the plant height, number of leaves, leaf area, yield attributes, and root yield in radish were significantly affected due to the residual effect of vermicompost in onion-radish cropping system.

TABLE 2: Effect of treatments on the yield attributes of mung bean

	Number of pods per plant		Number of grains per pod		Test weight (g)	
	2007	2008	2007	2008	2007	2008
Organic materials and biofertilizers						
Control	10.3	10.8	6.8	6.9	39.5	39.8
FYM	11.6	12.3	7.3	7.5	42.5	42.7
VC	12.9	13.7	7.4	7.7	43.1	43.3
FYM + CR	12.8	14.8	7.7	7.8	44.6	45.0
VC + CR	14.3	15.7	7.8	7.9	44.9	45.4
FYM + CR + B	13.6	16.0	7.8	7.9	45.3	45.4
VC + CR + B	15.1	16.8	8.0	8.1	45.6	46.1
SEM±	0.76	0.93	0.33	0.31	1.53	1.68
LSD (p = 0.05)	2.34	2.82	1.02	0.96	4.71	5.18

SEM±, standard error of the mean; LSD, least significant difference.

TABLE 3: Effect of treatments on the seed and stover of mung bean

	Grain yield (t ha^{-1})		Stover yield (t ha^{-1})	
	2007	2008	2007	2008
Organic materials and biofertilizers				
Control	0.68	0.69	2.69	2.70
FYM	0.74	0.76	2.93	3.02
VC	0.76	0.79	3.09	3.20
FYM + CR	0.92	0.96	3.30	3.44
VC + CR	0.95	1.00	3.47	3.63
FYM + CR + B	0.97	1.02	3.55	3.64
VC + CR + B	1.00	1.04	3.70	3.84
SEM±	0.050	0.051	0.20	0.28
LSD (p = 0.05)	0.154	0.157	0.62	0.86

SEM±, standard error of the mean; LSD, least significant difference.

6.3.2 GRAIN AND STOVER YIELDS

In terms of statistical significance (p = 0.05), FYM had no significant effect on the grain yield of mung bean, whereas FYM + CR, VC + CR, FYM + CR + B, and VC + CR + B, being at par, significantly increased the grain yield of mung bean over the control, FYM, and VC alone in the first year and over the control and FYM alone in the second year (Table 3). Weather conditions during the second year of study were more favorable than those of the first year. Mean monthly maximum and minimum temperatures were relatively low during the second year as compared to the first year. Total rainfall during the crop growth period of mung bean was about five times more in the second year than in the first year. All these favorable weather conditions resulted in higher yield during the second year as compared to the first year. On the other hand, the superiority of vermicompost was attributed to its slow and steady decomposition, which probably released the nutrients slowly and in higher quantity compared to other organic materials. Poonam et al. (2007) reported that seed inoculation with *Rhizobium* recorded an increase in yield by 12% to 16%.

Conjunctive use of *Rhizobium* with PSB and plant growth-promoting rhizobacteria revealed synergistic effect on the symbiotic parameters and grain yield of mung bean. In both years, stover yield of mung bean was not significantly affected by FYM and VC and FYM + CR, whereas other combinations of organic manures and biofertilizers significantly increased the stover yield of mung bean over the control. There was no significant difference between different combinations of organic manures and biofertilizers in both years of study. The findings are in corroboration with those reported by John De Britto and Sorna Girija (2006).

6.3.3 NPK CONCENTRATION

In 2007 as well as in 2008, all combinations of organic manures and biofertilizers significantly increased nitrogen concentration in mung bean grain over the control (Table 4). All the combinations of organic manures and biofertilizers were at par in respect of N concentration in mung bean grain in both years of study. In both years, FYM had no significant effect on N concentration in mung bean stover, whereas all other combinations of organic manures and biofertilizers significantly increased nitrogen concentration in mung bean stover over the control. There was no significant difference between the different combinations of organic manures and biofertilizers, and all the combinations, except FYM, significantly increased phosphorus concentration in mung bean grain over the control in both years. In both years, phosphorus concentration of mung bean was not significantly affected by FYM, VC, and FYM + CR, whereas other combinations of organic manures and biofertilizers significantly increased P concentration of stover over the control. There was no significant difference between different combinations of organic manures and biofertilizers in both years of study. In both years, the combinations of FYM, VC, FYM + CR, and VC + CR were at par and significantly increased K concentration in mung bean over the control. Similarly, FYM + CR + B and VC + CR + B, being at par, significantly increased K concentration in mung bean grain over FYM. In 2007 as well as in 2008, FYM had no significant effect on potassium concentration in mung bean stover, whereas VC, FYM + CR, VC + CR, and FYM + CR + B, being at par, significantly increased

P concentration over the control. The combination of VC + CR + B was at par with VC + CR and FYM + CR + B but significantly superior to VC and FYM + CR combinations. This is in conformity with the result obtained by applying seaweed extract as a biostimulant in organic farming of green gram (Zodape et al. 2010).

TABLE 4: Effect of treatments on nitrogen, phosphorus, and potassium concentrations (%) in mung bean

	Nitrogen				Phosphorus				Potassium			
	Grain		Stover		Grain		Stover		Grain		Stover	
	2007	2008	2007	2008	2007	2008	2007	2008	2007	2008	2007	2008
Organic materials and biofertilizers												
Control	2.91	2.92	1.29	1.28	0.316	0.316	0.08	0.08	0.77	0.76	0.38	0.37
FYM	3.31	3.24	1.50	1.52	0.342	0.351	0.10	0.10	0.83	0.84	0.42	0.43
VC	3.26	3.29	1.55	1.57	0.350	0.357	0.11	0.12	0.84	0.85	0.44	0.46
FYM + CR	3.27	3.29	1.54	1.57	0.348	0.356	0.11	0.12	0.85	0.87	0.44	0.46
VC + CR	3.32	3.38	1.59	1.62	0.356	0.363	0.12	0.13	0.87	0.89	0.47	0.50
FYM + CR + B	3.36	3.41	1.62	1.65	0.358	0.363	0.12	0.14	0.89	0.91	0.47	0.51
VC + CR + B	3.43	3.50	1.65	1.68	0.369	0.378	0.13	0.16	0.91	0.94	0.50	0.53
SEM±	0.10	0.10	0.08	0.09	0.01	0.01	0.01	0.02	0.02	0.02	0.02	0.02
LSD (p = 0.05)	0.31	0.32	0.25	0.27	0.03	0.04	0.03	0.06	0.06	0.06	0.05	0.07

SEM±, standard error of the mean; LSD, least significant difference.

6.3.4 NPK UPTAKE BY GRAIN AND STOVER

In both years, FYM and VC had no significant effect on nitrogen uptake over the control, whereas other combinations of organic manures and bio-fertilizers resulted in significantly (p = 0.05) higher nitrogen uptake by mung bean grain than the control (Table 5). In 2007 and 2008, FYM had no significant effect on nitrogen uptake by mung bean stover over the control, whereas VC, FYM + CR, VC + CR, and FYM + CR + B, being at par, significantly increased nitrogen uptake by mung bean stover over the control. The combination of VC + CR + B was at par with FYM + CR, VC + CR, and FYM + CR + B but significantly superior to VC and FYM alone. In 2007 and 2008, FYM had no significant effect on nitrogen uptake by mung bean, whereas VC, FYM + CR, and VC + CR, being at par, significantly increased nitrogen uptake by mung bean over the control. Similarly, FYM + CR + B and VC + CR + B were at par and significantly increased N uptake by mung bean over FYM and VC alone.

TABLE 5: Effect of treatments on nitrogen uptake (kg ha^{-1}) by mung bean

	2006			2007		
	Grain	Straw	Total	Grain	Straw	Total
Organic materials and biofertilizers						
Control	19.8	34.7	54.5	20.1	34.6	54.7
FYM	23.8	44.0	67.8	24.6	45.9	70.5
VC	24.8	47.9	72.7	26.0	50.2	76.2
FYM + CR	30.1	50.8	80.9	31.6	54.0	85.6
VC + CR	31.5	55.2	86.7	33.8	58.8	92.6
FYM + CR + B	32.6	57.5	90.1	34.8	60.1	94.9
VC + CR + B	34.3	61.1	95.4	36.4	64.5	100.9
SEM±	2.13	3.70	5.62	2.72	4.42	5.88
LSD (p = 0.05)	6.56	11.40	17.32	8.38	13.62	18.13

SEM±, standard error of the mean; LSD, least significant difference.

In both years, FYM and VC did not affect P uptake by mung bean grain significantly, whereas FYM + CR, VC + CR, FYM + CR + B, and VC + CR + B, being at par, significantly increased phosphorus uptake by mung bean grain over the control (Table 6). Also, FYM + CR + B and VC + CR + B were also significantly superior to FYM and VC alone. The combinations of FYM, VC, and FYM + CR in both years and VC + CR in only the first year had no significant effect on P uptake by mung bean stover, whereas FYM + CR + B and VC + CR + B in both years and VC + CR in the second year significantly increased phosphorus uptake by mung bean stover over the control. In both years, FYM and VC had no significant effect on P uptake by mung bean, whereas all other combinations of organic manures and biofertilizers significantly increased phosphorus uptake by mung bean over the control, the difference between different combinations being not significant.

TABLE 6: Effect of treatments on phosphorus uptake (kg ha^{-1}) by mung bean

	2006			2007		
	Grain	Straw	Total	Grain	Straw	Total
Organic materials and biofertilizers						
Control	2.2	2.2	4.4	2.2	2.2	4.4
FYM	2.5	2.9	5.6	2.7	3.0	5.7
VC	2.7	3.4	6.1	2.8	3.4	6.6
FYM + CR	3.2	3.6	6.8	3.4	4.1	7.5
VC + CR	3.4	4.2	7.6	3.6	4.7	8.3
FYM + CR + B	3.5	4.3	7.8	3.7	5.1	8.8
VC + CR + B	3.7	4.8	8.5	3.9	6.1	10.0
SEM±	0.21	0.65	0.71	0.25	0.79	0.87
LSD (p = 0.05)	0.65	2.01	2.19	0.78	2.43	2.68

SEM±, standard error of the mean; LSD, least significant difference.

In 2007 as well as in 2008, FYM and VC had no significant effect on K uptake by mung bean, whereas FYM + CR, VC + CR, FYM + CR + B, and VC + CR + B significantly increased potassium uptake by mung bean over the control (Table 7). There was no significant difference between different combinations of organic manures and biofertilizers in the second year, whereas in the first year, FYM + CR, VC + CR, FYM + CR + B, and VC + CR + B were at par but significantly superior to FYM alone. In both years, there was no significant effect of FYM, VC, and FYM + CR on potassium uptake by mung bean stover, whereas other combinations of organic manures and biofertilizers resulted in significantly higher potassium uptake than the control. The differences between different combinations of organic manures and biofertilizers were not significant in both years of study. In 2007, FYM and VC had no significant effect on K uptake by mung bean, whereas FYM + CR, VC + CR, and FYM + CR + B, being at par, significantly increased potassium uptake by mung bean over the control. The combination of VC + CR + B was at par with VC + CR and FYM + CR + B but significantly superior to FYM, VC, and FYM + CR. In 2008, FYM and VC also had no significant effect on K uptake by mung bean, whereas other combinations of organic manures and biofertilizers resulted significantly higher K uptake than the control. There was no significant difference between different combinations of organic manures and biofertilizers during this year.

TABLE 7: Effect of treatments on potassium uptake (kg ha^{-1}) by mung bean

	2006		2007			
	Grain	Straw	Grain	Straw	Grain	Straw
Organic materials and biofertilizers						
Control	5.2	10.2	15.4	5.2	10.0	15.2
FYM	6.1	12.3	18.4	6.4	13.0	19.4
VC	6.4	13.6	20.0	6.7	14.7	21.4
FYM + CR	7.8	14.5	22.3	8.4	15.8	24.2
VC + CR	8.3	16.3	24.6	8.9	18.2	27.1
FYM + CR + B	8.6	16.7	25.3	9.3	18.6	27.9
VC + CR + B	9.1	18.5	27.6	9.8	20.4	30.2
SEM±	0.54	1.89	1.70	0.69	2.23	2.15
LSD (p = 0.05)	1.66	5.82	5.24	2.12	6.87	6.61

SEM±, standard error of the mean; LSD, least significant difference.

6.3.5 MICRONUTRIENTS CONCENTRATION IN GRAIN

In 2007, FYM, VC, and FYM + CR, being at par, significantly (p = 0.05) increased Zn concentration of mung bean over the control (Table 8). Similarly, VC + CR, FYM, CR + B, and VC + CR + B were at par and significantly increased zinc concentration of mung bean grain over FYM. In

2008, FYM, VC, and FYM + CR, being at par, significantly increased Zn concentration of mung bean grain. Similarly, VC + CR was at par with FYM + CR + B but significantly superior to FYM and VC. Similarly, VC + CR + B was at par with FYM + CR + B but significantly superior to FYM, VC, FYM + CR, and VC + CR.

TABLE 8: Effect of treatments on micronutrient concentration (ppm) by mung bean grain

	Zinc		Copper		Iron		Manganese	
	2007	2008	2007	2008	2007	2008	2007	2008
Organic materials and biofertilizers								
Control	15.4	15.2	15.8	15.7	58.1	58.1	55.3	55.1
FYM	16.2	16.4	17.0	17.3	59.4	59.8	57.1	57.3
VC	16.4	16.7	17.4	17.6	60.2	60.8	57.5	57.7
FYM + CR	16.5	16.6	17.5	17.7	60.2	60.7	57.5	57.7
VC + CR	16.8	17.1	18.0	18.5	61.4	62.3	57.8	58.2
FYM + CR + B	16.9	17.5	18.3	18.7	61.4	62.2	58.0	58.7
VC + CR + B	17.1	17.8	19.1	19.7	62.8	63.7	58.5	59.0
SEM±	0.14	0.15	0.13	0.14	0.35	0.33	0.30	0.37
LSD (p = 0.05)	0.43	0.46	0.41	0.43	1.08	1.02	0.92	1.11

SEM±, standard error of the mean; LSD, least significant difference.

The copper concentration in mung bean grain with different combinations of organic manures and biofertilizers were found in the following increasing order: control < FYM < VC = FYM + CR < VC + CR = FYM + CR + B < VC + CR + B. In 2008, FYM, VC, and FYM + CR were at par and significantly increased Cu concentration by mung bean grain over the control. Similarly, VC + CR and FYM + CR + B were at par and resulted in significantly higher Cu concentration in mung bean grain than FYM, VC, and FYM + CR. The combination of VC + CR + B was significantly superior to other combinations of organic manures and biofertilizers.

In both years, FYM, VC, and FYM + CR, being at par, significantly increased iron concentration by mung bean over the control. Similarly, VC + CR and FYM + CR + B were at par and significantly increased Fe concentration of mung bean over FYM, VC, and FYM + CR. The combination of VC + CR + B was significantly superior to other combinations of organic manures and biofertilizers.

In the first year, FYM + VC, FYM + CR, VC + CR, and FYM + CR + B, being at par, significantly increased manganese concentration of mung bean grain over the control, whereas the combination of VC + CR + B was at par with VC + CR and FYM + CR + B but significantly superior to FYM, VC, and FYM + CR in respect to Mn concentration in mung bean grain. In the second year, FYM, VC, FYM + CR, and VC + CR were at par and significantly increased iron concentration in mung bean grain over the control. The combination of VC + CR + B was at par with VC, FYM + CR, VC + CR, and FYM + CR + B but significantly superior to FYM.

TABLE 9: Effect of treatments on protein content (%) and protein yield (kg ha^{-1}) of mung bean

	Protein content (%)		Protein yield(kg ha^{-1})	
	2007	2008	2007	2008
Organic materials and biofertilizers				
Control	18.2	18.3	124	126
FYM	20.1	20.3	149	154
VC	20.4	20.6	155	163
FYM + CR	20.4	20.6	188	198
VC + CR	20.8	21.1	198	211
FYM + CR + B	21.0	21.3	204	217
VC + CR + B	21.4	21.9	214	228
SEM±	0.51	0.56	21.2	23.10
LSD (p = 0.05)	1.56	1.74	65.47	71.05

SEM±, standard error of the mean; LSD, least significant difference.

6.3.6 PROTEIN

In both years, all the combinations of organic manures and biofertilizers were at par and significantly (p = 0.05) increased protein concentration in mung bean grain over the control (Table 9). In 2007, FYM, VC, and FYM + CR had no significant effect on the protein yield of mung bean, whereas VC + CR, FYM + CR + B, and VC + CR + B were at par with FYM, VC, and FYM + CR and increased the protein yield of mung bean over the control significantly. In 2008, FYM and VC had no significant effect on protein yield of mung bean, whereas FYM + CR, VC + CR, FYM + CR + B, and VC + CR + B being at par and significantly increased the protein yield of mung bean over the control. Dhaliwal et al. (2007) reported that the N and protein contents of seed in mung bean were significantly higher with both RFD (recommended fertilizer dose + residue incorporation over the chemical fertilizer treatments, being statistically on par with each other).

TABLE 10: Effect of treatments on economics ($\times 10^3$Rs ha^{-1}) of cultivation of mung bean

	Gross return		Cost of cultivation		Net return	
	2007	2008	2007	2008	2007	2008
Organic materials and biofertilizers						
Control	23.4	26.2	5.16	5.27	18.2	20.9
FYM	25.4	28.9	5.16	5.27	20.3	23.6
VC	26.2	30.1	5.16	5.27	21.0	24.8
FYM + CR	31.3	36.1	11.16	11.27	20.1	24.8
VC + CR	32.4	37.6	11.16	11.27	21.2	26.4
FYM + CR + B	33.0	38.3	11.34	11.45	21.7	26.9
VC + CR + B	34.1	39.2	11.34	11.45	22.8	27.8
SEM±	0.50	0.53	-	-	0.39	0.44
LSD (p = 0.05)	1.55	1.63	-	-	1.17	1.32

SEM±, standard error of the mean; LSD, least significant difference; Rs, Indian Rupees (Rs 1 = $0.0224).

6.3.7 ECONOMICS

In both years, FYM and VC were at par and significantly (p = 0.05) increased the gross income from mung bean over the control. Similarly, FYM + CR and VC + CR being at par and significantly increased the gross income of mung bean over FYM and VC alone (Table 10). The combinations of FYM + CR + B and VC + CR + B were at par with VC + CR but significantly superior to FYM, VC, and FYM + CR in respect of the gross income from mung bean.

The costs of cultivation of 2 years are given in Table 10. The cost of cultivation of a particular treatment did not vary in three replications; hence, data on cost of cultivation were not analyzed statistically. The cost of cultivation varied from Rs 5,160 to 5,270 to Rs 1,1340 to 1,1450. The incorporation of CR increased the cost of cultivation by 116% in the first year and 114% in the second year, whereas inoculation of biofertilizers increased the cost of cultivation by 1.6% in both years.

In the first year, FYM, VC, FYM + CR, and VC + CR were at par and significantly increased the net income of mung bean over the control. The combination of FYM + CR + B was at par with VC + CR and VC but significantly superior to FYM and FYM + CR. Similarly, VC + CR + B was at par with FYM + CR + B but significantly superior to FYM, VC, FYM + CR, and VC + CR. In the second year, FYM, VC, and FYM + CR were at par and significantly increased the net return of mung bean over the control. The combination of FYM + CR + B was at par with VC + CR and FYM + CR but significantly increased the net return of mung bean over FYM, VC, and FYM + CR. Similarly, VC + CR + B was at par with FYM + CR + B and significantly increased the net return of mung bean over FYM, VC, FYM + CR, and VC + CR. Naeem et al. (2006) reported the maximum net benefit of mung bean obtained from the treatment, where poultry manure was applied.

6.4 CONCLUSIONS

Organic farming may not lead to higher production and income in the short run as its returns are of long term nature. It is initially a soil-building process. Organic farming systems ensure built-in capacity to maintain and increase soil health and fertility leading to sustained increase in yield and production and low variability of crops which result to the stabilization and high jump in income and sustainability in agriculture. Findings of this study provided a sound base to believe that FYM + CR, VC + CR, FYM +

CR + B, and VC + CR + B, being at par, significantly increased the grain yield of mung bean over the control, but incorporation crop residue plus inoculation of seeds with biofertilizers (CC + PSB + Rhizobium) resulted the most economical treatment with respect to increasing net profit. This was because of the low price of biofertilizers compared with crop residue. Both of FYM + CR + B and VC + CR + B combinations resulted in improved grain quality and nutrient uptake by grain. The present study thus indicates that a combination of FYM + CR + biofertilizers or VC + RR + biofertilizers holds a promise for organic farming of mung bean.

REFERENCES

1. Cochran WG, Cox GM (1957) Experimental design. Wiley, New York. p 611
2. Dhaliwal R, Kler DS, Saini KS (2007) Effect of planting methods, farmyard manure and crop residue management on yield, yield contributing characters and correlations in mungbean under mungbean –Durum wheat ecosystem. Journal of Research 44(1):9-11
3. Duxbury KG, Gupta BR (2000) Effect of farmyard manure, chemical and biofertilizers on yiield and quality of rice (Oryza sativa L.) and soil properties. J Indian Soc Soil Sci 48(4):773-780
4. Haq A (1989) Studies on the yield and related morphological characters of some new mungbean genotypes in irrigated environment. M.Sc. Thesis. University of Agriculture, Faisalabad.
5. Jackson ML (1973) Soil chemical analysis. Prentice Hall of India Pvt Ltd., New Delhi.
6. Jamaval JS (2006) Effect of integrated nutrient management in maize (Zea mays) on succeeding winter crops under rainfed conditions. Indian J Agron 51(1):14-16
7. Jasdan M, Hutchaon SS (1996) Utilization of green manure for raising soil fertility in China. Soil Sci 135:65-69
8. John De Britto A, Sorna Girija L (2006) Investigation on the effect of organic and inorganic farming methods on black gram and green gram. Indian J Agric Res 40(3):204-207
9. Juliano BO (1985) Polysaccharides, protein and lipids of rice. In: Rice: chemistry and technology. American Association of Cereal Chemists, St. Paul. p 98
10. Kopke MM (1995) Studies on intercropping and biological N-fixation with Azospirillum in pearlmillet. M.Sc, Thesis, IARI.
11. Kumar R, Yadav NSP (2006) Improving the productivity and sustainability of rice (Oryza sativa)-wheat (Triticum aestivum) system through mungbean (Vigna radiata) residue management. In: Proceedings of the national symposium on conservation agriculture and environment. Banaras Hindu University, Varanasi, India.
12. Ladha S, Rolalndonu R, Moriu T (2000) Some aspects of the proper use chemical fertilizers combined with organic manures in rice production. Soils and Fertilizers in Taiwan, pp 13-3

13. Lianzheng S, Yixian U (1994) Some aspects of the proper use of chemical fertilizers combined with organic manures in rice production. Soils and Fertilizers in Taiwan. pp 13-3

14. Mandal S, Mandal M, Das M (2009) Simulation of inoleacetic acid production in Rhizobium isolate of Vigna mungo by root nodule phenolic acids. Arch Microbiol 191:389-393 PubMed Abstract | Publisher Full Text

15. Naeem M, Iqbal J, Ahmad Bakhsh MAA (2006) Comparative study of inorganic fertilizers and organic manures on yield and yield components of mungbean (Vigna radiat L.). Journal of Agriculture & Social Sciences 2(4):227-229 PubMed Abstract

16. Partap T (2006) The India organic pathway: making way for itself. Occasional paper 1. Bangalore: International Competence Center for Organic Agriculture (ICCOA).

17. Poonam S, Inderjit S, Veena K, Bains TS, Singh S (2007) Genotypic variability and association studies for growth, symbiotic parameters and grain yield in summer mungbean (Vigna radiate L.). Crop Improv 34(2):166-169

18. Prasad R (2005) Organic farming vis-à-vis modern agriculture. Curr Sci 89:252-254

19. Prasad R, Shivay YS, Kumar D, Sharma SN (2006) Learning by doing exercises in soil fertility. Division of Agronomy, Indian Agricultural Research Institute, New Delhi, India.

20. Reddy KC, Reddy KM (2005) Differential levels of vermicompost and nitrogen on growth and yield of onion (Allium cepa L.) Radish (Raphanus sativus L) cropping system. J Res ANGRAU 33(1):11-17

21. Sharma SK, Sharma SN (2004) Integrated nutrient management for sustainabilityof rice-wheat cropping system. Indian J Agric Sci 72:573-576

22. Sharma SN, Prasad R (1999) Effect of Sesbania green-manuring and mungbean residue incorporation on productivity and nitrogen uptake of a rice wheat cropping system. Bioresour Technol 87:171-175

23. Singh ID, Paroda NC (1994) Harvest index in cereals. Agron J 63:224-226

24. Srinivas M, Shaik M (2002) Performance of green gram and response functions as influenced by different levels of nitrogen and phosphorus. Crop Res Hisar India 24:458-462

25. Zodape ST, Mukhopadhyay S, Eswaran K, Reddy MP, Chikara J (2010) Enhanced yield and nutritional quality in green gram (Phaseolus radiate L) treated with sea-weed (Kappaphycus alvarezii) extract. J Sci Ind Res 69:468-471

CHAPTER 7

EVALUATION OF BIOFERTILIZERS IN IRRIGATED RICE: EFFECTS ON GRAIN YIELD AT DIFFERENT FERTILIZER RATE

NIÑO PAUL MEYNARD BANAYO, POMPE C. STA. CRUZ, EDNA A. AGUILAR, RODRIGO B. BADAYOS, AND STEPHAN M. HAEFELE

7.1 INTRODUCTION

Biofertilizers are becoming increasingly popular in many countries and for many crops. They are defined as products containing active or latent strains of soil microorganisms, either bacteria alone or in combination with algae or fungi that increase the plant availability and uptake of mineral nutrients [1]. In general, they contain free-living organisms associated with root surfaces but they may also include endophytes, microorganisms that are able to colonize the intercellular or even intracellular spaces of plant tissues without causing apparent damage to the host plant. The concept of biofertilizers was developed based on the observation that these microorganisms can have a beneficial effect on plant and crop growth (e.g., [2]). Consequently, a range of plant growth-promoting rhizobacteria (PGPR) has been identified and well characterized. Direct beneficial effects can occur when the microorganisms provide the plants with useful products. The best known case of this are microorganisms that can directly obtain N

This chapter was originally published under the Creative Commons Attribution License. Banayo NPM, Cruz PCS, Aguilar EA, Badayos RB, and Haefele SM. Evaluation of Biofertilizers in Irrigated Rice: Effects on Grain Yield at Different Fertilizer Rate. Agriculture 2,1 (2012).doi:10.3390/agriculture2010073.

from the atmosphere and convert this into organic forms usable by plants. Such biological nitrogen fixers (BNF) include members of the genus *Rhizobium, Azospirillum*, and blue-green algae. Rhizobia are symbiotically associated with legumes and nitrogen fixation occurs within root or stem nodules where the bacterium resides [3]. The genus *Azospirillum* also has several N-fixing species, which are rhizobacteria associated with monocots and dicots such as grasses, wheat, maize and *Brassica chinensis* L. [4,5]. *Azospirillum* strains have been isolated from rice repeatedly, and recently the strain *Azospirillum* sp. B510 has been sequenced [6,7]. Considerable N fixation by *Azotobacter* spp. and *Azospirillum* spp. in the rice crop rhizosphere was reported repeatedly [6,8], but others [9] questioned such high amounts of non-symbiotic N fixation in agriculture. Instead, it was hypothesized that the beneficial effect of *Azospirillum inoculums* may not derive from its N-fixing properties but from its stimulating effect on root development [2], probably often triggered by phytohormones [10]. This view was confirmed by [11], who concluded that the main effect of *Azospirillum* spp. is the stimulation of the density and length of root hairs, the rate of appearance of lateral roots, and the root surface area. Phytohormone production and a beneficial effect on plant growth were also shown for a range of other microorganisms [12,13].

Another important genus for biofertilizer producers is *Trichoderma*, a fungus present in nearly all soils. *Trichoderma* spp. thrive in the rhizosphere and can also attack and parasitize other fungi. *Trichoderma* spp. have been known for decades to increase plant growth and crop yield [14,15,16], to improve crop nutrition and fertilizer uptake [16,17], to speed up plant growth and enhance plant greenness [18], as well as to control numerous plant pathogens [19,20,21]. A part of these effects may also be related to the fact that some *Trichoderma* spp. seem to hasten the mineralization of organic materials [22], thus probably releasing nutrients from soil organic matter. Positive effects on plant nutrition were also described for other organisms, and many soil bacteria may enhance the mineral uptake of the plant, as for example by the increased solubility of phosphate in the soil solution [23].

There is a wide range of reports on the effect of biofertilizer application in crops grown in non-flooded soils (unlike lowland rice), and the technology for *Rhizobium* inoculation of leguminose plants is well established.

A review on results from *Azospirillum* inoculation experiments across the world and covering 20 years was conducted by [11]. They found a success rate of 60–70% with statistically significant yield increases on the order of 5–30%. However, the vast majority of these trials were on wheat, maize, sorghum, or millet, and only one of the experiments included in the analysis was on rice. Consequently, results from biofertilizer use in rice are still rare. Some reports from groups promoting the use of biofertilizers indicated considerable yield increases upon their use. *Trichoderma harzianum*, used as a coating agent for rice seed, was reported to result in a 15–20% yield increase compared with rice plants receiving full inorganic fertilizer rates only [22]. As already mentioned above [8], reported enhanced growth and development of rice and maize after the use of biofertilizer containing *Azospirillum* spp, and asserted the biofertilizer would provide 30–50% of the crop's N requirement. Similarly, [6] claimed that the inoculation of rice seedlings with *Azotobacter* spp. and *Azospirillum* spp. was able to substitute for the application of inorganic N fertilizer, and that this technology enabled rice yields of 3.9 to 6.4 t·ha^{-1} (yield increases in comparison with the control were about 2–3 t·ha^{-1}). Another study tested the effect of rice root inoculation with *Azospirillum* spp. under different N fertility levels, and found a more pronounced yield response at lower levels of inorganic N fertilization [24]. Generally, rice yield increases in this study were lower, and ranged around 0.5 t·ha^{-1}. A yield-increasing effect on rice by inoculation with *Azospirillum* sp. strain B510 was also shown by [25] but the experiment was conducted in pots only.

Based on these reports, it can be assumed that biofertilizers could offer an opportunity for rice farmers to increase yields, productivity, and resource use efficiency. And, the increasing availability of biofertilizers in many countries and regions and the sometimes aggressive marketing brings ever more farmers into contact with this technology. However, rice farmers get little advice on biofertilizers and their use from research or extension because so little is known on their usefulness in rice. Necessary would be recommendations describing under which conditions biofertilizers are effective, what their effect on the crop is, and how they should best be used. To start addressing these issues, we conducted this study, testing different biofertilizers in an irrigated lowland rice system in the Philippines during four seasons. The objectives of the study were (1) to evaluate the effects of different biofertilizers on irrigated rice grain yield, (2) to investigate possible interactions of the effect of these biofertilizers with different inorganic fertilizer rates, and (3) to determine, based on the results, whether biofertilizers are a possible option to improve the productivity of rice production and under which conditions they give good results.

TABLE 1: Average top-soil characteristics (0–15 cm depth) for all experimental seasons and both experimental sites.

Site		UPLB		IRRI	
Soil type		Anthraquic Gleysols		Anthraquic Gleysols	
		2008/2009 DS	2009 WS	2010 WS	2010/2011 DS
pH (1:1)	-	6.9	6.8	6.9	6.5
Total organic C	g kg^{-1}	18.6	15.9	16.2	15.0
Total soil N	g kg^{-1}	2.7	1.6	1.5	1.5
Olsen P	mg kg^{-1}	55	40	35	30
Avail K	cmol kg^{-1}	-	-	1.26	1.32
Exch K	cmol kg^{-1}	1.50	1.06	1.50	1.50
Exch Ca	cmol kg^{-1}	-	-	18.9	18.1
Exch Mg	cmol kg^{-1}	-	-	13.5	13.3
Exch Na	cmol kg^{-1}	-	-	1.01	1.00
CEC	cmol kg^{-1}	-	-	33.6	33.0
Clay	g kg^{-1}	-	-	441	445
Silt	g kg^{-1}	-	-	332	355
Sand	g kg^{-1}	-	-	227	200

7.2 MATERIALS AND METHODS

7.2.1 SITE DESCRIPTION

The experiments were conducted during two dry seasons (DS) and two wet seasons (WS). In the 2008/09 DS and the 2009 WS, an experimental site at the Central Experimental Station of the University of the Philippines at

Los Baños (CES-UPLB) was used, whereas the experiment in the 2010 WS and the 2010/11 DS was conducted at the Experimental Station of the International Rice Research Institute (IRRI) in Los Baños (ES-IRRI). Both experimental sites were located in close vicinity (about 1 km apart) in Laguna Province, Philippines (14°11' North, 121°15' East, 21 masl), in a typical lowland rice production area with the dominant soil type "anthraquic Gleysols" [26]. Detailed soil characteristics were analyzed only for the field at ES-IRRI (Table 1) but the soil at CES-UPLB was similar. The soil at both sites had a fine texture (clayey loam) and a high cation exchange capacity (CEC). Topsoil pH values at CES-UPLB in the 2009 DS and WS were 6.9 and 6.8, respectively, while pH values of 6.9 (2010 WS) and 6.5 (2011 DS) were observed at the ES-IRRI site. The soil organic carbon concentrations at both farms were relatively high, ranging between 1.5% and 1.9%. Related to this, organic N concentrations were also high at both farms (0.15–0.27%). The high soil organic matter content also caused high P availability as indicated by high Olsen P values, which were far above the critical low level of 10–15 $mg \cdot kg^{-1}$ [27]. Similarly, the exchangeable K was adequate for both experimental sites at the start of the cropping seasons [27].

TABLE 2: Inorganic fertilizer treatments in all four experimental seasons as ratio of the recommended rate (RR) and as actual nutrients applied in the dry and wet season.

Fertilizer Rate	Unit	Dry Season	Wet Season
0% RR	$N-P_2O_5-K_2O$ in $kg \cdot ha^{-1}$	0-0-0	0-0-0
25% RR	$N-P_2O_5-K_2O$ in $kg \cdot ha^{-1}$	30-15-15	22.5-7.5-7.5
50% RR	$N-P_2O_5-K_2O$ in $kg \cdot ha^{-1}$	60-30-30	45-15-15
100% RR	$N-P_2O_5-K_2O$ in $kg \cdot ha^{-1}$	120-60-60	90-30-30
or			
0% RR	N-P-K in $kg \cdot ha^{-1}$	0-0-0	0-0-0
25% RR	N-P-K in $kg \cdot ha^{-1}$	30-7-13	22.5-3-6
50% RR	N-P-K in $kg \cdot ha^{-1}$	60-13-25	45-7-13
100% RR	N-P-K in $kg \cdot ha^{-1}$	120-26-50	90-13-25

7.2.2 EXPERIMENTAL TREATMENTS AND DESIGN

In all four seasons, the experiment was a two-factor experiment arranged in a randomized complete block design (RCBD) with three replications. Main plots were assigned to four different fertilizer levels: i) the full recommended rate (100% RR) of inorganic fertilizer; ii) 50% RR, 25% of RR, and the Control treatment in which no inorganic fertilizer was applied. However, the recommended rate changed between seasons and was 120 kg N ha^{-1}, 60 kg P$_2$O$_5$ ha^{-1}, and 60 kg K$_2$O ha^{-1} in the DS, and 90 kg N ha^{-1}, and 30 kg P$_2$O$_5$ ha^{-1}, 30 kg K$_2$O ha^{-1} in the WS. The exact N, P, and K amounts applied are given in Table 2.

Subplots (30 m^2 each) were assigned to the different biofertilizers tested in the experiment. Three different biofertilizers available in the Philippines were used, and an overview of their characteristics is given in Table 3. The products were Bio-N® (BN), BioGroe® (BG), and BioSpark® (BS; the same product was called BioCon in 2009). In addition, a Control treatment was used in which no biofertilizer was applied. Thus, the total number of treatment combinations tested was 16.

BN was developed in the early 1980s by Dr. M Umali-Garcia [28]. According to the distributor (BIOTECH, UPLB), it contains *Azospirillum lipoferum* and *A. brasilense*, isolated from *Saccharum spontaneum* (local name is Talahib). BN is available in dry powder form in a 200-gram package, which can be used for seed inoculation, direct broadcasting on seeds, or mixed with water as a root dip. The BN product has a shelf-life of 3 months and the package we used was well before its expiry date. BN is specifically targeted at rice and corn.

The second product tested was BG, developed by Dr. ES Paterno of BIOTECH at UPLB. It contains unknown plant growth-promoting bacteria (rhizobacteria) that influence root growth by producing plant hormones and providing nutrients in soluble form [28].

The last product tested was BS, developed by Dr. VC Cuevas. According to personal information from her, it contains three different species of *Trichoderma* isolated from Philippine forest soils (including *Trichoderma harzianum*), and is mass-produced using a pure organic carrier [29]. The product can be used for seed coating or for soil application in the seedbed.

TABLE 3: Characteristics of the three biofertilizers used and tested.

Product ID	BN	BG	BS
Product name	Bio-N®	BioGroe®	BioSpark®
Active ingredient	*Azospirillum lipoferum, A. brasilense*	Plant growth-promoting rhizobacteria (not defined)	*Trichoderma parceramosum, T. pseudokoningii*, and UV-irradiated strain of *T. harzianum*
Active organism	Bacteria	Bacteria	Fungus
Product type	Dry powder in200-g pack	Dry powder in100-g pack	Dry powder in250-g pack
Carrier medium	Sterile charcoal/soil mixture	Sterile charcoal/soil mixture	Dry organic medium (rice hull)
Producer declared cell number	108 cfu g^{-1}	-	109 cfu g^{-1}
Shelf life	3 months	6 months	24 months
Product amount recommended and used (for 1 ha)	1000 g 40 kg^{-1} seed	400 g 40 kg^{-1} seed	200 g 40 kg^{-1} seed
2011 biofertilizer costs needed for 40 kg seed	US$6.82	US$3.64	US$6.36
Elemental contents *			
N %	0.13	0.34	1.27
P %	0.091	0.063	0.687
K %	0.22	0.24	0.72
Supplier	BioTech UPLB	BioTech UPLB	BioSpark Corp.

* *Source: Analytical Service Laboratory, GQNPC, IRRI.*

7.2.3 CROP ESTABLISHMENT AND MANAGEMENT

In all experiments, rice variety PSB Rc18, a modern-type variety with 120 days duration, was used. Seed for the BN and Control treatments was soaked for 24 h, incubated for another 24 h, and sown using the modified dapog (mat) method. BN was prepared in a slurry solution and applied by dipping the roots of the seedlings into the slurry, 1 h before transplanting in the field. For the BG and BS treatments, seeds were initially also soaked for 24 h. The biofertilizers BG and BS were then applied by mixing the seeds with the biofertilizer product, thus coating the seeds. BG and BS were applied at 400 g 40 kg^{-1} seed and 200 g 40 kg^{-1} seed, respectively. The seed-biofertilizer mixture was then incubated for 10 hours in an open jute sac to allow cooling, followed by 14 hours incubation in the closed sack like the control. Seeds were sown using the modified dapog method. In all treatments, 14-day-old seedlings were transplanted at 2–3 seedlings per hill with a planting distance of 20 cm × 20 cm. Missing hills were re-planted within 7 days after transplanting (DAT).

Inorganic fertilizers used for the fertilizer treatments were urea (46-0-0 N-P$_2$O$_5$-K$_2$O) and compound (14-14-14 N-P$_2$O$_5$-K$_2$O) fertilizer. Compound fertilizer was applied basal just before transplanting according to the treatment. The remaining N was applied in equal splits at 10 DAT and at 55 DAT. A water depth of 3–5 cm was aimed for at every irrigation from early tillering until 1–2 weeks before physiological maturity. To control insect pests and diseases in the 2010 WS and 2010/11 DS, granular Furadan was applied 20 DAT at a rate of 33 kg·ha^{-1} and Hopcin was applied at a rate of 0.8 L·ha^{-1} at flowering. Molluscicide was applied right after transplanting to control golden apple snails in the field. Post-emergence herbicide was applied once at the 2-3-leaf stage of emerging weeds. Hand-weeding was done thereafter as needed. Application rates were based on the recommended rate of the specific pesticides that were used.

7.2.4 SAMPLING AND STATISTICAL ANALYSIS

Grain yields were determined in the study for a 5-m^2 (2.5 m × 2.0 m) designated sampling area, which was strategically located at the center of each subplot, leaving at least two border rows. Grain moisture content was determined immediately after threshing (Riceter grain moisture meter, Kett Electric Laboratory, Tokyo, Japan) and all grain yields are reported at 14% moisture content. The data gathered in the study were statistically analyzed using the procedures described by [30]. Analysis of variance was conducted using SAS (Version 9.0) and treatment means were compared by the least significant difference (LSD) and were considered significant at $p \leq 0.05$.

7.3 RESULTS AND DISCUSSION

In all four seasons and across the biofertilizer treatments, grain yield increased with increasing amounts of applied fertilizer (Table 4, Figure 1). However, this increase was not always statistically significant and the yield increase varied considerably between seasons. Overall, the lowest grain yields occurred in the 2009 WS, ranging only from 1.9 to 2.7 t·ha^{-1}. Generally, low yields in that season were due to a typhoon that caused considerable damage through flooding of the experimental field and lodging of the crop. For this reason, the crop was harvested prematurely by about 1 week, which further reduced attainable yields.

Grain yields in the other three experimental seasons were similar and ranged from 4.0 to 5.2 t·ha^{-1} in the 2009 DS, from 3.4 to 5.1 t·ha^{-1} in the 2010 WS, and from 3.8 to 5.6 t·ha^{-1} in the 2010/11 DS. These ranges already indicate a relatively low yield increase due to fertilizer application in the 2008/09 DS (up to 1.2 t·ha^{-1} for the full fertilizer rate of 120-60-60 kg N-P_2O_5-K_2O ha^{-1}) and the 2009 WS (up to 0.8 t·ha^{-1} for the full fertilizer rate of 90-30-30 kg N-P_2O_5-K_2O ha^{-1}). A higher response to inorganic fertilizer was achieved in the 2010 WS (up to 1.7 t·ha^{-1} for the full fertilizer rate of 90-30-30 kg N-P_2O_5-K_2O ha^{-1}) and the 2010/11 DS (up to 1.8 t·ha^{-1} for the full fertilizer rate of 120-60-60 kg N-P_2O_5-K_2O ha^{-1}).

The effects of biofertilizer treatments on grain yield, depending on the inorganic fertilizer treatment, are shown in Table 4 and Figure 1. Significant yield increases due to biofertilizer use were observed in all experimental seasons with the exception of the 2008/09 DS. In the 2010/11 DS, no significant difference between the three biofertilizers tested was detected, but all three achieved better yields than the Control. The biofertilizer achieving the highest average grain yields across all four inorganic fertilizer treatments and in all four seasons was BN. Statistically significant interactions between biofertilizer treatment and inorganic fertilizer treatment could not be detected in any season (at $p \leq 0.05$), suggesting that the effect of the biofertilizer was independent of the inorganic fertilizer rate.

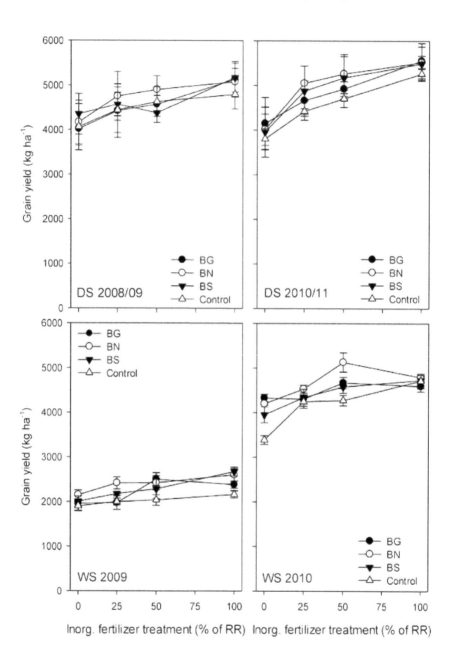

FIGURE 1: Grain yield of PSB Rc18 as affected by inorganic fertilizer rates and biofertilizer treatments. Shown are the results of all four seasons and bars represent the standard error of the mean.

However, there was a trend of higher yield increases due to biofertilizer use at low to medium inorganic fertilizer rates (Table 4, Figure 1). This trend was most obvious for the BN biofertilizer whereas the performance of the BS and BG biofertilizers was less consistent.

TABLE 4: Grain yield of the variety PSB Rc18 as affected by inorganic fertilizer level and biofertilizer treatments in all four experimental seasons and both sites.

Season	Biofertilizer treatment [***]	Inorganic fertilizer treatment [**]				
		0% RR	25% RR	50% RR	100% RR	Mean [*]
		Grain yield (kg·ha^{-1})				
2008/09 DS	BG	4016	4421	4569	5134	4508 a
	BN	4163	4753	4900	5081	4683 a
	BS	4351	4569	4375	5173	4610 a
	Control	4062	4440	4630	4799	4534 a
	Mean [*]	4158 c	4548 b	4617 b	5034 a	
2009 WS	BG	1963	1975	2502	2383	2206 bc
	BN	2149	2417	2420	2604	2398 a
	BS	2005	2179	2287	2674	2286 ab
	Control	1902	2000	2038	2165	2026 c
	Mean [*]	2005 c	2143 bc	2038 ab	2456 a	
2010 WS	BG	4326	4303	4670	4596	4482 ab
	BN	4197	4529	5131	4794	4663 a
	BS	3952	4336	4578	4732	4399 bc
	Control	3389	4245	4274	4716	4219 c
	Mean [*]	3965 c	4353 b	4659 a	4710 a	
2010/11 DS	BG	4145	4665	4926	5556	4825 a
	BN	4009	5049	5262	5519	4960 a
	BS	3955	4876	5175	5492	4861 a
	Control	3801	4420	4707	5265	4548 b
	Mean [*]	3977 c	4751 b	5014 b	5458 a	

*In each season, mean values in a column or row followed by the same letter are not significantly different at the 5% level of significance according to LSD; ** RR: Recommended rate: 120-60-60 kg N-P$_2$O$_5$-K$_2$O ha^{-1} in the DS; 90-30-30 kg N-P$_2$O$_5$-K$_2$O ha^{-1} in the WS; *** Biofertilizer treatments are described in detail in the text.*

The grain yield increase due to biofertilizer only (0% RR inorganic fertilizer treatment) usually ranged from 200 to 300 kg grain ha^{-1} for the best biofertilizers with the exception of the 2010 WS, when the BN treatment had an almost 800 kg·ha^{-1} better grain yield than the Control. In relative terms (Table 5), the seasonal yield increase across fertilizer treatments was between 5% and 18% for the BN biofertilizer (up to 24% for individual treatment combinations), between 3% and 13% for the BS biofertilizer (up to 24% for individual treatment combinations), and between 1% and 9% for the BG biofertilizer (up to 28% for individual treatment combinations). For the calculation of the relative yield increase, only average values could be compared and no statistical analysis could be conducted.

The effect of biofertilizer on the agronomic efficiency of N fertilizer (AEN) is shown in Table 6. For these calculations, the yield of each treatment was compared with the grain yield baseline (the Control treatment in which no biofertilizer and no inorganic fertilizer were used) and the yield increase was divided by the N rate applied. Again, only average values could be compared and no statistical analysis was possible. The results (Table 6) indicate considerably higher overall AEN values in the 2010 WS and the 2010/11 DS. Also, the AEN values are generally higher at low N rates and decrease with higher N application rates. The biggest AEN increase caused by biofertilizer occurred at the lowest N fertilizer rate (25% RR treatment), and, among the different biofertilizers tested, the BN biofertilizer resulted in the highest and most consistent AENs.

In our experiments, the selected biofertilizers were used as recommended by the producers but we could not check the viability or the contents of the products. Thus, we did not verify whether the biofertilizers contained the declared organisms (Table 6; the contents of BG remained unidentified) or the required number of living cells in the inoculate. The importance of quality control and regulation for biofertilizer production was emphasized by [31], who also pointed out that the frequent absence of such mechanisms can cause non-functional products. Maintenance of high standards for *Azospirillum* inoculants with proven efficient strains and cell numbers on the order of 1×10^9 to 1×10^{10} colony-forming units (cfu) g^{-1} or mL^{-1} was also requested by [11]. But, the fact that the products in our study caused a significant effect on grain yield in three out of four seasons (only two out of four seasons for BG) indicated that the biofertilizers tested had sufficient active ingredients and that the producers maintained a good quality over the four seasons (or 2.5 years). Theoretically, the effect of the biofertilizers could also have been caused by non-living ingredients but the applied amount was so small that even micronutrients could not explain the observed effects. Also, no micronutrient deficiencies are known from either of the two experimental sites.

TABLE 5: Relative yield increase over the Control treatments with the same inorganic fertilizer rate for all biofertilizers tested, in all seasons and at both experimental sites.

Season	Biofertilizer treatment [***]	Inorganic fertilizer treatment [**]				
		0% RR	25% RR	50% RR	100% RR	Mean
Relative yield increase (%) [*]						
2008/09 DS	BG	-1	0	-1	7	1
	BN	2	7	6	6	5
	BS	7	3	-6	8	3
	Control	-	-	-	-	-
2009 WS	BG	3	-1	23	10	9
	BN	13	21	19	20	18
	BS	5	9	12	24	13
	Control	-	-	-	-	-
2010 WS	BG	28	1	9	-3	8
	BN	24	7	20	2	12
	BS	17	2	7	0	6
	Control	-	-	-	-	-
2010/11 DS	BG	9	6	5	6	6
	BN	5	14	12	5	9
	BS	4	10	10	4	7
	Control	-	-	-	-	-

*The relative yield increase was calculated for treatment means and in comparison to the control without biofertilizer use but within the same inorganic fertilizer treatment; ** RR: Recommended rate: 120-60-60 kg $N-P_2O_5-K_2O$ ha^{-1} in the DS; 90-30-30 kg $N-P_2O_5-K_2O$ ha^{-1} in the WS; *** Biofertilizer treatments are described in detail in the text.*

TABLE 6: Estimated agronomic efficiency (AEN) of applied N depending on the inorganic fertilizer treatment and the biofertilizer used.

Season	Biofertilizer treatment **	Inorganic fertilizer treatment *			
		0% RR	25% RR	50% RR	100% RR
		Reference grain yield (kg·ha⁻¹)	AEN *** (kg grain yield increase kg⁻¹ N applied)		
2008/09 DS	BG		12	8	9
	BN		23	14	8
	BS		17	5	9
	Control	4062	13	9	6
2009 WS	BG		3	13	5
	BN		23	12	8
	BS		12	9	9
	Control	1902	4	3	3
2010 WS	BG		41	28	13
	BN		51	39	16
	BS		42	26	15
	Control	3389	38	20	15
2010/11 DS	BG		29	19	15
	BN		42	24	14
	BS		36	23	14
	Control	3801	21	15	12

* RR: Recommended rate: 120-60-60 kg N-P_2O_5-K_2O ha⁻¹ in the DS; 90-30-30 kg N-P_2O_5-K_2O ha⁻¹ in the WS; ** Biofertilizer treatments are described in detail in the text; *** For the estimation of AEN in each experimental season, the grain yield of the treatment without inorganic fertilizer and biofertilizer (0% RR and Control) was used as reference.

The general effect of inorganic fertilizer was as expected, and grain yields increased continuously with increasing fertilizer rates (Table 4). However, the response to inorganic fertilizer was low in the 2008/09 DS and the 2009 WS, as also indicated by the low AEN (Table 5). Good and economic values for AEN are usually 15–20 kg grain yield per kg N applied, and, at AEN < 10, inorganic fertilizer use may give negative economic returns depending on the input and output prices [32,33]. Low response in the 2009 WS can be explained by the negative effects of a typhoon and the early harvest. The low response in the 2008/09 DS could be due to the combination of a very fertile soil (high grain yield in the 0% RR treatment) and a limited yield potential in that season (low maximum yields in the 100% RR treatment).

The tested biofertilizers did increase grain yield significantly, and especially the BN biofertilizer did so consistently. Even in seasons in which no significant effect could be detected due to the yield variability between plots, the grain yield with biofertilizer was usually better than without. The seasonal yield increase across fertilizer treatments was between 5% and 18% for the BN biofertilizer (up to 24% for individual treatments; Table 5), which is within the 5–30% range reported for *Azospirillum inoculums* and non-rice crops by [4,11]. Similarly, the here-observed yield increase for the *Trichoderma*-based BS (3–13%) was close to the 15–20% rice yield increase described by [22]. The trend of yield increases between the different inorganic fertilizer treatments was not so clear across seasons but yield increases were often lower at higher inorganic fertilizer rates (Figure 1), which was also reported by [24]. Absolute grain yield increases due to biofertilizer were usually below 0.5 t·ha^{-1} (Table 1, Figure 1), corresponding to an estimated additional N uptake of less than 7.5 kg N ha^{-1} (based on 0.5% N in straw, 1.0% N in grain, and harvest index 0.5). Both values are far below grain yield increases and additional N uptake reported by [6] and [8], but similar to the rice grain yield increases reported by [24].

The calculated AEN values (Table 6) suggested higher N use efficiency for treatments with biofertilizer use. Increased nutrient uptake and fertilizer use efficiency were also reported for *Trichoderma* spp. [16,17,34] and for *Azospirillum* spp. [11]. But, the results could be explained in several ways. One possibility is that the biofertilizer stimulated root growth and thereby increased the uptake of indigenous N from the soil (the higher AEN would then be only an artifact of the calculation method). Second, the increased root growth could reduce N fertilizer losses, and the third option could be biological N fixation (which could explain the superior performance of the BN biofertilizer, supposedly containing organisms capable of biological N fixation). But, our experiment cannot answer the question of which process or combination of processes is at work here, if that is possible at all under field conditions [9].

7.4 SUMMARY AND CONCLUSIONS

The study was conducted to evaluate the effect of different biofertilizers on the grain yield of lowland rice, and investigate possible interaction effects with different inorganic fertilizer amounts. The results showed significant yield increases for all products tested in some seasons but the most consistent results were achieved by the *Azospirillum*-based biofertilizer. In most cases, the observed grain yield increases were not huge (0.2 to 0.5 $t \cdot ha^{-1}$) but could provide substantial income gains given the relatively low costs of all biofertilizers tested. The positive effect of the tested biofertilizers was not limited to low rates of inorganic fertilizers and some effect was still observed at grain yields up to 5 $t \cdot ha^{-1}$. However, the trends in our results seem to indicate that the use of biofertilizers might be most helpful in low- to medium-input systems. The results achieved can already be used to develop better advice for farmers on biofertilizer use in lowland rice, but several important questions remain. In particular, biofertilizers need to be evaluated under conditions with abiotic stresses typical for most low- to medium-input systems (e.g., under drought or low soil fertility) and with a range of germplasm because their effect might depend also on the variety used. More upstream-oriented research would be needed to better understand the actual mechanisms involved, which in turn could also contribute to making the best use of biofertilizers in rice-based systems.

REFERENCES

1. Vessey, J.K. Plant growth promoting rhizobacteria as biofertilizers. Plant Soil 2003, 255, 571–586, doi:10.1023/A:1026037216893.
2. Davidson, J. Plant beneficial bacteria. Biotechnol. 1988, 6, 282–286, doi:10.1038/nbt0388-282.
3. Dela Cruz, R.E. State of the art in biotechnology: Crop production. In Biotechnology for Agriculture, Forestry and the Environment; PCARRD: Los Baños, Laguna, Philippines, 1993.

4.   Bashan, Y.; Levanony, H. Current status of Azospirillum inoculation technology: Azospirillum as a challenge for agriculture. Can. J. Microbiol. 1990, 36, 591–608, doi:10.1139/m90-105.

5.   Singh, S.; Rekha, P.D.; Arun, A.B.; Hameed, A.; Singh, S.; Shen, F-T.; Young, C-C. Glutamate wastewater as a culture medium for Azospirillum rugosum production and its impact on plant growth. Biol. Fert. Soils 2011, 47, 419–426, doi:10.1007/s00374-011-0547-3.

6.   Razie, F.; Anas, I. Effect of Azotobacter and Azospirillum on growth and yield of rice grown on tidal swamp rice fields in south Kalimantan. Jurnal Tanah dan Lingkungan 2008, 10, 41–45.

7.   Kaneko, T.; Minamisawa, K.; Isawa, T.; Nakatsukasa, H.; Mitsui, H.; Kawaharada, Y.; Nakamura, Y.; Watanabe, A.; Kawashima, K.; Ono, A.; et al. Complete genomic structure of the cultivated rice endophyte Azospirillum sp. B510. DNA Res. 2010, 17, 37–50, doi:10.1093/dnares/dsp026.

8.   Sison, M.L.Q. Available biotechnologies and products. Presented at the workshop on promoting popular awareness and appreciation of biotechnolog. Cagayan de Oro City, Philippines, 16 February 1999.

9.   Giller, K.E.; Merckx, R. Exploring the boundaries of N2-fixation in cereals and grasses: An hypothetical and experimental framework. Symbiosis 2003, 35, 3–17.

10. Ereful, N.C.; Paterno, E.S. Assessment of cytokinin production in some plant growth-promoting bacteria. Asia Life Sci. 2007, 16, 137–152.

11. Okon, Y.; Labandera-Gonzales, C.A. Agronomic applications of Azospirillum: An evaluation of 20 years worldwide field inoculation. Soil Biol. Biochem. 1994, 26, 1591–1601, doi:10.1016/0038-0717(94)90311-5.

12. Fernando, L.M.; Merca, F.E.; Paterno, E.S. Isolation and partial structure elucidation of gibberellin produced by plant growth promoting bacteria (PGPB) and its effect on the growth of hybrid rice (Oryza sativa L.). Philipp. J. Crop Sci. 2010, 35, 12–22.

13. Difuntorum-Tamabalo, D.; Paterno, E.S.; Barraquio, W.; Duka, I.M. Identification of an indole-3-acetic acid-producing plant growth-promoting bacterium (PGPB) isolated from the roots of Centrosema pubescens Benth. Philipp. Agr. Sci. 2006, 89, 149–156.

14. Lindsey, D.L.; Baker, R. Effect of certain fungi on dwarf tomatoes grown under gnotobiotic conditions. Phytopathology 1967, 57, 1262–1263.

15. Chang, Y-C.; Chang, Y-C.; Baker, R.; Kleifeld, O.; Chet, I. Increased growth of plants in the presence of the biological control agent Trichoderma harzianum. Plant Diseases 1986, 70, 145–148, doi:10.1094/PD-70-145.

16. Harman, G.E. Myths and dogmas of biocontrol. Changes in perceptions derived from research on Trichoderma harzianum T-22. Plant Diseases 2000, 84, 377–393, doi:10.1094/PDIS.2000.84.4.377.

17. Yedidia, I.; Srivastva, A.K.; Kapulnik, Y.; Chet, I. Effect of Trichoderma harzianum on microelement concentrations and increased growth of cucumber plants. Plant Soil 2001, 235, 235–242, doi:10.1023/A:1011990013955.

18. Harman, G.E. Overview of mechanisms and uses of Trichoderma spp. Phytopathology 2006, 96, 190–194, doi:10.1094/PHYTO-96-0190.

19. Weindling, R. Trichoderma lignorum as a parasite of other soil fungi. Phytopathology 1932, 22, 837–845.

20. Shoresh, M.; Harman, G.E. The molecular basis of shoot responses of maize seedlings to Trichoderma harzianum T22 inoculation of the root: A proteomic approach. Plant Physiol. 2008, 147, 2147–2163, doi:10.1104/pp.108.123810.

21. Cuevas, V.C.; Sinohin, A.M.; Orajay, J.I. Performance of selected Philippine species of Trichoderma as biocontrol agents of damping off pathogens and as growth enhancer of vegetables in farmer's field. Philipp. Agr. Sci. 2005, 88, 63–71.

22. Cuevas, V.C. Rapid composting for intensive rice land use. In Innovation for Rural Development; SEAMEO-SEARCA: Los Baños, Philippines, 1991; Volume 1, pp. 5–10.

23. Goldstein, A.H.; Liu, S.T. Molecular cloning and regulation of a mineral phosphate solubilizing gene from Erwinia herbicola. Nat. Biotech. 1987, 5, 72–74, doi:10.1038/nbt0187-72.

24. Rajabamamohan, R.V.; Nayak, D.N.; Charyulu, P.B.B.N.; Adhy, T.K. Yield response of rice to root inoculation with Azospirillum. J. Agr. Sci. 1983, 100, 689–691, doi:10.1017/S0021859600035462.

25. Isawa, T.; Yasuda, M.; Awazaki, H.; Minamisawa, K.; Shinozaki, S.; Nakashita, H. Azospirillum sp. strain B510 enhances rice growth and yield. Microbes Environ. 2010, 25, 58–61, doi:10.1264/jsme2.ME09174.

26. World Reference Base for Soil Resources; Food and Agriculture Organization of the United Nations: Rome, Italy, 2006; p. 128.

27. Fairhurst, T.H.; Witt, C.; Buresh, R.J.; Dobermann, A. A Practical Guide to Nutrient Management, 2nd ed.; International Rice Research Institute, International Plant Nutrition Institute, and the International Potash Institute: Singapore, 2007; p. 89.

28. Paredes, J.C.; Go, I. Putting biofertilizers to good use. BioLife 2008, 5, 4–10.

29. Cuevas, V.C. Soil inoculation with Trichoderma pseudokoningii rifai enhances yield of rice. Philip. J. Sci. 2006, 135, 31–37.

30. Gomez, K.A.; Gomez, A.A. Statistical Procedures for Agricultural Research, 2nd ed.; John Wiley & Sons: New York, NY, USA, 1984; p. 680.

31. Reddy, L.N.; Giller, K.E. How effective are effective micro-organisms? LEISA Magazine 2008, 24, 18–19.

32. Haefele, S.M.; Sipaseuth, N.; Phengsouvanna, V.; Dounphady, K.; Vongsouthi, S. Agro-economic evaluation of fertilizer recommendations for rainfed lowland rice. Field Crop. Res. 2010, 119, 215–224, doi:10.1016/j.fcr.2010.07.002.

33. Witt, C.; Buresh, R.J.; Peng, S.; Balasubramanian, V.; Dobermann, A. Nutrient management. In Rice: A Practical Guide to Nutrient Management; Fairhurst, T.H., Witt, C., Buresh, R.J., Dobermann, A., Eds.; International Rice Research Institute, International Plant Nutrition Institute, and the International Potash Institute: Singapore, 2007; pp. 1–45.

34. Harman, G.E. Microbial tools to improve crop performance and profitability and to control plant diseases. In Proceedings of International Symposium on Biological Control of Plant Diseases for the New Century—Mode of Action and Application Technology, Taichung City, Taiwan, 12–13 November 2001.

CHAPTER 8

EFFECTS OF FERTILIZER AND PLANT DENSITY ON YIELD AND QUALITY OF ANISE (*PIMPINELLA ANISUM* L.)

MAHDI FARAVANI, BEHJAT SALARI, MOSTAFA HEIDARI, MOHAMMAD TAGHI KASHKI, AND BARAT ALI GHOLAMI

8.1 INTRODUCTION

Anise (*Pimpinella anisum* L.) is one of the most important annual medicinal plants from Apiaceae family in the world. It is a dainty, white-flowered plant, about 44 cm high, native of Iran, Turkey, Egypt, Greece, Crete and Asia Minor. Its active substances are used in various pharmaceutical and food industries (Nabizadeh et al., 2012). It is widely distributed and mainly cultivated for the seeds. The seeds of anise contain an essential oil (1 to 4%) and active substances like anethole used in various pharmaceutical and food industries (Klaus et al., 2009). It is a folkloric remedy for seizures in traditional medicine (Pourgholami et al., 1999; Staub, 2008; Özel, 2009; Aćimović, 2013).

Fertilizers have played an effective role in increasing crop yield, and the indiscriminate use of chemical fertilizers can cause environmental and

*This chapter was originally published under the Creative Commons Attribution License. Faravani M, Salari B, Heidari M, Kashki MT, and Gholami BA. Effects of Fertilizer and Plant Density on Yield and Quality of Anise (*Pimpinella anisum L.*). Journal of Agricultural Sciences **58**,3 (2013). DOI: 10.2298/ JAS1303209F.*

human health problems, depletion of non-renewable resources, and make plants less resistant to pests and diseases (Jevdjović and Maletić, 2006; Mal et al., 2013). Environmental problems related to the use of synthetic fertilizers and to organic waste management have led to an increased interest in the use of organic materials as an alternative source of nutrients for crops, but this is also associated with N_2O emissions (Aguilera et al., 2013).

The vermicompost is the product of composting using various worms, usually red wigglers, white worms, and other earthworms to create a heterogeneous mixture of decomposing vegetable or food waste, bedding materials, and vermicast. The vermicompost is a management practice that may contribute to sustainable agroecosystems by making them less dependent on inorganic fertilizers (Amossé et al., 2013).

The results from a study on anise (*Pimpinella anisum* L.) in Turkey showed that plant height, number of branches per plant, number of umbellets per plant, number of seeds per umbellet, seed yield and thousand seed weight were affected positively by organic fertilizer and organic–inorganic fertilizer combination application (Doğramaci and Arabaci, 2010). Nabizadeh et al. (2012) reported that the highest seed yield and the highest oil content were achieved (1,286 and 179 kg/ha) by fertilizer treatment using biological Nitroxin (3 l/ha) and chemical nitrogen (60 ha/kg) with the density of 25 plants/m² respectively.

Plant spacing is an important factor in determining the microenvironment in the anise field and can lead to a higher yield in the crop by favorably affecting the absorption of nutrients and exposure of the plant to the light. The current study was carried out to determine the effects of different plant densities and type of fertilizers on the yield of seed and its main components—the weight of 1000 seeds, the number of umbels per plant, the number of umbellets per umbel, the number of seeds per umbellet and yield of essential oil.

8.2 MATERIAL AND METHODS

This study was conducted at Khorasan Agriculture and Natural Resources Center with geographic coordinates of latitudes 36°13' N and longitudes

59°25' E at an average altitude of 985 m above sea level in 2011–2012. According to De Martonne climate classification, the areas of this province are located in arid or semi-arid climate. While, according to Emberger climate classification, most of areas of this province are located in cold-dry climate. In order to determine the physical and chemical properties of soil, soil sampling was done before planting. The results of soil analysis showed that the electrical conductivity of the saturation extract was 1.62 dS m^{-1} and pH value was 8. Seed planting was done in 36 plots with dimensions 3 m × 2 m with row spacing of 50 cm and seed planting depth of 3 cm on 21 April 2012.

The experiment was conducted by using the split-plot design which involved two experimental factors. The main factor was fertilization, applied in four treatments: control, vermicompost – 5 t/ha, cow manure – 25 t/ha, and full chemical fertilizer (NPK) – 60 kg/ha (with the same rate of each nutrient). The second factor was plant density, applied at three levels: 17, 25, and 50 plants/m^2. Levels of fertilizers and plant densities with the whole plots were arranged in a randomized complete block design. The vermicompost used in this study was produced by the activity of *Eisenia foetida* worm on manure. Irrigation was performed weekly. Thinning and weed control were done at growth stages of 4–6 leaves in order to get the desirable plant density and the best use of space and nutrient requirements.

Plants were harvested when leaf color changed to yellow. Different plant morphological characters as the number of umbels per plant, number of umbellets per umbel, plant height, internode distance, the number of lateral branches, grain number, grain weight and yield components were measured from five randomly selected plants. A ruler was applied to measure the length and the width of the canopy. Canopy area (S) was calculated by using the following formula: S = 3.14 × (x + y)/2, where: x is length of canopy, and y is canopy width.

The seed essential oil was measured by Clevenger hydro-distillation method. Calculations and statistical analysis were done by using SAS statistical software. Duncan's multiple range tests were used for means comparisons.

8.3 RESULTS AND DISCUSSION

An analysis of variance showed that the anise plant height was not affected by applied treatments ($p<0.05$) but plant height was significantly ($p<0.05$) changed by the interaction between plant density and fertilizer (Table 1). The maximum plant height (42.83 cm) was observed in combination of 50 plants per square meter and application of vermicompost (Table 2). The maximum cover (1,442 cm^2) was obtained when using vermicompost ($p<0.05$) but it was not affected by plant density. Lateral branches were significantly ($p<0.05$) affected by plant density; as plant density increased, plants lateral branches decreased. The maximum number of branches was observed in plant density of 25 plants/m^2.

Thousand kernel weight was not affected significantly ($p<0.05$) by applied treatments. However, the highest thousand kernel weight with 3.5 g was observed in plant density of 25 plants/m^2. Number of umbels per plant and number of umbellets per umbel were affected ($p<0.05$) by fertilizers, but the effect of plant density and its interaction with fertilizer were not significant ($p<0.05$). The maximum number of umbels per plant – 30.56 and 31.56 were found in the application of vermicompost and cow manure, respectively. No significant difference ($p<0.05$) was observed among the applications of the treatments, but their interaction was significant ($p<0.01$). The maximum number of seeds (8.66 seeds per umbellet) was obtained in application of cow manure and using plant density of 50 plants/m^2. Both seed yield and biological yield were affected by plant density ($p<0.05$) and no difference was observed in the fertilizer consumption. When a density of 50 plants per square meter was used, the highest seed yield (552.26 kg/ha) and biological yield (876.67 kg/ha) were obtained in the field. The lowest yield was observed at plant density of 17 plants/m^2 (Table 2).

Results indicated significant differences ($p<0.05$) for harvest index in different levels of fertilizers without any differences in the level of plant density. The fertilizer caused a greater decrease in harvest index (HI) through an increased biological yield or vegetative growth than seed yield. The highest harvest index (23%) was observed in the control treatment.

The integrated use of plant density and fertilizer (cow manure) increased the percentage of essential oil to 6.18% in 25 plants per square meter and the essential oil yield to 20.19 kg/ha in 25 plants per square meter (Table 2). The application of organic fertilizer did not indicate any significant effect on the essential oil, but its interaction with plant density showed a significant (p<0.01) difference. Choosing an optimum plant density and fertilizer may improve soil physical and chemical properties, nutrient availability that might increase the quality of yield. It was also observed that at the same oil yield level of essential oil of *Artemisia annua*, increased 1.5, 2 and 2.5 fold when the plant densities increased about 2, 4, and 8 fold (Ram et al., 1997).

Khorshidi (2009) in his research on Foeniculum vulgare Mill. var. "Soroksary" reported that the highest essential oil percentage (3.53%) was obtained when applying the lowest densities of planting. The highest percentage of anethole (83.07%), estragol (3.47%), fenchone (8.04%), p-cymene (4.45%), α-terpinene (0.54%), sabinene (0.51%), and α-pinene (0.48%) were obtained with the space between plants of 25, 10, 20, 20, 15, 20, and 25 cm, respectively.

The yield and essential oil of aniseed are affected by the genotype, the ecological conditions and especially by agricultural practices, such as the sowing date, fertilizer and water application, and plant density (Nabizadeh et al., 2012).

8.4 CONCLUSION

The highest yields of seeds and essential oil of anise were obtained from the application of vermicompost and optimum plant density of 25 and 50 plants per square meter. Seedlings grown at the highest planting density (50 plants/m^2) and treated with vermicompost (5 t/ha) had better development than all groups without vermicompost, even at the lowest density. The application of organic fertilizers has great potential to accelerate growth development and increase germination, growth, flowering and fruit production when applying an optimum plant density.

TABLE 1: Mean squares from analysis of variance for morphological characters of anise at different plant densities and fertilizer treatments.

Source of variation	Degree of freedom	Biological yield	Seed yield	Harvest index (HI)	1,000 kernel weight	No. of seeds per umbellet	No. of umbellets per umbel	No. of umbels per plant	Canopy area	No. of lateral branches	Plant height	Essential oil yield	Essential oil percentage
Replication	2	31,743	7,952.33	0.00004	0.25	1.58	16.30**	23.36	223.74	0.0007	1.27	45.24	0.49
Fertilizer (F)	3	33,773ns	14,041.92ns	0.007*	1.62ns	7.87ns	6.89*	148.66*	75,898.77**	0.091ns	22.97ns	21.18ns	0.25ns
Main plot	6	24,258	14,855.35	0.001	0.65	2.99	0.91	16.36	5,376.40	0.082	10.71	14.91	0.25
Plant density (PD)	2	149,791,102**	411.75	0.004ns	0.25ns	0.25ns	0.53ns	4.96ns	17,548.11ns	0.63*	0.82ns	534.31**	0.44*
Interaction FxPD	6	82,018*	9,710.72ns	0.003ns	0.54ns	3.87*	1.75ns	24.40ns	65,811.91**	0.21ns	8.18*	49.32ns	2.30**
Subplot	16	23,643	13,269.49	0.001	0.68	1.01	3.44	24.61	5,188.55	0.15	2.97	26.61	5.73
CV (%)		8.40	32.43	17.43	24.74	12.99	17.83	18.00	5.43	12.50	4.25	37.85	16.30

*and ** means significant difference at $p < 0.05$ and $p < 0.01$ respectively.

TABLE 2: Multiple comparisons of measured characters of anise at different plant densities and fertilizer treatments.

Treatments	Biomass (kg/ha)	Seed yield (kg/ha)	Harvest index (HI)	1,000 kernels weight (g)	No. of seeds per umbellet	No. of umbellets per umbel	No. of umbels per plant	Canopy area (cm²)	No. of lateral branches	Plant height (cm)	Essential oil yield (kg/ha)	Essential oil (%)
Plant density (plants/m²)												
50	3,059.00a	552.26a	0.18a	3.25a	7.91a	10.20a	27.58a	1,306b	3.14ab	40.73a	20.19a	3.58b
25	1,552.08a	328.59b	0.21a	3.50a	7.66a	10.41a	28.16a	1,302b	3.31a	40.69a	13.08ab	4.40a
17	876.66c	184.66b	0.21a	3.25a	7.66a	10.62a	26.91a	1,370a	2.86b	40.26a	6.85b	3.71b
Fertilizers												
Control	1,765.82a	403.68a	0.23a	3.88a	6.44b	9.44c	24.22b	1,316b	3.00a	40.00b	14.93a	3.78a
Vermicompost	1,912.96a	319.40a	0.17b	3.00a	8.00ab	10.77ab	30.56a	1,442a	3.24a	42.92a	14.93a	3.69a
Cow manure	1,815.85a	326.81a	0.19b	3.00a	8.66a	11.44a	31.56a	1,218c	3.11a	9.92ab	12.66a	3.92a
NPK	1,822.37a	370.80a	0.20ab	3.44a	7.88ab	10.00bc	23.89b	1,329b	3.07a	39.40b	11.97a	4.10a

Means followed by the same letter are not significantly different at p<0.05 according to Duncan's multiple range test.

REFERENCES

1. Aćimović, M.G. (2013): The influence of fertilization on yield of caraway, anise and coriander in organic agriculture. Journal of Agricultural Sciences (Belgrade) 58(2):85-94.
2. Aguilera, E., Lassaletta, L., Sanz-Cobena, A., Garnier, J., Vallejo, A. (2013): The potential of organic fertilizers and water management to reduce N2O emissions in Mediterranean climate cropping systems. A review. Agriculture, Ecosystems and Environment 164:32-52.
3. Amossé, J., Bettarel, Y., Bouvier, C., Bouvier, T., Tran Duc, T., Doan Thu, T., Jouquet, P. (2013): The flows of nitrogen, bacteria and viruses from the soil to water compartments are influenced by earthworm activity and organic fertilization (compost vs. vermicompost). Soil Biology and Biochemistry 66:197-203.
4. Doğramaci, S., Arabaci, O. (2010): The effect of the organic and inorganic fertilizer applications on yield and yield components of anise (*Pimpinella anisum* L.) type and ecotypes. Journal of Adnan Menderes University, Agricultural Faculty 7(2):103-109.
5. Klaus, A., Beatović, D., Nikšić, M., Jelačić, S., Petrović, T. (2009): Antibacterial activity oils from Serbia against the Listeria monocytogenes. Journal of Agricultural Sciences (Belgrade) 54(2):95-104.
6. Khorshidi, J., Tabatabaie, M.F., Omidbaigi, R., Sefidkon, F. (2009): Effect of densities of planting on yield and essential oil components of fennel (Foeniculum vulgare Mill. var. Soroksary). Journal of Agricultural Science 1(1):152-157.
7. Jevdjović, R., Maletić, R. (2006): Effects of application of certain types of fertilizers on anise seed yield and quality. Journal of Agricultural Sciences (Belgrade) 52(2):117-122.
8. Mal, S., Chattopadhyay, G.N., Chakrabarti, K. (2013): Compost quality assessment for successful organic waste recycling. Ecoscan 3(spec. issue):199-203.
9. Nabizadeh, E., Habibi, H., Hosainpour, M. (2012): The effect of fertilizers and biological nitrogen and planting density on yield quality and quantity *Pimpinella anisum* L. European Journal of Experimental Biology 2:1326-1336.
10. Özel, A. (2009): Anise (*Pimpinella anisum*): changes in yields and component composition on harvesting at different stages of plant maturity. Experimental Agriculture 45:117-126.
11. Pourgholami, M.H., Majzoob, S., Javadi, M., Kamalinejad, M., Fanaee, G.H.R., Sayyah, M. (1999): The fruit essential oil of *Pimpinella anisum* exerts anticonvulsant effects in mice. Journal of Ethnopharmacology 66:211-215.
12. Ram, M., Gupta, M., Dwivedi, S., Kumar, S. (1997): Effect of plant density on the yields of artemisinin and essential oil in Artemisia annua cropped under low input cost management in north-central India. Planta Medica 63:372-374.
13. Staub, J. (2008): 75 exceptional herbs for your garden. Gibbs Smith, Layton, Utah, USA.

MICROBIAL DIVERSITY OF VERMICOMPOST BACTERIA THAT EXHIBIT USEFUL AGRICULTURAL TRAITS AND WASTE MANAGEMENT POTENTIAL

JAYAKUMAR PATHMA and NATARAJAN SAKTHIVEL

9.1 INTRODUCTION

Soil, is the soul of infinite life that promotes diverse microflora. Soil bacteria viz., *Bacillus, Pseudomonas* and *Streptomyces* etc., are prolific producers of secondary metabolites which act against numerous co-existing phytopathogeic fungi and human pathogenic bacteria (Pathma et al. 2011b). Earthworms are popularly known as the "farmer's friend" or "nature's plowman." Earthworm influences microbial community, physical and chemical properties of soil. They breakdown large soil particles and leaf litter and thereby increase the availability of organic matter for microbial degradation and transforms organic wastes into valuable vermicomposts by grinding and digesting them with the help of aerobic and anaerobic microbes (Maboeta and Van Rensburg 2003). Earthworms ac-

This chapter was originally published under the Creative Commons Attribution License. Pathma J and Sakthivel N. Microbial Diversity of Vermicompost Bacteria that Exhibit Useful Agricultural Traits and Waste Management Potential. SpringerPlus 1,26 (2012). doi:10.1186/2193-1801-1-26.

tivity is found to enhance the beneficial microflora and suppress harmful pathogenic microbes. Soil wormcasts are rich source of micro and macro-nutrients, and microbial enzymes (Lavelle and Martin 1992). Vermicom-posting is an efficient nutrient recycling process that involves harnessing earthworms as versatile natural bioreactors for organic matter decomposi-tion. Due to richness in nutrient availability and microbial activity ver-micomposts increase soil fertility, enhance plant growth and suppress the population of plant pathogens and pests. This review paper describes the bacterial biodiversity and nutrient status of vermicomposts and their im-portance in agriculture and waste management.

9.2 EARTHWORMS

Earthworms are capable of transforming garbage into 'gold'. Charles Dar-win described earthworms as the 'unheralded soldiers of mankind,' and Aristotle called them as the 'intestine of earth', as they could digest a wide variety of organic materials (Darwin and Seward 1903; Martin 1976). Soil volume, microflora and fauna influenced by earthworms have been termed as "drilosphere" and the soil volume includes the external structures pro-duced by earthworms such as surface and below ground casts, burrows, middens, diapause chambers as well as the earthworms' body surface and internal gut associated structures in contact with the soil (Lavelle et al. 1989; Brown et al. 2000). Earthworms play an essential role in carbon turnover, soil formation, participates in cellulose degradation and humus accumulation. Earthworm activity profoundly affects the physical, chemi-cal and biological properties of soil. Earthworms are voracious feeders of organic wastes and they utilize only a small portion of these wastes for their growth and excrete a large proportion of wastes consumed in a half digested form (Edwards and Lofty 1977; Kale and Bano 1986; Jambhekar 1992). Earthworms intestine contains a wide range of microorganisms, enzymes and hormones which aid in rapid decomposition of half-digested material transforming them into vermicompost in a short time (neary 4–8 weeks) (Ghosh et al. 1999; Nagavallemma et al. 2004) compared to tra-ditional composting process which takes the advantage of microbes alone and thereby requires a prolonged period (nearly 20 weeks) for compost production (Bernal et al. 1998; Sánchez-Monedero et al. 2001). As the organic matter passes through the gizzard of the earthworm it is grounded into a fine powder after which the digestive enzymes, microorganisms and other fermenting substances act on them further aiding their breakdown within the gut, and finally passes out in the form of "casts" which are later acted upon by earthworm gut associated microbes converting them into mature product, the "vermicomposts" (Dominguez and Edwards 2004).

Earthworms, grouped under phylum annelida are long, narrow, cylindrical, bilaterally symmetrical, segmented soil dwelling invertebrates with a glistening dark brown body covered with delicate cuticle. They are hermaphrodites and weigh over 1,400–1,500 mg after 8–10 weeks. Their body contains 65% protein (70–80% high quality 'lysine rich protein' on a dry weight basis), 14% fats, 14% carbohydrates, and 3% ash. Their life span varies between 3–7 years depending upon the species and ecological situation. The gut of earthworm is a straight tube starting from mouth followed by a muscular pharynx, oesophagus, thin walled crop, muscular gizzard, foregut, midgut, hindgut, associated digestive glands, and ending with anus. The gut consisted of mucus containing protein and polysaccharides, organic and mineral matter, amino acids and microbial symbionts viz., bacteria, protozoa and microfungi. The increased organic carbon, total organic carbon and nitrogen and moisture content in the earthworm gut provide an optimal environment for the activation of dormant microbes and germination of endospores etc. A wide array of digestive enzymes such as amylase, cellulase, protease, lipase, chitinase and urease were reported from earthworm's alimentary canal. The gut microbes were found to be responsible for the cellulase and mannose activities (Munnoli et al. 2010). Earthworms comminutes the substrate, thereby increases the surface area for microbial degradation constituting to the active phase of vermicomposting. As this crushed organic matter passes through the gut it get mixed up with the gut associated microbes and the digestive enzymes and finally leaves the gut in partially digested form as "casts" after which the microbes takes up the process of decomposition contributing to the maturation phase (Lazcano et al. 2008).

Association of earthworms with microbes is found to be complex. Certain groups of microbes were found to be a part of earthworm's diet which is evidenced by the destruction of certain microbes as they pass through the earthworms digestive system. Few yeasts, protozoa and certain groups of fungi such as F*usarium oxysporum, Alternaria solani,* and microfungi were digested by the earthworms, *Drawida calebi, Lumbricus terrestris* and *Eisenia foetida. Bacillus cereus* var *mycoides* were reported to decrease during gut passage while *Escherichia coli* and *Serratia marcessens* were completely eliminated during passage through earthworm gut (Edwards and Fletcher, 1988).

Earthworms are classified into epigeic, anecic and endogeic species based on definite ecological and trophic functions (Brown 1995; Bhatnagar and Palta 1996) (Table 1). Epigeic earthworms are smaller in size, with uniformly pigmented body, short life cycle, high reproduction rate and regeneration. They dwell in superficial soil surface within litters, feeds on the surface litter and mineralize them. They are phytophagous and rarely ingest soil. They contain an active gizzard which aids in rapid conversion of organic matter into vermicomposts. In addition epigeic earthworms are efficient bio-degraders and nutrient releasers, tolerant to disturbances, aids in litter comminution and early decomposition and hence can be efficiently used for vermicomposting. Epigeic earthworms includes *Eisenia foetida, Lumbricus rubellus, L. castaneus, L. festivus, Eiseniella tetraedra, Bimastus minusculus, B. eiseni, Dendrodrilus rubidus, Dendrobaena veneta, D. octaedra*. Endogeics earthworms are small to large sized worms, with weakly pigmented body, life cycle of medium duration, moderately tolerant to disturbance, forms extensive horizontal burrows and they are geophagous feeding on particulate organic matter and soil. They bring about pronounced changes in soil physical structure and can efficiently utilize energy from poor soils, hence can be used for soil improvements. Endogeics include *Aporrectodea caliginosa, A. trapezoides, A. rosea, Millsonia anomala, Octolasion cyaneum, O. lacteum, Pontoscolex corethrurus, Allolobophora chlorotica* and *Aminthas sp.* They are further classified into polyhumic endogeic which are small sized, rich soil feeding earthworms, dwelling in top soil (A1); mesohumic endogeic which are medium sized worms, dwelling in A and B horizon, feeding on bulk (A1) soil; and oligohumic endogeic which are very large worms, dwelling in B and C horizons, feeding on poor, deep soil. Aneceics are larger, dorsally pigmented worms, with low reproductive rate, sensitive to disturbance, nocturnal, phytogeophagous, bury the surface litter, forms middens and extensive, deep, permanent vertical burrows, and live in them. Formation of vertical burrows affects airwater relationship and movement from deep layers to surface helps in efficient mixing of nutrients. *Lumbricus terrestris, L. polyphemus* and *Aporrectodea longa* are examples of aneceics earthworms (Kooch and Jalilvand 2008). Epigeics and aneceics are harnessed largely for vermicomposting (Asha et al. 2008). Epigeics namely *Eisenia foetida* (Hartenstein et al. 1979), *Eudrilus eugeniae* (Kale and Bano 1988), *Perionyx excavatus* (Sinha et al. 2002; Suthar and Singh 2008) and *Eisenia anderi* (Munnoli et al. 2010) have been used in converting organic wastes into vermicompost. These surface dwellers capable of working on litter layers converting them into manure are of no significant value in modifying the soil structure. In contrast, anecics such as *Lampito mauritii* are efficient creators of an effective drilosphere as well as excellent compost producers (Ismail 1997). Earthworms thus act as natural bio-reactors, altering the nature of the organic waste by fragmenting them.

Earthworm activity engineers the soil by forming extensive burrows which loosen the soil and makes it porous. These pores improve aeration, water absorption, drainage and easy root penetration. Soil aggregates formed by earthworms and associated microbes, in the casts and burrow walls play an indispensible role in soil air ecosystem. These aggregates are mineral granules bonded in a way to resist erosion and to avoid soil compaction both in wet and dry condition. Earthworms speed up soil reclamation and make them productive by restoring beneficial microflora (Nakamura 1996). Thus degraded unproductive soils and land degraded by mining could be engineered physically, chemically and biologically and made productive by earthworms. Hence earthworms are termed as ecosystem engineers (Brown et al. 2000; Munnoli et al. 2010).

9.3 VERMICOMPOSTING

Vermicomposting is a non-thermophilic biological oxidation process in which organic material are converted into vermicompost which is a peat like material, exhibiting high porosity, aeration, drainage, water holding capacity and rich microbial activities (Edwards 1998; Atiyeh et al. 2000b; Arancon et al. 2004a), through the interactions between earthworms and associated microbes. Vermiculture is a cost-effective tool for environmentally sound waste management (Banu et al. 2001; Asha et al. 2008). Earthworms are the crucial drivers of the process, as they aerate, condition and fragment the substrate and thereby drastically alter the microbial activity and their biodegradation potential (Fracchia et al. 2006; Lazcano et al. 2008). Several enzymes, intestinal mucus and antibiotics in earthworm's intestinal tract play an important role in the breakdown of organic macromolecules. Biodegradable organic wastes such as crop residues, municipal, hospital and industrial wastes pose major problems in disposal and treatment. Release of unprocessed animal manures into agricultural fields contaminates ground water causing public health risk. Vermicomposting is the best alternative to conventional composting and differs from it in several ways (Gandhi et al. 1997). Vermicomposting hastens the decomposition process by 2–5 times, thereby quickens the conversion of wastes into valuable biofertilizer and produces much more homogenous materials compared to thermophilic composting (Bhatnagar and Palta 1996; Atiyeh et al. 2000a). Distinct differences exist between the microbial communities found in vermicomposts and composts and hence the nature of the microbial processes is quite different in vermicomposting and composting (Subler et al. 1998). The active phase of composting is the thermophilic stage characterized by thermophilic bacterial community where intensive decomposition takes place followed by a mesophilic maturation phase (Lazcano et al. 2008; Vivas et al. 2009). Vermicomposting is a mesophilic process characterized by mesophilic bacteria and fungi (Benitez et al. 1999). Vermicomposting comprises of an active stage during which earthworms and associated microbes jointly process the substrate and the maturation phase that involves the action of associated microbes and occurs once the worm's moves to the fresher layers of undigested waste or when the product is removed from the vermireacter. The duration of the active phase depends on the species and density of the earthworms involved (Ndegwa et al. 2000; Lazcano et al. 2008; Aira et al. 2011). A wide range of oganic wastes viz., horticultural residues from processed potatoes (Edwards 1988); mushroom wastes (Edwards 1988; Tajbakhsh et al. 2008); horse wastes (Hartenstein et al. 1979; Edwards et al. 1998); pig wastes (Chan and Griffiths 1988; Reeh 1992); brewery wastes (Butt 1993); sericulture wastes (Gunathilagraj and Ravignanam 1996); municipal sewage

sludge (Mitchell et al. 1980; Dominguez et al. 2000); agricultural residues (Bansal and Kapoor 2000); weeds (Gajalakshmi et al. 2001); cattle
dung (Gunadi et al. 2002); industrial refuse such as paper wastes (Butt
1993; Elvira et al. 1995; Gajalakshmi et al. 2002); sludge from paper mills
and dairy plants (Elvira et al. 1997; Banu et al. 2001); domestic kitchen
wastes (Sinha et al. 2002); urban residues and animal wastes (Edwards et
al. 1985; Edwards 1988) can be vermicomposted (Sharma et al. 2005).

Effects of vermicomposting on pH, electrical conductivity (EC), C:N
ratio and other nutrients have been documented. Earthworm activity
reduced pH and C:N ratio in manure (Gandhi et al. 1997; Atiyeh et al.
2000b). Chemical analysis showed vermicompost had a lower pH, EC, organic carbon (OC) (Nardi et al. 1983; Albanell et al. 1988; Mitchell 1997),
C:N ratio (Riffaldi and Levi-Minzi 1983; Albanell et al. 1988), nitrogen
and potassium and higher amounts of total phosphorous and micronutrients compared to the parent material (Hashemimajd et al. 2004). Slightly
decreased pH values of vermicompost compared to traditional compost
might be attributed due to mineralization of N and P, microbial decomposition of organic materials into intermediate organic acids, fulvic acids,
humic acids (Lazcano et al. 2008; Albanell et al. 1988; Chan and Griffiths
1988; Subler et al. 1998) and concomitant production of CO_2 (Elvira et al.
1998; Garg et al. 2006). Vermicomposting of paper mill and dairy sludge
resulted in 1.2–1.7 fold loss of organic carbon as CO_2 (Elvira et al. 1998).
In contrast to the parent material used, vermicomposts contain higher humic acid substances (Albanell et al. 1988). Humic acid substances occur
naturally in mature animal manure, sewage sludge or paper-mill sludge,
but vermicomposting drastically increases the rate of production and their
amount from 40–60 percent compared to traditional composting. The enhancement in humification processes is by fragmentation and size reduction of organic matter, increased microbial activity within earthworm intestine and soil aeration by earthworm feeding and movement (Dominguez
and Edwards, 2004). EC indicates the salinity of the organic amendment.
Minor production of soluble metabolites such as ammonium and precipitation of dissolved salts during vermicomposting lead to lower EC values.
Compared to the parent material used, vermicomposts contain less soluble
salts and greater cation exchange capacity (Holtzclaw and Sposito 1979;
Albanell et al. 1988). C:N ratio is an indicator of the degree of decompo-

sition. During the process of biooxidation, CO_2 and N is lost and loss of N takes place at a comparatively lower rate. Comparison of compost and vermicompost showed that vermicompost had significantly less C:N ratios as they underwent intense decomposition (Lazcano et al. 2008).

Vermicomposting of cow manure using earthworm species *E. andrei* (Atiyeh et al. 2000b) and *E. foetida* (Hand et al. 1988) favored nitrification, resulting in the rapid conversion of ammonium-nitrogen to nitrate-nitrogen. Vermicomposting increased the concentration of nitrate-nitrogen to 28 fold after 17 weeks, while in conventional compost there was only 3-fold increase (Subler et al. 1998; Atiyeh et al. 2000a). Increase in ash concentration during vermicomposting suggests that vermicomposting accelerates the rate of mineralization (Albanell et al. 1988). Mineralization is the process in which the chemical compounds in the organic matter decompose or oxidise into forms that could be easily assimilated by the plants. Increase in ash content increases the rate of mineralization. Ash is an alkaline substance which hinders the formation of H_2S as well as improves the availibility of O_2 and thereby renders composts odorless. Thus vermicomposting increases the ash content and accelerates the rate of mineralization which is essential to make nutrients available to plants. The observed increase of total phosphorous (TP) in vermicompost is probably due to mineralization and mobilization of phosphorus resulting from the enhanced phosphatase activity by microorganisms in the gut epithelium of the earthworms (Zhang et al. 2000; Garg et al. 2006). Vermicomposts showed a significant increase in exchangeable Ca^{2+}, Mg^{2+} and K^+ compared to fresh sludge indicating the conversion of nutrients to plant-available forms during passage through the earthworm gut (Garg et al. 2006; Yasir et al. 2009a). Vermicomposts contain higher nutrient concentrations, but less likely to produce salinity, than composts. Additionally, vermicomposts possess outstanding biological properties and have microbial populations significantly larger and more diverse compared to conventional composts (Edwards 1998). Soil supplemented with vermicompost showed better plant growth compared to soil treated with inorganic fertilizers or cattle manure (Kalembasa 1996; Subler et al. 1998).

TABLE 1: Ecological categories and niches of earthworms and their characteristic features and beneficial traits

Species	Ecological category	Ecological niche	Characteristic features	Beneficial trait
Eisenia foetida,	Epigeics	Superficial soil layers, leaf litter, compost	Smaller in size, body uniformly pigmented, active gizzard, short life cycle, high reproduction rate and regeneration, tolerant to disturbance, phytophagous	Efficient biodegraders and nutrient releasers, efficient compost producers, aids in litter comminution and early decomposition
Lumbricus rubellus,				
L. castaneus,				
L. festivus,				
Eiseniella tetraedra,				
Bimastus minusculus,				
B. eiseni,				
Dendrodrilus rubidus,				
Dendrobaena veneta,				
D. octaedra				
Aporrectodea caliginosa,	Endogeics	Topsoil or subsoil	Small to large sized worms, weakly pigmented, life cycle of medium duration, moderately tolerant to disturbance, geophagous	Brings about pronounced changes in soil physical structure, can efficiently utilize energy from poor soils hence can be used for soil improvements
A. trapezoides,				
A. rosea,				
Millsonia anomala,				

TABLE 1: *Cont.*

Species	Ecological category	Ecological niche	Characteristic features	Beneficial trait
Octolasion cyaneum,	Polyhumic endogeic	Top soil (A1)	Small size, un-pigmented, forms horizontal burrows, rich soil feeder	
O. lacteum,				
Pontoscolex corethrurus,	Mesohumic endogeic	A and B horizon	Medium size, unpigmented, forms extensive horizontal burrows, bulk (A1) soil feeder	
Allolobophora chlorotica,				
Aminthas sp.	Oligohumic endogeic	B and C horizon	Very large in size, unpigmented, forms extensive horizontal burrows, feeds on poor, deep soils	
L. terrestris,	Anecics	Permanent deep burrows in soil	Large in size, dorsally pigmented, forms extensive, deep, vertical perma-nent burrows, low reproductive rate, sensitive to distur-bance, phytogeopha-gous, nocturnal	Forms vertical burrows affect-ing air-water relationship and movement from deep layers to surface helps in efficient mixing of nutrients
L. polyphemus,				
A. longa				

9.4 DIVERSITY OF BACTERIA ASSOCIATED WITH EARTHWORMS

Earthworm's ability to increase plant nutrient availability is likely to be dependent on the activity of earthworm gut microflora. Earthworms indi-rectly influence the dynamics of soil chemical processes, by comminuting the litter and affecting the activity of the soil micro-flora (Petersen and Luxton 1982; Lee 1985; Edwards and Bohlen 1996). Interactions between earthworms and microorganisms seem to be complex. Earthworms ingest plant growth-promoting rhizospheric bacteria such as *Pseudomonas, Rhizobium, Bacillus, Azosprillium, Azotobacter,* etc. along with rhizospheric soil, and they might get activated or increased due to the ideal micro-environment of the gut. Therefore earthworm activity increases the population of plant growth-promoting rhizobacteria (PGPR) (Sinha et al. 2010). This specific group of bacteria stimulates plant growth directly by solubilization of nutrients (Ayyadurai et al. 2007; Ravindra et al. 2008), production of growth hormone, 1-aminocyclopropane-1-carboxylate (ACC) deaminase (Correa et al. 2004), nitrogen fixation (Han et al. 2005), and indirectly by suppressing fungal pathogens. Antibiotics, fluorescent pigments, siderophores and fungal cell-wall degrading enzymes namely chitinases and glucanases (Han et al. 2005; Sunish et al. 2005; Ravindra et al. 2008; Jha et al. 2009; Pathma et al. 2010; Pathma et al. 2011a, b) produced by bacteria mediate the fungal growth-suppression. Earthworms are reported to have association with such free living soil bacteria and constitute the drilosphere (Ismail 1995). Earthworm microbes mineralize the organic matter and also facilitate the chelation of metal ions (Pizl and Novokova 1993; Canellas et al. 2002). Gut of earthworms *L. terrestris, Allolobophora caliginosa* and *Allolobophora terrestris* were reported to contain higher number of aerobes compared to soil (Parle 1963). Earthworms increased the number of microorganisms in soil as much as five times (Edwards and Lofty 1977) and the number of bacteria and 'actinomycetes' contained in the ingested material increased upto 1,000 fold while passing through their gut (Edwards and Fletcher 1988). Similar increase was observed in plate counts of total bacteria, proteolytic bacteria and actinomycetes by passage through earthworms gut (Parle 1963; Daniel and Anderson 1992; Pedersen and Hendriksen 1993; Devliegher and Verstraete 1995). Similarly microbial biomass either decreased (Bohlen and Edwards 1995; Devliegher and Verstraete 1995), or increased (Scheu 1992) or remained unchanged (Daniel and Anderson 1992) after passage through the earthworm gut. An oxalate-degrading bacterium *Pseudomonas oxalaticus* was isolated from intestine of *Pheretima* species (Khambata and Bhat 1953) and an actinomycete *Streptomyces lipmanii* was identified in the gut of *Eisenia lucens* (Contreras 1980). Scanning electron micrographs provided evidence for endogenous microflora in guts of earthworms, *L. terrestris* and *Octolasion cyaneum* (Jolly et al. 1993). Gut of *E. foetida* contained various anaerobic N_2-fixing bacteria such as *Clostridium butyricum, C. beijerinckii* and *C. paraputrificum* (Citernesi et al. 1977). Alimentary canal of *Lumbricus rubellus* and *Octolasium lacteum* were found to contain more numbers of aerobes and anerobes (Karsten and Drake 1995) and culturable denitrifiers (Karsten and Drake 1997). List of vermicompost bacteria and their beneficial traits is presented in Table 2.

Earthworms harbor 'nitrogen-fixing' and 'decomposer microbes' in their gut and excrete them along with nutrients in their excreta (Singleton et al. 2003). Earthworms stimulate and accelerate microbial activities by increasing the population of soil microorganisms (Binet et al. 1998), microbial numbers and biomass (Edwards and Bohlen 1996), by improving aeration through burrowing actions. Vermicomposting modified the original microbial community of the waste in a diverse way. Actinobacteria and Gammaproteobacteria were abundant in vermicompost, while conventional compost contained more Alphaproteobacteria and Bacteriodetes, the bacterial phylogenetic groups typical of non-cured compost (Vivas et al. 2009). Total bacterial counts exceeded $10^{-10}/$ g of vermicompost and it included nitrobacter, azotobacter, rhizobium, phosphate solubilizers and actinomycetes (Suhane 2007). Molecular and culture-dependent analyses of bacterial community of vermicompost showed the presence of α-Proteobacteria, β-Proteobacteria, γ-Proteobacteria, Actinobacteria, Planctomycetes, Firmicutes and Bacteroidetes (Yasir et al. 2009a). Several findings showed considerable increase in total viable counts of actinomycetes and bacteria in the worm treated compost (Parthasarathi and Ranganathan 1998; Haritha Devi et al. 2009). The increase of microbial population may be due to the congenial condition for the growth of microbes within the digestive tract of earthworm and by the ingestion of nutrient rich organic wastes which provide energy and also act as a substrate for the growth of microorganisms (Tiwari et al. 1989). The differences in microbial species, numbers and activity between the earthworm alimentary canal or burrow and bulk soil indirectly support the hypothesis that the bacterial community structures of these habitats are different from those of the soil. Specific phylogenetic groups of bacteria such as Aeromonas hydrophila in *E. foetida* (Toyota and Kimura 2000), fluorescent pseudomonads in *L. terrestris* (Devliegher and Verstraete 1997), and Actinobacteria in *L. rubellus* (Kristufek et al. 1993) have been found in higher numbers in earthworm guts, casts, or burrows.

TABLE 2: Biodiversity of vermicompost bacteria and their beneficial traits

Vermicompost earthworm	Names of bacteria	Beneficial traits	References
Pheretima sp.	*Pseudomonas oxalaticus*	Oxalate degradation	Khambata and Bhat, 1953
Unspecified	*Rhizobium trifolii*	Nitrogen fixation and growth of leguminous plants	Buckalew et al. 1982
Lumbricus rubellus	*R. japonicum, P. putida*	Plant growth promotion	Madsen and Alexander 1982
L. terrestris	*Bradyrhizobium japonicum*	Improved distribution of nodules on soybean roots	Rouelle, 1983
Aporrectodea trapezoids,	*P. corrugata 214OR*	Suppress *Gaeumannomyces graminis* var. Tritd in wheat	Doube et al. 1994
A. rosea			
A. trapezoids,	*R. meliloti L5-30R*	Increased root nodulation and nitrogen fixation in legumes	Stephens et al. 1994b
Microscolex dubius			
Eisenia foetida	*Bacillus spp., B. megaterium,*	Antimicrobial activity against *Enterococcus faecalisDSM* 2570, *Staphylococcus aureus* DSM 1104	Vaz-Moreira et al. 2008
B. pumilus, B. subtilis			
L. terrestris	*Fluorescent pseudomonads,*	Suppress *Fusarium oxysporum* f. sp. asparagi and *F. proliferatum* in asparagus, *Verticillium dahlia* in eggplant and *F. oxysporum* f. sp. *lycopersici* Race 1 in tomato	Elmer, 2009
Filamentous actinomycetes			
Eudrilus sp.	*Free-living N_2 fixers,*	Plant growth promotion by nitrification, phosphate solubilisation and plant disease suppression	Gopal et al. 2009
Azospirillum, Azotobacter,			
Autotrophic Nitrosomonas,			
Nitrobacter, Ammonifying			

TABLE 2: *Cont.*

Vermicompost earthworm	Names of bacteria	Beneficial traits	References
bacteria, Phosphate solubilizers,			
Fluorescent pseudo-monads			
E. foetida	*Proteobacteria, Bacteroidetes,*	Antifungal activity against *Colletotrichum coccodes, R. solani, P. ultimum, P. capsici* and *F. moliniforme*	Yasir et al. 2009a
Verrucomicrobia, Actinobacteria,			
Firmicutes			
Unspecified	*Eiseniicola composti YC06271T*	Antagonistic activity against *F. moniliforme*	Yasir et al. 2009b

Enzymatic activity characterization and quantification has a direct correlation with type and population of microbes and reflects the dynamics of the composting process in terms of the decomposition of organic matter and nitrogen transformations and provide information about the maturity of the compost (Tiquia 2005). Wormcasts contain higher activities of cellulase, amylase, invertase, protease, peroxidase, urease, phosphatase and dehydrogenase (Sharpley and Syers 1976; Edwards and Bohlen 1996). Dehydrogenase is an intracellular enzyme related to the oxidative phosphorylation process (Trevors 1984) and is an indicator of microbial activity in soil and other biological ecosystems (Garcia et al. 1997). The maximum enzyme activities (cellulase, amylase, invertase, protease and urease) were observed during 21–35 days in vermicomposting and on 42–49 days in conventional composting. Also, microbial numbers and their extracellular enzyme profiles were more abundant in vermicompost produced from fruitpulp, vegetable waste, groundnut husk and cowdung compared to the normal compost of the same parental origin (Haritha Devi et al. 2009). *Pseudomonas, Paenibacillus, Azoarcus, Burkholderia, Spiroplasm, Acaligenes,* and *Acidobacterium*, the potential degraders of several categories of organics are seen associated with the earthworm's intestine and vermicasts (Singleton et al. 2003). *Firmicutes* viz., *Bacillus benzoevorans,*

B. cereus, B. licheniformis, B. megaterium, B. pumilus, B. subtilis, B. macroides; Actinobacteria namely *Cellulosimicrobium cellulans, Microbacterium spp., M. oxydans*; Proteobacteria such as *Pseudomonas spp., P. libaniensis*; ungrouped genotypes *Sphingomonas* sp., *Kocuria palustris* and yeasts namely *Geotrichum* spp. and *Williopsis californica* were reported from vermicomposts (Vaz-Moreira et al. 2008). Pinel et al. (2008) reported the presence of a novel nephridial symbiont, Verminephrobacter eiseniae from *E. foetida. Ochrobactrum sp., Massilia sp., Leifsonia sp.* and bacteria belonging to families Aeromonadaceae, Comamonadaceae, Enterobacteriaceae, Flavobacteriaceae, Moraxellaceae, Pseudomonadaceae, Sphingobacteriaceae, Actinobacteria and Microbacteriaceae were reported to occur in earthworms alimentary canal (Byzov et al. 2009). The microbial flora of earthworm gut and cast are potentially active and can digest a wide range of organic materials and polysaccharides including cellulose, sugars, chitin, lignin, starch and polylactic acids Zhang et al. (2000; Aira et al. 2007; Vivas et al. 2009). Single-strand conformation polymorphism (SSCP) profiles on the diversity of eight bacterial groups viz., Alphaproteobacteria, Betaproteobacteria, Bacteroidetes, Gammaproteobacteria, Deltaproteobacteria, Verrucomicrobia, Planctomycetes, and Firmicutes from fresh soil, gut, and casts of the earthworms *L. terrestris* and *Aporrectodea caliginosa* showed the presence of Bacteroidetes, Alphaproteobacteria, Betaproteobacteria and representatives of classes Flavobacteria, Sphingobacteria (Bacteroidetes) and *Pseudomonas* spp. in the worm casts in addition to unclassified *Sphingomonadaceae* (Alphaproteobacteria) and *Alcaligenes* spp. (Betaproteobacteria) (Nechitaylo et al. 2010).

9.5 ROLE OF VERMICOMPOST IN SOIL FERTILITY

Vermicomposts can significantly influence the growth and productivity of plants (Kale et al. 1992; Kalembasa 1996; Edwards 1988; Sinha et al. 2009) due to their micro and macro elements, vitamins, enzymes and hormones (Makulec 2002). Vermicomposts contain nutrients such as nitrates, exchangeable phosphorus, soluble potassium, calcium, and magnesium in plant available forms (Orozco et al. 1996; Edwards 1998) and have large particular surface area that provides many microsites for microbial activity and for the strong retention of nutrients (Shi-wei and Fu-zhen 1991). Uptake of nitrogen (N), phosphorus (P), potassium (K) and magnesium (Mg) by rice (*Oryza sativa*) plant was highest when fertilizer was applied in combination with vermicompost (Jadhav et al. 1997). N uptake by ridge gourd (*Luffa acutangula*) was higher when the fertilizer mix contained 50% vermicompost (Sreenivas et al. 2000). Apart from providing mineralogical nutrients, vermicomposts also contribute to the biological fertility by adding beneficial microbes to soil. Mucus, excreted through the earthworm's digestive canal, stimulates antagonism and competition between diverse microbial populations resulting in the production of some antibiotics and hormone-like biochemicals, boosting plant growth (Edwards and Bohlen 1996). In addition, mucus accelerates and enhances decomposition of organic matter composing stabilized humic substances which embody water-soluble phytohormonal elements (Edwards and Arancon 2004) and plant-available nutrients at high levels (Atiyeh et al. 2000c). Adding vermicasts to soil improves soil structure, fertility, plant growth and suppresses diseases caused by soil-borne plant pathogens, increasing crop yield (Chaoui et al. 2002; Scheuerell et al. 2005; Singh et al. 2008). Kale (1995) reported the nutrient status of vermicomposts with organic carbon 9.15-17.98%, total nitrogen 0.5-1.5%, available phosphorus 0.1-0.3%, available potassium 0.15%, calcium and magnesium 22.70-70 mg/100 g, copper 2–9.3 ppm, zinc 5.7-11.5 ppm and available sulphur 128–548 ppm.

Effects of a variety of vermicomposts on a wide array of field crops (Chan and Griffiths 1988; Arancon et al. 2004b), vegetable plants (Edwards and Burrows 1988; Wilson and Carlile 1989;Subler et al. 1998; Atiyeh et al. 2000b), ornamental and flowering plants (Edwards and Burrows 1988; Atiyeh et al. 2000c) under greenhouse and field conditions have been documented. Vermicomposts are used as alternative potting media due to their low-cost, excellent nutrient status and physiochemical characters. Considerable improvements in plant growth recorded after amending soils with vermicomposts have been attributed to the physico-chemical and biological properties of vermicomposts.

Vermicompost addition favorably affects soil pH, microbial population and soil enzyme activities (Maheswarappa et al. 1999) and also reduces the proportion of water-soluble chemical, which cause possible environmental contamination (Mitchell and Edwards 1997). Vermicompost addition increases the macropore space ranging from 50–500 μm, resulting in improved air-water relationship in the soil, favourably affecting plant growth (Marinari et al. 2000). Evaluation of various organic and inorganic amendments on growth of raspberry proves that vermicompost has beneficial buffering capability and ameliorate the damage caused by excess of nutrients which may otherwise cause phytotoxicity (Subler et al. 1998). Thus, vermicompost acts a soil conditioner (Albanell et al. 1988) and a slow-release fertilizer (Atiyeh et al. 2000a). During vermicomposting the heavy metals forms complex, aggregates with humic acids and other polymerized organic fractions resulting in lower availability of heavy metals to the plant, which are otherwise phytotoxic (Dominguez and Edwards 2004). Soil amended with vermicompost produced better quality fruits and vegetables with less content of heavy metals or nitrate, than soil fertilized with mineral fertilizers (Kolodziej and Kostecka 1994).

9.6 ROLE OF VERMICOMPOST BACTERIA IN BIOMEDICAL WASTE MANAGEMENT

The importance of sewage sludge, biosolids and biomedical waste management by safe, cheap and easy methods need no further emphasis. All these wastes are infectious and have to be disinfected before being disposed into the environment. Biosolids also contain an array of pathogenic microorganisms (Hassen et al. 2001). Biocomposting of wastes bring about biological transformation and stabilization of organic matter and effectively reduces potential risks of pathogens (Burge et al. 1987; Gliotti et al. 1997; Masciandaro et al. 2000). Vermicomposting does not involves a thermophilic phase which might increase the risk of using this technology for management of infectious wastes, but surprisingly vermicomposting resulted into a noticeable reduction in the pathogen indicators such as fecal coliform, Salmonella sp., enteric virus and helminth ova in the biosolids (Eastman 1999; Sidhu et al. 2001). Vermicomposting of biosolids resulted in reduction of faecal coliforms and *Salmonella* sp. from 39,000 MPN/g to 0 MPN/g and < 3 MPN to < 1MPN/g respectively (Dominguez and Edwards 2004). Vermicomposting of municipal sewage sludge with

L. mauritii eliminated *Salmonella* and *Escherichia* sp., and the earthworm gut analysis also proved that *Salmonella* sp. ranging 15–17×10^3 CFU/g and *Escherichia* sp. ranging 10–14×10^2 CFU/g were completely eliminated in the gut after 70 days of vermicomposting period (Ganesh Kumar and Sekaran 2005). Activities by earthworms on sludge reduced levels of pathogens and odors of putrefaction and accelerated sludge stabilization (Mitchell 1978; Brown and Mitchell 1981; Hartenstein 1983). The reduction or removal of these enteric bacterial populations at the end of vermicomposting period, correlates with the findings that earthworm's diet include microorganisms and earthworms ability to selectively digest them (Bohlen and Edwards 1995; Edwards and Bohlen 1996). Apart from solid waste management, earthworms are also used in sewage water treatment. Earthworms promote the growth of 'beneficial decomposer bacteria' in wastewater and acts as aerators, grinders, crushers, chemical degraders, and biological stimulators (Dash 1978; Sinha et al. 2002). Earthworms also granulate the clay particles and increase the hydraulic conductivity and natural aeration and further grind the silt and sand particles and increase the total specific surface area and thereby enhance adsorption of the organic and inorganic matter from the wastewater. In addition, earthworms body acts as a 'biofilter' and remove the biological oxygen demand (BOD), chemical oxygen demand (COD), total dissolved solids (TDS) and total suspended solids (TSS) from wastewater by 90%, 80–90%, 90–92% and 90–95% respectively by 'ingestion' and biodegradation of organic wastes, heavy metals, and solids from wastewater and by their 'absorption' through body walls (Sinha et al. 2008).

Reports reveal that vermicomposting converts the infected biomedical waste containing various pathogens viz., *Staphylococcus aureus, Proteus vulgaris, Pseudomonas pyocyaneae* and *Escherichia coli* to an innocuous waste containing commensals like *Citrobactor freundii* and aerobic spore bearing microorganism usually found in the soil and alimentary canal of earthworms (Umesh et al. 2006). Vermicomposting plays a vital role for safe management of biomedical wastes and solid wastes generated from wastewater treatment plants and its bioconversion into valuable composts free from enteric bacterial populations. Depending on the earthworm species, vermicomposting was known to reduce the level of different pathogens such as *Salmonella enteriditis, Escherichia coli,* total and faecal coliforms, helminth ova and human viruses in different types of waste. Direct means of reduction in these microbial numbers during gut passage might be due to the digestive enzymes and mechanical grinding, while indirect means of pathogen removal might be due to promotion of aerobic conditions which could bring down the load of coliforms (Monroy et al. 2009; Edwards 2011; Aira et al. 2011).

9.7 ROLE OF VERMICOMPOST IN PLANT GROWTH PROMOTION

Use of vermicomposts as biofertilizers has been increasing recently due to its extraordinary nutrient status, and enhanced microbial and antagonistic activity. Vermicompost produced from different parent material such as food waste, cattle manure, pig manure, etc., when used as a media supplement, enhanced seedling growth and development, and increased productivity of a wide variety of crops (Edwards and Burrows 1988; Wilson and Carlile 1989; Buckerfield and Webster 1998; Edwards 1998; Subler et al. 1998; Atiyeh et al. 2000c). Vermicompost addition to soil-less bedding plant media enhanced germination, growth, flowering and fruiting of a wide range of green house vegetables and ornamentals (Atiyeh et al. 2000a, b, c), marigolds (Atiyeh et al. 2001), pepper (Arancon et al. 2003a), strawberries (Arancon et al. 2004b) and petunias (Chamani et al. 2008). Vermicompost application in the ratio of 20:1 resulted in a significant and consistent increase in plant growth in both field and greenhouse conditions (Edwards et al. 2004), thus providing a substantial evidence that biological growth promoting factors play a key role in seed germination and plant growth (Edwards and Burrows 1988; Edwards 1998). Investigations revealed that plant hormones and plant-growth regulating substances (PGRs) such as auxins, gibberellins, cytokinins, ethylene and abscisic acid are produced by microorganisms (Barea et al. 1976; Arshad and Frankenberger 1993).

Several researchers have documented the presence of plant growth regulators such as auxins, gibberellins, cytokinins of microbial origin (Krishnamoorthy and Vajranabhiah 1986; Grappelli et al. 1987; Tomati et al. 1988; Muscolo et al. 1999) and humic acids (Senesi et al. X1992;

Masciandaro et al. 1997; Atiyeh et al. 2002) in vermicompost in appreciable quantities. Cytokinins produced by *Bacillus* and *Arthrobacter* spp. in soils increase the vigour of seedlings (Inbal and Feldman 1982; Jagnow 1987). Microbially produced gibberellins influence plant growth and development (Mahmoud et al. 1984; Arshad and Frankenberger 1993) and auxins produced by *Azospirillum brasilense* affects the growth of plants belonging to paoceae (Barbieri et al. 1988). Extensive investigations on the biological activities of humic substances showed that they also posses plant growth stimulating property (Chen and Aviad 1990). Humic substances increased the dry matter yields of corn and oat seedlings (Lee and Bartlett 1976; Albuzio et al. 1994); number and length of tobacco roots (Mylonas and Mccants 1980); dry weights of roots, shoots and number of nodules of groundnut, soyabean and clover plants (Tan and Tantiwiramanond 1983) and vegetative growth of chicory plants (Valdrighi et al. 1996) and induced root and shoot formation in plant tissue culture (Goenadi and Sudharama 1995). High levels of humus have been reported from vermicomposts originating from food wastes, animal manure, sewage and paper mill sludges (Atiyeh et al. 2002; Canellas et al. 2002; Arancon et al. 2003c). The humic and fulvic acid in the humus dissolves insoluble minerals in the organic matter and makes them readily available to plants and in addition they also help plants to overcome stress and stimulates plant growth (Sinha et al. 2010). Studies on biological activities of vermicompost derived humic substances, revealed that they had similar growth-promoting hormonal effect (Dell'Agnola and Nardi 1987; Nardi et al. 1988; Muscolo et al. 1993). The humic materials extracted from vermicomposts have been reported to produce auxin-like cell growth and nitrate metabolism in carrots (*Daucus carota*) (Muscolo et al. 1996). Humates obtained from pig manure vermicompost increased growth of tomato (Atiyeh et al. 2002) and those obtained from cattle, food and paper waste vermicompost increased the growth of strawberries and peppers (Arancon et al. 2003a).

Earthworms produce plant growth regulators (Gavrilov 1963). Since earthworms increase the microbial activity by several folds they are considered as important agents which enhance the production of plant growth regulators (Nielson 1965; Graff and Makeschin 1980; Dell'Agnola and Nardi 1987; Grappelli et al. 1987; Tomati et al. 1987, 1988; Edwards and

Burrows 1988; Nardi et al. 1988; Edwards 1998). Plant growth stimulating substances of microbial origin were isolated from tissues of *Aporrectodea longa, L. terrestris* and *Dendrobaena rubidus* and indole like substances were detected from the tissue extracts of *A. caliginosa, L. rubellus* and *E. foetida* which increased the growth of peas (Nielson 1965) and dry matter production of rye grass (Graff and Makeschin 1980). A. trapezoids aided in the dispersal of Rhizobium through soil resulting in increased root colonization and nodulation of leguminous plants (Bernard et al. 1994). Use of earthworm casts in plant propagation promoted root initiation, increased root numbers and biomass. The hormone-like effect produced by earthworm casts on plant metabolism, growth and development causing dwarfing, stimulation of rooting, internode elongation and precociousness of flowering was attributed to the fact of presence of microbial metabolites (Tomati et al. 1987; Edwards 1998). Earthworm casts stimulated growth of ornamental plants and carpophore formation in *Agaricus bisporus* when used as casing layer in mushroom cultivation (Tomati et al. 1987). Aqueous extracts of vermicompost produced growth comparable to the use of hormones such as auxins, gibberellins and cytokinins on Petunia, Begonia and Coleus, providing solid evidence that vermicompost is a rich source of plant growth regulating substances (Grappelli et al. 1987; Tomati et al. 1987, 1988). Addition of vermicompost at very low levels to the growth media dramatically increased the growth of hardy ornamentals *Chamaecyparis lawsonian, Elaeagnus pungens, Pyracantha spp., Viburnum bodnantense, Cotoneaster conspicus* and *Cupressocyparis leylandi.* Cucumber (Hahn and Bopp 1968), dwarf maize (Sembdner et al. 1976) and coleus bioassays (Edwards et al. 2004) evidenced that vermicompost contained appreciable amounts of cytokinins, gibberellins and auxins respectively. Maize seedlings dipped in vermicompost water showed marked difference in plumule length compared to normal water indicating that plant growth promoting hormones are present in vermicompost (Nagavallemma et al. 2004). Comparative studies on the impact of vermiwash and urea solution on seed germination, root and shoot length in *Cyamopsis tertagonoloba* proved that vermiwash contained hormone like substances (Suthar 2010). High performance liquid chromatography (HPLC) and gas chromatography-mass spectroscopy (GC-MS) analyses of aqueous extracts of cattle waste derived vermicompost showed presence of significant amounts indole-acetic-acid (IAA), gibberellins and cytokinins (Edwards et al. 2004).

Earthworm gut associated microbes enrich vermicomposts with highly water-soluble and light-sensitive plant growth hormones, which gets absorbed on humic acid substances in vermicompost making them extremely stable and helps them persist longer in soils thereby influencing plant growth (Atiyeh et al. 2002; Arancon et al. 2003c). This is confirmed by presence of exchangeable auxin group in the macrostructure of humic acid extract from vermicompost (Canellas et al. 2002). Apart from the rich nutritional status and ready nutrient availability, presence of humic acids and plant growth regulating substances makes vermicompost a biofertilizer which increases germination, growth, flowering and fruiting in a wide range of crops. Vermicompost substitution in a relatively small proportion (10–20%) to the potting mixture increased dry matter production and tomato growth significantly (Subler et al. 1998). Soil amended with 20% vermicompost was more suitable for tomato seedling production (Valenzuela et al. 1997). Similarly vermicompost addition upto 50% in the medium resulted in enhanced growth of *Chamaecyparis lawsoniana* (Lawson's Cypress), *Juniperus communis* (Juniper) and *Elaeagnus pungens* (Silverberry) rooted liners (Bachman and Edgar Davice 2000).

Vermicompost application increased plant spread (10.7%), leaf area (23.1%), dry matter (20.7%) and increased total strawberry fruit yield (32.7%) (Singh et al. 2008). Substitution of vermicompost drastically reduced the incidence of physiological disorders like albinism (16.1–4.5%), fruit malformation (11.5–4.0%) and occurrence of grey mould (10.4–2.1%) in strawberry indicating its significance in reducing nutrient-related disorders and Botrytis rot, thereby increasing the marketable fruit yield upto 58.6% with better quality parameters. Fruit harvested from plant receiving vermicompost were firmer, had higher total soluble solids (TSS), ascorbic acid content and attractive colour. All these parameters appeared to be dose dependent and best results were achieved at 7.5 t ha^{-1} (Singh et al. 2008). Vermicompost application showed significant increase in germination percent (93%), growth and yield of mung bean (*Vigna radiata*) compared to the control (Karmegam et al. 1999). Similarly, the fresh and dry matter yields of cowpea (*Vigna unguiculata*) were higher in soil amended with vermicompost than with biodigested slurry, (Karmegam and Daniel 2000). Combined application of vermicompost with N fertilizer gave higher dry matter (16.2 g plant^{-1}) and grain yield (3.6 t ha^{-1}) of wheat (*Triticum aestivum*) and higher dry matter yield (0.66 g plant^{-1}) of the following coriander (*Coriandrum sativum*) crop in wheat-coriander cropping system (Desai et al. 1999). Vermicompost application produced herbage yields of coriander cultivars comparable to those obtained with chemical fertilizers (Vadiraj et al. 1998). Yield of pea (*Pisum sativum*) increased with the application of vermicompost (10 t ha^{-1}) along with recommended NPK (Meena et al. 2007). Vermicompost application to sorghum (*Sorghum bicolor*) (Patil and Sheelavantar 2000), sunflower (*Helianthus annuus*) (Devi et al. 1998), tomato (*Lycopersicon esculentum*) (Nagavallemma et al. 2004), eggplant (*Solanum melangona*) (Guerrero and Guerrero, 2006), okra (*Abelmoschus esculentus*) (Gupta et al. 2008), hyacinth bean (*Lablab purpureas*) (Karmegam and Daniel 2008), grapes (Buckerfield and Webster 1998) and cherry (Webster 2005) showed a positive result. Vermicompost amendment at the rate of 10 t ha^{-1} along with 50% of recommended dose of NPK fertilizer increased the number and fresh weight of flowers per plant, flower diameter and yield, while at the rate of 15 t ha^{-1} along with 50% of recommended dose of NPK increased vase life of *Chrysanthemum chinensis* (Nethra et al. 1999). Red Clover and cucumber grown in soil amended with vermicompost showed an increase in mineral contents viz., Ca, Mg, Cu, Mn and Zn in their shoot tissues (Sainz et al. 1998). Vermicomposted cow manure stimulated the growth of lettuce and tomato plants while the unprocessed parent material did not (Atiyeh et al. 2000b). Similarly, vermicomposted duck wastes resulted in better growth of tomatoes, lettuce, and peppers than the unprocessed wastes (Wilson and Carlile 1989). The enhancement in plant growth might be attributed to the fact that processed waste had improved physicochemical characteristics and nutrients, in forms readily available to the plant as well as the presence of plant growth promoting and antagonistic disease suppressing beneficial bacteria.

9.8 ROLE OF VERMICOMPOST IN PLANT DISEASE MANAGEMENT

9.8.1 PLANT PATHOGEN CONTROL

Soils with low organic matter and microbial activity are conducive to plant root diseases (Stone et al. 2004) and addition of organic amendments can effectively suppress plant disease (Raguchander et al. 1998; Blok et al. 2000; Lazarovits et al. 2000). Several researchers reported the disease suppressive properties of thermophilic compost (Hoitink et al. 1997; Goldstein 1998; Pitt et al. 1998) on a wide range of phytopathogens viz., Rhizoctonia (Kuter et al. 1983), Phytopthora (Hoitink and Kuter 1986; Pitt et al. 1998), *Plasmidiophora brassicae* and *Gaeumannomyces graminis* (Pitt et al. 1998) and Fusarium (Kannangowa et al. 2000; Cotxarrera et al. 2002). Microbial antagonism might be one of the possible reasons for disease suppression as organic amendments enhances the microbial population and diversity. Traditional thermophilic composts promote only selected microbes while non-thermophilic vermicomposts are rich sources of microbial diversity and activity and harbour a wide variety of antagonistic bacteria thus acts as effective biocontrol agents aiding in suppression of diseases caused by soil-borne phytopathogenic fungi (Chaoui et al. 2002; Scheuerell et al. 2005; Singh et al. 2008).

Earthworm feeding reduces the survival of plant pathogens such as *Fusarium* sp. and *Verticillium dahliae* (Yeates 1981; Moody et al. 1996) and increases the densities of antagonistic fluorescent pseudomonads and filamentous actinomycetes while population densities of *Bacilli* and *Trichoderma* spp. remains unaltered (Elmer 2009). Earthworm activities reduce root diseases of cereals caused by *Rhizoctonia* (Doube et al. 1994). It has been proved that earthworms decreased the incidence of field diseases of clover, grains, and grapes incited by *Rhizoctonia* spp. (Stephens et al. 1994a; Stephens and Davoren 1997) and *Gaeumannomyces* spp. (Clapperton et al. 2001). Earthworms *Aporrectodea trapezoides* and *Aporrectodea rosea* act as vectors of *Pseudomonas corrugata* 214OR, a biocontrol agent for wheat take-all caused by *G. graminis* var. *tritd* (Doube et al. 1994). Greenhouse studies on augmentation of pathogen infested soils with *L. terrestris* showed a significant reduction of disease caused by *Fusarium oxysporum* f. sp. *asparagi* and *F. prolifera-tum* on susceptible cultivars of asparagus (*Asparagus officinalis*), *Verticillium dahliae* on eggplant (*Solanum melongena*) and *F. oxysporum* f. sp. *lycopersici* race 1 on tomato. Plant weights increased by 60-80% and disease severity reduced by 50-70% when soils were augmented with earthworms. Incorporation of soil with vermicompost effectively suppressed *R. solani* in wheat (Stephens et al. 1993), *Phytophthora nicotianae* (Nakamura 1996; Szczech 1999; Szczech and Smolinska 2001) and Fusarium in tomatoes (Nakamura 1996; Szczech 1999), *Plasmodiophora brassicae* in tomatoes and cabbage (Nakamura 1996), Pythium and Rhizoctonia (root rot) in cucumber and radish (Simsek Ersahin et al. 2009), *Botrytis cineria* (Singh et al. 2008) and *Verticillium* (Chaoui et al. 2002) in strawberry and *Sphaerotheca fulginae* in grapes (Edwards et al. 2004). Vermicompost application drastically reduced the incidence of 'Powdery Mildew', 'Color Rot' and 'Yellow Vein Mosaic' in Lady's finger (Abelmoschus esculentus) (Agarwal et al. 2010). Substitution of vermicompost in the growth media reduced the fungal diseases caused by *R. solani, P. drechsleri* and *F. oxysporum* in gerbera (Rodriguez et al. 2000). Amendment of vermicompost at low rates (10-30%) in horticulture bedding media resulted in significant suppression of Pythium and Rhizoctonia under green house conditions (Edwards et al. 2004). Research findings proved that vermicompost when added to container media significantly reduced the infection of tomato plants by *P. nicotianae* var. *nicotianae* and *F. oxysporum* sp. *lycopersici* (Szcech et al. 1993; Szczech 1999). Club-rot of cabbage caused by *P. brassicae* was inhibited by dipping cabbage roots into a mixture of clay and vermicompost (Szcech et al. 1993). Potato plants treated with vermicompost were less susceptible to *P. infestans* than plants treated with inorganic fertilizers (Kostecka et al. 1996a). Aqueous extracts of vermicompost inhibited mycelial growth of *B. cineria, Sclerotinia sclerotiorum, Corticium rolfsii, R. solani* and *F. oxysporum* (Nakasone et al. 1999), effectively controlled powdery mildew of barley (Weltzien 1989) and affected the development of powdery mildews on balsam (*Impatiens balsamina*) and pea (*Pisum sativum*) caused by *Erysiphe cichoracearum* and *E. pisi*, respectively in field conditions (Singh et al. 2003).

9.8.2 MECHANISMS THAT MEDIATE PATHOGEN SUPPRESSION

Two possible mechanisms of pathogen suppression have been described, one depends on systemic plant resistance and the other is mediated by microbial competition, antibiosis and hyperparasitism (Hoitink and Grebus 1997). The microbially mediated suppression is again classified into two mechanisms viz., 'general suppression' where a wide range of microbes suppress the pathogens such as *Pythium* and *Phytopthora* (Chen et al. 1987) and 'specific suppression' where a narrow range of organisms facilitates suppression, for instance disease caused by *Rhizoctonia* (Hoitink et al. 1997). The disease suppressive effect of vermicompost against fusarium wilt of tomato clearly depicted that fungus inhibition was purely biotic and no chemical factors played any role, since the experiments with heat-sterilized vermicompost failed to control the disease (Szczech 1999). Experiments on suppression of damping-off caused by *R. solani*, in vermicompost amended nurseries of white pumpkin proved that vermicompost suppressed the disease in a dosage and temperature dependent manner (Rivera et al. 2004). Earthworm castings are rich in nutrients (Lunt and Jacobson 1944; Parle 1963) and calcium humate, a binding agent (Edwards 1998) that reduces desiccation of individual castings and favors the incubation and proliferation of beneficial microbes, such as *Trichoderma* spp. (Tiunov and Scheu 2000), *Pseudomonas* spp. (Schmidt et al. 1997), and mycorrhizal spores (Gange 1993; Doube et al. 1995). Earthworm activity increased the communities of Gram-negative bacteria (Clapperton et al. 2001; Elmer 2009). Vermicompost associated chitinolytic bacterial communities viz., *Nocardioides oleivorans,* several species of *Streptomyces* and *Staphylococcus epidermidis* showed inhibitory effects against plant phytopathogens such as, *R. solani, Colletotrichum coccodes, Pythium ultimum, P. capsici* and *Fusarium moniliforme* (Yasir et al. 2009a).

9.9 ROLE OF VERMICOMPOST IN ARTHROPOD PEST CONTROL

Addition of organic amendments helped in suppression of various insect pests such as European corn borer (Phelan et al. 1996), other corn insect pests (Biradar et al. 1998), aphids and scale insects (Culliney and Pimentel 1986; Costello and Altiei 1995; Huelsman et al. 2000) and brinjal shoot and fruit borer (Sudhakar et al. 1998). Several reports also evidenced that vermicompost addition decreased the incidence of *Spodoptera litura, Helicoverpa armigera,* leaf miner (*Apoaerema modicella*), jassids (*Empoasca kerri),* aphids (*Aphis craccivora*) and spider mites on groundnuts (Rao et al. 2001; Rao 2002, 2003) and psyllids (*Heteropsylla cubana*) on a tropical leguminous tree (*Leucaena leucocephala*) (Biradar et al. 1998). Vermicompost amendment decreased the incidence of sucking pests under field conditions (Ramesh 2000) and suppressed the damage caused by of two-spotted spider mite (*Tetranychus* spp.), aphid (*Myzus persicae*) (Edwards et al. 2007) and mealy bug (*Pseudococcus* spp.) under green house conditions (Arancon et al. 2007). Vermicompost substitution to soil less plant growth medium MetroMix 360 (MM360) at a rate less then 50% reduced the damage caused by infestation of pepper seedlings by *M. persicae* and *Pseudococcus* spp. and tomato seedlings by *Pseudococcus* spp., cabbage seedlings by *M. persicae* and cabbage white caterpillars (*Pieris brassicae* L.) (Arancon et al. 2005). Greenhouse cage experiments conducted on tomatoes and cucumber seedlings infested with *M. persicae*, citrus mealybug (*Planococcus citri*), two spotted spider mite (*Tetranychus urticae*); striped cucumber beetles (*Acalymna vittatum*) attacking cucumbers and tobacco hornworms (*Manduca sexta*) attacking tomatoes proved that treatment of infested plants with aqueous extracts of vermicompost suppressed pest establishment, and their rates of reproduction. Vermicompost teas at higher dose also brought about pest mortality (Edwards et al. 2010b). Suppression of aphid population gains importance since they are key vectors in transmission of plant viruses. Addition of solid vermicompost reduced damage by *A. vittatum* and spotted cucumber beetles (*Diabotrica undecimpunctata*) on cucumbers and larval hornworms (*Manduca quinquemaculata*) on tomatoes in both greenhouse and field experiments (Yardim et al. 2006). Combined application of vermicompost and vermiwash spray to chilli (*Capiscum annum*) significantly reduced the incidence of 'Thrips' (*Scirtothrips dorsalis*) and 'Mites' (*Polyphagotarsonemus latus*) (Saumaya et al. 2007).

9.9.1 MECHANISMS THAT MEDIATE PEST CONTROL

Plants grown in inorganic fertilizers are more prone to pest attack than those grown on organic fertilizers (Culliney and Pimentel 1986; Yardim and Edwards 2003; Phelan 2004). Inorganic nitrogen fertilization improves the nutritional quality and palatability of the host plants, inhibits the raise of secondary metabolite concentrations (Fragoyiannis et al. 2001; Herms 2002), enhances the fecundity of insects dieting on them, attracts more individuals for oviposition (Bentz et al. 1995) and increases the population growth rates of insects (Culliney and Pimentel 1986; Jannsson and Smilowitz 1986). Though organic fertilizer has an enhanced nutritional composition they release nutrients at a slower rate (Patriquin et al. 1995) hence plants grown with organic fertilizers possess decreased N levels (Steffen et al. 1995) and have higher phenol content (Asami et al. 2003) resulting in resistance of these plants to pest attack. Similarly vermicomposts exhibit a slow, balanced nutritional release pattern, particularly in release of plant available N, soluble K, exchangeable Ca, Mg and P (Edwards and Fletcher 1988; Edwards 1998). Vermicomposts are rich in humic acid and phenolic compounds. Phenolic compounds act as feeding deterrents and hence significantly affect pest attacks (Kurowska et al. 1990; Summers and Felton 1994; QiTian 2004; Hawida et al. 2007; Koul 2008; Mahanil et al. 2008; Bhonwong et al. 2009). Soil containing earthworms contained polychlorinated phenols and their metabolites (Knuutinen et al. 1990). An endogenous phenoloxidase present in *L. rubellus* bioactivate compounds to form toxic phenols viz., p-nitrophenol (Park et al. 1996). Monomeric phenols could be absorbed by humic acids in the gut of earthworms (Vinken et al. 2005). Uptake of soluble phenolic compounds from vermicompost, by the plant tissues makes them unpalatable thereby affecting pest rates of reproduction and survival (Edwards et al. 2010a; Edwards et al. 2010b).

9.10 ROLE OF VERMICOMPOST IN NEMATODE CONTROL

It has been well documented that addition of organic amendments decreases the populations of plant parasitic nematodes (Addabdo 1995; Sipes et al. 1999; Akhtar and Malik 2000). Vermicompost amendments appreciably suppress plant parasitic nematodes under field conditions (Arancon et al.

2003b). Vermicomposts also suppressed the attack of *Meloidogyne incognita* on tobacco, pepper, strawberry and tomato (Swathi et al. 1998; Edwards et al. 2007; Arancon et al. 2002; Morra et al. 1998) and decreased the numbers of galls and egg masses of *Meloidogyne javanica* (Ribeiro et al. 1998).

9.10.1 MECHANISMS THAT MEDIATE NEMATODE CONTROL

There are several feasible mechanisms that attribute to the suppression of plant parasitic nematodes by vermicompost application and it involves both biotic and abiotic factors. Organic matter addition to the soil stimulates the population of bacterial and fungal antagonists of nematodes (e.g., *Pasteuria penetrans, Pseudomonas* spp. and chitinolytic bacteria, *Trichoderma* spp.), and other typical nematode predators including nematophagous mites viz., *Hypoaspis calcuttaensis* (Bilgrami 1996), Collembola and other arthropods which selectively feeds on plant parasitic nematodes. (Thoden et al. 2011). Vermicompost amendment promoted fungi capable of trapping nematode and destroying nematode cysts (Kerry 1988) and increased the population of plant growth-promoting rhizobacteria which produce enzymes toxic to plant parasitic nematodes (Siddiqui and Mahmood 1999). Vermicompost addition to soils planted with tomatoes, peppers, strawberry and grapes showed a significant reduction of plant parasitic nematodes and increased the population of fungivorous and bacterivorous nematodes compared to inorganic fertilizer treated plots (Arancon et al. 2002). In addition, few abiotic factors viz., nematicidal compounds such as hydrogen sulphide, ammonia, nitrates, and organic acids released during vermicomposting, as well as low C/N ratios of the compost cause direct adverse effects while changes in soil physiochemical characterists viz., bulk density, porosity, water holding capacity, pH, EC, CEC and nutrition posses indirect adverse effects on plant parasitic nematodes (Rodriguez-Kabana 1986; Thoden et al. 2011).

9.11 CONCLUSION

Vermicomposting is a cost-effective and eco-friendly waste management technology which takes the previlige of both earthworms and the associated microbes and has many advantages over traditional thermophilic composting. Vermicomposts are excellent sources of biofertilizers and their addition improves the physiochemical and biological properties of agricultural soil. Vermicomposting amplifies the diversity and population of beneficial microbial communities. Although there are some reports indicating that few harmful microbes such as spores of Pythium and Fusarium are dispersed by earthworms (Edwards and Fletcher 1988), the presence and amplification of antagonistic disease-suppressing and other plant growth-promoting beneficial bacteria during vermicomposting out weigh these harmful effects (Edwards and Fletcher 1988; Gammack et al. 1992; Brown 1995). Vermicomposts with excellent physio-chemical properties and buffering ability, fortified with all nutrients in plant available forms, antagonistic and plant growth-promoting bacteria are fantabulous organic amendments that act as a panacea for soil reclamation, enhancement of soil fertility, plant growth, and control of pathogens, pests and nematodes for sustainable agriculture.

REFERENCES

1. Addabdo TD (1995) The nematicidal effect of organic amendments: a review of the literature 1982–1994. Nematol Mediterranea 23:299-305
2. Aira M, Monroy F, Dominguez J (2007) Earthworms strongly modify microbial biomass and activity triggering enzymatic activities during vermicomposting independently of the application rates of pig slurry. Sci Total Environ 385:252-261
3. Aira M, Gómez-Brandón M, González-Porto P, Domínguez J (2011) Selective reduction of the pathogenic load of cow manure in an industrial-scale continuous-feeding vermireactor. Bioresource Technol 102:9633-9637
4. Agarwal S, Sinha RK, Sharma J, et al. (2010) Ver-miculture for sustainable horticulture: Agronomic impact studies of earthworms, cow dung compost and vermi-compost vis-à-vis chemical fertilizers on growth and yield of lady's finger (Abelmoschus esculentus). In: Sinha RK (ed) Special Issue on 'Vermiculture Technology', International Journal of Environmental Engineering, Inderscience Publishers, Geneva, Switzerland.
5. Akhtar M, Malik A (2000) Role of organic amendments and soil organisms in the biological control of plant parasitic nematodes: a review. Bioresour Technol 74:35-47
6. Albanell E, Plaixats J, Cabrero T (1988) Chemical changes during vermicomposting (Eisenia fetida) of sheep manure mixed with cotton industrial wastes. Biol Fertil Soils 6:266-269
7. Albuzio A, Concheri G, Nardi S, Dell'Agnola G (1994) Effect of humic fractions of different molecular size on the development of oat seedlings grown in varied nutritional conditions. In: Senesi N, Miano TM (eds) Humic substances in the Global En-

vironment and Implications on Human Health, Elsevier, Amsterdam, Netherlands. pp 199-204

8.   Arancon NQ, Edwards CA, Atiyeh R, Metzger JD (2004) Effects of vermicomposts produced from food waste on the growth and yields of greenhouse peppers. Bioresour Technol 93:139-144

9.   Arancon NQ, Edwards CA, Bierman P, Metzger JD, Lee S, Welch C (2003) Effects of vermicomposts to tomatoes and peppers grown in the field and strawberries under high plastic tunnels. Pedobiologia 47:731-735

10. Arancon NQ, Edwards CA, Bierman P, Welch C, Metzger JD (2004) The influence of vermicompost applications to strawberries: Part 1. Effects on growth and yield. Bioresour Technol 93:145-153

11. Arancon NQ, Edwards CA, Lee S (2002) Management of plant parasitic nematode populations by use of vermicomposts. Proceedings Brighton Crop Protection Conference – Pests and Diseases. 705-716

12. Arancon NQ, Edwards CA, Yardim EN, Oliver TJ, Byrne RJ, Keeney G (2007) Suppression of two-spotted spider mite (Tetranychus urticae), mealy bug (Pseudococcus sp) and aphid (Myzus persicae) populations and damage by vermicomposts. Crop Prot 26:29-39

13. Arancon NQ, Galvis PA, Edwards CA (2005) Suppression of insect pest populations and damage to plants by vermicomposts. Bioresour Technol 96:1137-1142

14. Arancon NQ, Galvis P, Edwards CA, Yardim E (2003) The trophic diversity of nematode communities in soils treated with vermicomposts. Pedobiologia 47:736-740

15. Arancon NQ, Lee S, Edwards CA, Atiyeh RM (2003) Effects of humic acids and aqueous extracts derived from cattle, food and paper-waste vermicomposts on growth of greenhouse plants. Pedobiologia 47:744-781

16. Arshad M, Frankenberger WT Jr (1993) Microbial production of plant growth regulators. In: Metting FB Jr (ed) Soil Microbial Ecology: Applications in Agricultural and Environmental Management, Marcell Dekker, New York. pp 307-347

17. Asami DK, Hang YJ, Barnett DM, Mitchell AE (2003) Comparison of the total phenolic and ascorbic acid content of freeze-dried and air-dried marionberry, strawberry and corn grown using conventional organic and sustainable agricultural practices. J Agric Food Chem 51:1237-1241

18. Asha A, Tripathi AK, Soni P (2008) Vermicomposting: A Better Option for Organic Solid Waste Management. J Hum Ecol 24:59-64

19. Atiyeh RM, Arancon NQ, Edwards CA, Metzger JD (2001) The influence of earthworm-processed pig manure on the growth and productivity of marigolds. Bioresour Technol 81:103-108

20. Atiyeh RM, Dominguez J, Subler S, Edwards CA (2000) Changes in biochemical properties of cow manure during processing by earthworms (Eisenia andrei, Bouché) and the effects on seedling growth. Pedobiologia 44:709-724

21. Atiyeh RM, Lee S, Edwards CA, Arancon NQ, Metzger JD (2002) The influence of humic acids derived from earthworm-processed organic wastes on plant growth. Bioresour Technol 84:7-14

22. Atiyeh RM, Subler S, Edwards CA, Bachman G, Metzger JD, Shuster W (2000) Effects of vermicomposts and composts on plant growth in horticulture container media and soil. Pedobiologia 44:579-590

23. Atiyeh RM, Arancon NQ, Edwards CA, Metzger JD (2000) Influence of earthworm-processed pig manure on the growth and yield of green house tomatoes. Bioresour Technol 75:175-180

24. Ayyadurai N, Ravindra Naik P, Sakthivel N (2007) Functional characterization of antagonistic fluorescent pseudomonads associated with rhizospheric soil of rice (Oryza sativa L.). J Microbiol Biotechnol 17:919-927

25. Bachman GR, Edgar Davice W (2000) Growth of magnolia virginiana liners in vermicompost-amended media. Proceeding of SNA Research Conference, Southern Nursery Association, Atlanta. pp 65-67

26. Bansal S, Kapoor KK (2000) Vermicomposting of crop residues and cattle dung with Eisenia foetida. Bioresour Technol 73:95-98

27. Banu JR, Logakanthi S, Vijayalakshmi GS (2001) Biomanagement of paper mill sludge using an indigenous (Lampito mauritii) and two exotic (Eudrilus eugineae and Eisenia foetida) earthworms. J Environ Biol 22:181-185

28. Barbieri P, Bernardi A, Galli E, Zanetti G (1988) Effects of inoculation with different strains of Azospirillum brasilense on wheat roots development. In: Klingmüller W (ed) Azospirillum IV, Genetics, Physiology, Ecology, SpringerVerlag, Berlin. pp 181-188

29. Barea JM, Navarro E, Montana E (1976) Production of plant growth regulators by rhizosphere phosphate solubilizing bacteria. J Appl Bacteriol 40:129-134

30. Benitez E, Nogales R, Elvira C, Masciandaro G, Ceccanti B (1999) Enzymes activities as indicators of the stabilization of sewage sludges composting by Eisenia foetida. Bioresour Technol 67:297-303

31. Bentz JA, Reeves J, Barbosa P, Francis B (1995) Nitrogen fertilizer effect on selection, acceptance and suitability of Euphorbia pulcherrima (Euphorbiaceae) as a host plant to Bemisia tabaci (Homoptera: Aleyrodidae). Environ Entomol 24:40-45

32. Bernal MP, Faredes C, Sanchez-Monedero MA, Cegarra J (1998) Maturity and stability parameters of composts prepared with a, wide range of organic wastes. Bioresour Technol 63:91-99

33. Bernard MD, Peter MS, Christopher WD, Maarten HR (1994) Interactions between earthworms, beneficial soil microorganisms and root pathogens. Appl Soil Ecol 1:3-10

34. Bhatnagar RK, Palta RK (1996) Earthworm-Vermiculture and Vermicomposting. Kalyani Publishers, New Delhi.

35. Bhonwong A, Stout MJ, Attajarusit J, Tantasawat P (2009) Defensive role of tomato polyphenoloxidases against cotton bollworm (Helicoverpa armigera) and beet army worm (Spodoptera exigua). J Chem Ecol 35:28-38

36. Bilgrami AL (1996) Evaluation of the predation abilities of the mite Hypoaspis calcuttaensis, predaceous on plant and soil nematodes. Fund Appl Nematol 20:96-98

37. Binet F, Fayolle L, Pussard M (1998) Significance of earthworms in stimulating soil microbial activity. Biol Fertil Soils 27:79-84

38. Biradar AP, Sunita ND, Teggel RG, Devaradavadgi SB (1998) Effect of vermicompost on the incidence of subabul psyllid. Insect- Environ 4:55-56

39. Blok WJ, Lamers JG, Termoshuizen AJ, Bollen GJ (2000) Control of soil-borne plant pathogens by incorporating fresh organic amendments followed by tarping. Phytopathology 90:253-259

40. Bohlen PJ, Edwards CA (1995) Earthworm effects on N dynamics and soil respiration in microcosms receiving organic and inorganic nutrients. Soil Biol Biochem 27:341-348

41. Brown BA, Mitchell MJ (1981) Role of the earthworm, Eisenia fetida, in affecting survival of Salmonella enteritidis ser. Typhimurium. Pedobiologia 22:434-438

42. Brown GG (1995) How do earthworms affect microfloral and faunal community diversity? Plant Soil 170:209-231

43. Brown GG, Barois I, Lavelle P (2000) Regulation of soil organic matter dynamics and microbial activity in the drilosphere and the role of interactions with other edaphic functional domains. Eur J Soil Biol 36:177-198

44. Buckalew DW, Riley RK, Yoder WA, Vail WJ (1982) Invertebrates as vectors of endomycorrhizal fungi and Rhizobium upon surface mine soils. West Virginia Acad Sci Proc 54:1

45. Buckerfield JC, Webster KA (1998) Worm-worked waste boosts grape yields: prospects for vermicompost use in vineyards. Australian and New Zealand Wine Industry Journal 13:73-76

46. Burge WD, Enkiri NK, Hussong D (1987) Salmonella regrowth in compost as influenced by substrate. Microbial Ecol 14:243-253

47. Butt KR (1993) Utilization of solid paper mill sludge and spent brewery yeast as a feed for soil dwelling earthworms. Bioresour Technol 44:105-107

48. Byzov BA, Nechitaylo TY, Bumazhkin BK, Kurakov AV, Golyshin PN, Zvyagintsev DG (2009) Culturable microorganisms from the earthworm digestive tract. Microbiology 78:360-368

49. Canellas LP, Olivares FL, Okorokova FAR (2002) Humic acids isolated from earthworm compost enhance root elongation, lateral root emergence and plasma membrane H+− ATPase activity in maize roots. Plant Physiol 130:1951-1957

50. Chamani E, Joyce DC, Reihanytabar A (2008) Vermicompost Effects on the Growth and Flowering of Petunia hybrida 'Dream Neon Rose'. American-Eurasian J Agric And Environ Sci 3:506-512

51. Chan LPS, Griffiths DA (1988) The vermicomposting of pretreated pig manure. Biol Wastes 24:57-69

52. Chaoui H, Edwards CA, Brickner M, Lee S, Arancon N (2002) Suppression of the plant diseases, Pythium (damping off), Rhizoctonia (root rot) and Verticillum (wilt) by vermicomposts. Proceedings of Brighton Crop Protection Conference – Pests and Diseases II(8B-3):711-716

53. Chen W, Hoitink HA, Schmitthenner AF, Touvinen O (1987) The role of microbial activity in suppression of damping off caused by Pythium ultimum. Phytopathology 78:314-322

54. Chen Y, Aviad T (1990) Effects of humic substances on plant growth. In: MacCarthy P, Clapp CE, Malcolm RL, Bloom PR (eds) Humic Substances in Soil and Crop Sciences, Selected Reading ASA and SSSA, Madison. pp 161-186

55. Citernesi U, Neglia R, Seritti A, Lepidi AA, Filippi C, Bagnoli G, Nuti MP, Galluzzi R (1977) Nitrogen fixation in the gastro-enteric cavity of soil animals. Soil Biol Biochem 9:71-72

56. Clapperton MJ, Lee NO, Binet F, Conner RL (2001) Earthworms indirectly reduce the effect of take-all (Gaeumannomyces graminis var. tritici) on soft white spring wheat (Triticium aestivum cv. Fielder). Soil Biol Biochem 33:1531-1538

57. Contreras E (1980) Studies on the intestinal actinomycete flora of Eisenia lucens (Annelida, Oligochaeta). Pedobiologia 20:411-416

58. Correa JD, Barrios ML, Galdona RP (2004) Screening for plant growth promoting rhizobacteria in Chamaecytisus proliferus (tagasaste), a forage tree-shrub legume endemic to the Canary Islands. Plant Soil 266:75-84

59. Costello MJ, Altiei MA (1995) Abundance, growth rate and parasitism of Brevicoryne brassicae and Myzus persicae (Homoptera: Aphididae) on broccoli grown in living mulches. Agric Ecosyst Environ 52:187-196

60. Cotxarrera L, Trillas-Gayl MI, Steinberg C, Alabouvette C (2002) Use of sewage sludge compost and Trichoderma asperellum isolates to suppress Fusarium wilt of tomato. Soil Biol Biochem 34:467-476

61. Culliney TW, Pimentel D (1986) Ecological effects of organic agricultural practices on insect populations. Agric Ecosyst Environ 15:253-256

62. Daniel O, Anderson JM (1992) Microbial biomass and activity in contrasting soil material after passage through the gut of earthworm Lumbricus rubellus Hoffmeister. Soil Biol Biochem 24:465-470

63. Darwin F, Seward AC (1903) More letters of Charles Darwin. In: John M (ed) A record of his work in series of hitherto unpublished letters, London. p 508

64. Dash MC (1978) Role of earthworms in the decomposer system. In: Singh JS, Gopal B (eds) Glimpses of ecology, India International Scientific Publication, New Delhi. pp 399-406

65. Dell'Agnola G, Nardi S (1987) Hormone-like effect and enhanced nitrate uptake induced by depolyconder humic fractions obtained from Allolobophora rosea and A. caliginosa faeces. Biol Fertil Soils 4:111-118

66. Desai VR, Sabale RN, Raundal PV (1999) Integrated nitrogen management in wheat-coriander cropping system. J Maharasthra Agric Univ 24:273-275

67. Devi D, Agarwal SK, Dayal D (1998) Response of sunflower (Helianthus annuus) to organic manures and fertilizers. Indian J Agron 43:469-473

68. Devliegher W, Verstraete W (1995) Lumbricus terrestris in a soil core experiment: nutrient-enrichment processes (NEP) and gut-associated processes (GAP) and their effect on microbial biomass and microbial activity. Soil Biol Biochem 27:1573-1580

69. Devliegher W, Verstraete W (1997) Microorganisms and soil physicochemical conditions in the drilosphere of Lumbricus terrestris. Soil Biol Biochem 29:1721-1729

70. Dominguez J, Edwards CA, Webster M (2000) Vermicomposting of sewage sludge: effects of bulking materials on the growth and reproduction of the earthworm Eisenia andrei. Pedobiologia 44:24-32

71. Dominguez J, Edwards CA (2004) Vermicomposting organic wastes: A review. In: Shakir Hanna SH, Mikhail WZA (eds) Soil Zoology for sustainable Development in the 21st century, Cairo. pp 369-395

72. Doube BM, Ryder MH, Davoren CW, Meyer T (1995) Earthworms: a down under delivery system service for biocontrol agents of root disease. Acta Zool Fennica 196:219-223

73. Doube BM, Stephens PM, Davorena CW, Ryderb MH (1994) Interactions between earthworms, beneficial soil microorganisms and root pathogens. Appl Soil Ecol 1:3-10

74. Eastman BR (1999) Achieving pathogen stabilization using vermicomposting. Bio-Cycle 40:62-64

75. Edwards CA (1988) Breakdown of animal, vegetable and industrial organic wastes by earthworms. In: Edwards CA, Neuhauser EF (eds) Earthworms in Waste and Environmental Management SPB, The Hague, Netherlands. pp 21-31

76. Edwards CA (1998) The use of earthworms in the breakdown and management of organic wastes. In: Edwards CA (ed) Earthworm Ecology, CRC Press, Boca Raton. pp 327-354

77. Edwards CA (2011) Human pathogen reduction during vermicomposting. In: Edwards CA, Arancon NQ, Sherman R (eds) Vermiculture technology: earthworms, organic wastes and environmental management, CRC Press, Boca Raton. pp 249-261

78. Edwards CA, Arancon NQ, Bennett MV, Askar A, Keeney G, Little B (2010) Suppression of green peach aphid (Myzus persicae) (Sulz.), citrus mealybug (Planococcus citri) (Risso), and two spotted spider mite (Tetranychus urticae) (Koch.) attacks on tomatoes and cucumbers by aqueous extracts from vermicomposts. Crop Prot 29:80-93

79. Edwards CA, Arancon NQ, Bennett MV, Askar A, Keeney G (2010) Effect of aqueous extracts from vermicomposts on attacks by cucumber beetles (Acalymna vittatum) (Fabr.) on cucumbers and tobacco hornworm (Manduca sexta) (L.) on tomatoes. Pedobiologia 53:141-148

80. Edwards CA, Arancon NQ, Emerson E, Pulliam R (2007) Supressing plant parasitic nematodes and arthropod pests with vermicompost teas. Biocycle. 38-39

81. Edwards CA, Arancon NQ (2004) Vermicomposts suppress plant pest and disease attacks. BioCycle 45:51-53

82. Edwards CA, Bohlen PJ (1996) Biology and Ecology of earthworms. Chapman and Hall, London. p 426

83. Edwards CA, Burrows I, Fletcher KE, Jones BA (1985) The use of earthworms for composting farm wastes. In: Gasser JKR (ed) Composting Agricultural and Other Wastes, Elsevier, London. pp 229-241

84. Edwards CA, Burrows I (1988) The potential of earthworm composts as plant growth media. In: Edwards CA, Neuhauser E (eds) Earthworms in Waste and Environmental Management, SPB Academic Press, The Hague. pp 21-32

85. Edwards CA, Dominguez J, Arancon NQ (2004) The influence of vermicomposts on pest and diseases. In: Shakir Hanna SH, Mikhail WZA (eds) Soil Zoology for Sustainable Development in the 21st centuary, Cairo. pp 397-418

86. Edwards CA, Dominguez J, Neuhauser EF (1998) Growth and reproduction of Perionyx excavatus (Perr.) (Megascolecidae) as factors in organic waste management. Biol Fertil Soils 27:155-161

87. Edwards CA, Fletcher KE (1988) Interaction between earthworms and microorganisms in organic matter breakdown. Agric Ecosyst Environ 20:235-249

88. Edwards CA, Lofty R (1977) The Biology of Earthworms. Chapmann and Hall, London.

89. Elmer WH (2009) Influence of earthworm activity on soil microbes and soilborne diseases of vegetables. Plant Dis 93:175-179

90. Elvira C, Dominguez J, Sampedro L, Mato S (1995) Vermicomposting for the pulp industry. Biocycle 36:62-63

91. Elvira C, Sampedro L, Benítez E, Nogales R (1998) Vermicomposting of sludges from paper mill and dairy industries with Eisenia andrei: a pilot-scale study. Bioresour Technol 63:205-211

92. Elvira C, Sampedro L, Dominguez J, Mato S (1997) Vermicomposting of wastewater sludge from paper-pulp industry with nitrogen rich materials. Soil Biol Biochem 9:759-762
93. Fracchia L, Dohrmann AB, Martinotti MG, Tebbe CC (2006) Bacterial diversity in a finished compost and vermicompost: differences revealed by cultivation independent analyses of PCR-amplified 16S rRNA genes. Appl Microbiol Biotechnol 71:942-952
94. Fragoyiannis DA, McKinlay RG, D'Mello JPF (2001) Interactions of aphids herbivory and nitrogen availability on the total foliar glycoalkoloid content of potato plants. J Chem Ecol 27:1749-1762
95. Gajalakshmi S, Ramasamy EV, Abbasi SA (2001) Assessment of sustainable vermiconversion of water hyacinth at different reactor efficiencies employing Eudrilus engeniae Kingburg. Bioresour Technol 80:131-135
96. Gajalakshmi S, Ramasamy EV, Abbasi SA (2002) Vermicomposting of paper waste with the anecic earthworm Lampito mauritii Kingburg. Indian J Chem Technol 9:306-311
97. Gammack SM, Paterson E, Kemp JS, Cresser MS, Killham K (1992) Factors affecting movement of microorganisms in soils. In: Stotzky G, Bolla LM (eds) Soil Biochemistry, 7, Marcel Dekker, New York. pp 263-305
98. Gandhi M, Sangwan V, Kapoor KK, Dilbaghi N (1997) Composting of household wastes with and without earthworms. Environ Ecol 15:432-434
99. Ganesh kumar A, Sekaran G (2005) Enteric pathogen modification by anaecic earthworm, Lampito Mauritii. J Appl Sci Environ Mgt 9:15-17
100. Gange AC (1993) Translocation of mycorrhizal fungi by earthworms during early succession. Soil Biol Biochem 25:1021-1026
101. Garcia C, Hernandez T, Costa F (1997) Potential use of dehydrogenase activity as an index of microbial activity in degraded soils. Commun Soil Sci Plant Anal 28:123-134
102. Garg P, Gupta A, Satya S (2006) Vermicomposting of different types of waste using Eisenia foetida: a comparative study. Bioresour Technol 97:391-395
103. Gavrilov K (1963) Earthworms, producers of biologically active substances. Zh Obshch Biol 24:149-154
104. Ghosh M, Chattopadhyay GN, Baral K (1999) Transformation of phosphorus during vermicomposting. Bioresour Technol 69:149-154
105. Gliotti C, Giusquiani PL, Businelli D, Machioni A (1997) Composition changes of dissolved organic matter in a soil amended with municipal waste compost. Soil Sci 162:919-926
106. Goenadi DH, Sudharama IM (1995) Shoot initiation by humic acids of selected tropical crops grown in tissue culture. Plant Cell Rep 15:59-62
107. Goldstein J (1998) Compost suppresses diseases in the lab and fields. BioCycle 39:62-64
108. Gopal M, Gupta A, Sunil E, Thomas VG (2009) Amplification of plant beneficial microbial communities during conversion of coconut leaf substrate to vermicompost by Eudrilus sp. Curr Microbiol 59:15-20

109. Graff O, Makeschin F (1980) Beeinlussung des Ertrags von Weidelgrass (Lolium muttiflorum) Ausscheidungen von Regenwurmen dreier verschiedener Arten. Pedobiologia 20:176-180

110. Grappelli A, Galli E, Tomati U (1987) Earthworm casting effect on Agaricus bisporus fructification. Agrochimica 2:457-462

111. Guerrero RD, Guerrero LA (2006) Response of eggplant (Solanum melongena) grown in plastic containers to vermicompost and chemical fertilizer. Asia Life Sciences 15:199-204

112. Gunadi B, Blount C, Edward CA (2002) The growth and fecundity of Eisenia foetida (Savigny) in cattle solids pre-composted for different periods. Pedobiologia 46:15-23

113. Gunathilagraj K, Ravignanam T (1996) Vermicomposting of sericulture wastes. Madras Agric J 83:455-457

114. Gupta AK, Pankaj PK, Upadhyava V (2008) Effect of vermicompost, farm yard manure, biofertilizer and chemical fertilizers (N, P, K) on growth, yield and quality of lady's finger (Abelmoschus esculentus). Pollution Research 27:65-68

115. Hahn H, Bopp M (1968) A cytokinin test with high specificity. Planta 83:115-118

116. Han J, Sun L, Dong X, Cai Z, Yang H, Wang Y, Song W (2005) Characterization of a novel plant growth-promoting bacteria strain Delftia tsuruhatensis HR4 both as a diazotroph and a potential biocontrol agent against various pathogens. Syst Appl Microbiol 28:66-76

117. Hand P, Hayes WA, Frankland JC, Satchell JE (1988) Vermicomposting of cow slurry. Pedobiologia 31:199-209

118. Haritha Devi S, Vijayalakshmi K, Pavana Jyotsna K, Shaheen SK, Jyothi K, Surekha Rani M (2009) Comparative assessment in enzyme activities and microbial populations during normal and vermicomposting. J Environ Biol 30:1013-1017

119. Hartenstein R, Neuhauser EF, Kaplan DL (1979) Reproductive potential of the earthworm Eisenia foetida. Oecologia 43:329-340

120. Hartenstein R (1983) Assimilation by earthworm Eisenia fetida. In: Satchell JE (ed) Earthworm ecology. From Darwin to vermiculture, Chapman and Hall, London. pp 297-308

121. Hashemimajd K, Kalbasi M, Golchin A, Shariatmadari H (2004) Comparison of vermicompost and composts as potting media for growth of tomatoes. J Plant Nutr 27:1107-1123

122. Hassen A, Belguith K, Jedidi N, Cherif A, Cherif M, Boudabous A (2001) Microbial characterization during composting of municipal solid waste. Bioresour Technol 80:217-225

123. Hawida S, Kapari L, Ossipov V, Ramtala MJ, Ruuhola T, Haukioja E (2007) Foliar phenolics are differently associated with Epirrita autmnata growth and immuno competence. J Chem Ecol 33:1013-1023

124. Herms DA (2002) Effects of fertilization on insect resistance of woody ornamental plants. Environ Entomol 31:923-933

125. Hoitink HA, Kuter GA (1986) Effects of composts in growth media on soil-bome pathogens. In: Chen Y, Avnimelech Y (eds) The role of organic matter in modern agriculture, Martinus Nijhoff Publishers, Dordrecht. pp 289-306

126. Hoitink HA, Stone AG, Han DY (1997) Suppression of plant diseases by compost. Hort Sci 32:184-187

127. Hoitink HA, Grebus ME (1997) Composts and Control of Plant Diseases. In: Hayes MHB, Wilson WS (eds) Humic Substances Peats and Sludges Health and Environmental Aspects, Royal Society of Chemistry, Cambridge. pp 359-366

128. Holtzclaw KM, Sposito G (1979) Analytical properties of the soluble metal-complexing fractions in sludge-soil mixtures. IV. Determination of carboxyl groups in fulvic acid. Soil Sci Soc Am J 43:318-323

129. Huelsman MF, Edwards CA, Lawrence JL, Clarke-Harris DO (2000) A study of the effect of soil nitrogen levels on the incidence of insect pests and predators in Jamaican sweet potato (Ipomoea batatus) and Callaloo (Amaranthus). Proc Brighton Pest Control Conference: Pests and Diseases 8D–13:895-900

130. Inbal E, Feldman M (1982) The response of a hormonal mutant of common wheat to bacteria of the Azospirillium. Israel J Bot 31:257-263

131. Ismail SA (1995) Earthworms in soil fertility management. In: Thampan PK (ed) Organic Agriculture, pp. 77-100

132. Ismail SA (1997) Vermicology: The biology of Earthworms. Orient Longman Limited, Chennai.

133. Jadhav AD, Talashilkar SC, Pawar AG (1997) Influence of the conjunctive use of FYM, vermicompost andurea on growth and nutrient uptake in rice. J Maharashtra Agric Univ 22:249-250

134. Jagnow G (1987) Inoculation of cereal crops and forage grasses with nitrogen-fixing rhizosphere bacteria: a possible cause of success and failure with regard to yield response - a review. Z Pflanzenernaehr Dueng Bodenkde 150:361-368

135. Jambhekar H (1992) Use of earthworm as a potential source of decompose organic wastes. Proc Nat Sem Org Fmg, Coimbatore. pp 52-53

136. Jannsson RK, Smilowitz Z (1986) Influence of nitrogen on population parameters of potato insects: abundance, population growth and within-plant distribution of the green peach aphid, Myzus persicae (Homoptera: Aphididae). Environ Entomol 15:49-55

137. Jha BK, Gandhi Pragash M, Cletus J, Raman G, Sakthivel N (2009) Simultaneous phosphate solubilization potential and antifungal activity of new fluorescent pseudomonad strains, Pseudomonas aeruginosa, P. plecoglossicida and P. mosselii. W J Microbiol Biotech 25:573-581

138. Jolly JM, Lappin-Scott HM, Anderson JM, Clegg CD (1993) Scanning electron microscopy of the gut microflora of two earthworms: Lumbricus terrestris and Octolasion cyaneum. Microbial Ecol 26:235-245

139. Kale RD, Bano K (1986) Field Trials with vermicompost (vee comp. E. 8. UAS) on organic fertilizers. In: Dass MC, Senapati BK, Mishra PC (eds) Proceedings of the national seminar on organic waste utilization, Sri Artatrana Ront, Burla. pp 151-157

140. Kale RD, Bano K (1988) Earthworm cultivation and culturing techniques for the production of vee COMP83E UAS. Mysore J Agric Sci 2:339-344

141. Kale RD, Mallesh BC, Bano K, Bagyaray DJ (1992) Influence of vermicompost application on the available macronutrients and selected microbial populations in paddy field. Soil Biol Biochem 24:1317-1320

142. Kale RD (1995) Vermicomposting has a bright scope. Indian Silk 34:6-9

143. Kalembasa D (1996) The influence of vermicomposts on yield and chemical composition of tomato. Zesz Probl Post Nauk Roln 437:249-252

144. Kannangowa T, Utkhede RS, Paul JW, Punja ZK (2000) Effect of mesophilic and thermophilic composts on suppression of Fusarium root and stem rot of greenhouse cucumber. Canad J Microbiol 46:1021-1022

145. Karmegam N, Alagermalai K, Daniel T (1999) Effect of vermicompost on the growth and yield of greengram (Phaseolus aureus Rob.). Trop Agric 76:143-146

146. Karmegam N, Daniel T (2000) Effect of biodigested slurry and vermicompost on the growth and yield of cowpea (Vigna unguiculata (L.). Environ Ecol 18:367-370

147. Karmegam N, Daniel T (2008) Effect of vermi-compost and chemical fertilizer on growth and yield of Hyacinth Bean (Lablab purpureas). Dynamic Soil, Dynamic Plant, Global Science Books 2:77-81

148. Karsten GR, Drake HL (1995) Comparative assessment of the aerobic and anaerobic microfloras of earthworm guts and forest soils. Appl Environ Microbiol 61:1039-1044

149. Karsten GR, Drake HL (1997) Denitrifying bacteria in the earthworm gastrointestinal tract and in vivo emission of nitrous oxide (N2O) by earthworms. Appl Environ Microbiol 63:1878-1882

150. Kerry B (1988) Fungal parasites of cyst nematodes. In: Edwards CA, Stinner BR, Stinner D, Rabatin S (eds) Biological Interactions in Soil, Elsevier, Amsterdam. pp 293-306

151. Khambata SR, Bhat JV (1953) Studies on a new oxalate-decomposing bacterium, Pseudomonas oxalaticus. J Bacteriol 66:505-507

152. Knuutinen J, Palm H, Hakala H, Haimi J, Huhta V, Salminen J (1990) Polychlorinated phenols and their metabolites in soil and earthworms of a saw mill environment. Chemosphere 20:609-623

153. Kolodziej M, Kostecka J (1994) Some qualitative features of the cucumbers and carrots cultivated on the vermicompost. Zeszyty Naukowe Akademii Rolniczej W Krakowie 292:89-94

154. Kooch Y, Jalilvand H (2008) Earthworm as ecosystem engineers and the most important detritivors in forst soils. Pak J Boil Sci 11:819-825

155. Kostecka J, Blazej JB, Kolodziej M (1996a) Investigations on application of vermicompost in potatoes farming in second year of experiment. Zeszyty Naukowe Akademii Rolniczej W Krakowie 310:69-77

156. Koul O (2008) Phytochemicals and insect control: an antifeedant approach. Crit Rev Plant Sci 27:1-24

157. Krishnamoorthy RV, Vajranabhiah SN (1986) Biological activity of earthworm casts: An assessment of plant growth promoter levels in casts. Proc Indian Acad Sci (Anim Sci) 95:341-35

158. Kristufek V, Ravasz K, Pizl V (1993) Actinomycete communities in earthworm guts and surrounding soil. Pedobiologia 37:379-384

159. Kurowska A, Gora J, Kalemba D (1990) Effects of plant phenols on insects. Pol Wiad Chem 44:399-409

160. Kuter GA, Nelson GB, Hoitink HA, Madden LV (1983) Fungal population in container media amended with composted hardwood bark suppressive and conductive to Rhizoctonia damping-off. Phytopathology 73:1450-1456

161. Lavelle E, Barois I, Martin A, Zaidi Z, Schaefer R (1989) Management of earthworm populations in agro-ecosystems: A possible way to maintain soil quality? In: Clarholm M, Bergstrom L (eds) Ecology of Arable Land: Perspectives and Challenges, Kluwer Academic Publishers, London. pp 109-122

162. Lavelle P, Martin A (1992) Small-scale and large-scale effects of endogeic earthworms on soil organic matter dynamics in soils of the humid tropics. Soil Biol Biochem 12:1491-1498

163. Lazarovits G, Tenuta M, Conn KL, Gullino ML, Katan J, Matta A (2000) Utilization of high nitrogen and swine manure amendments for control of soil-borne diseases: efficacy and mode of action. Acta Hortic 5:559-564

164. Lazcano C, Gomez-Brandon M, Dominguez J (2008) Comparison of the effectiveness of composting and vermicomposting for the biological stabilization of cattle manure. Chemosphere 72:1013-1019

165. Lee KE (1985) Earthworms: Their Ecology and Relationships with Soils and Land Use. Academic Press, Sydney.

166. Lee YS, Bartlett RJ (1976) Stimulation of plant growth by humic substances. Soil Sci Soc Am J 40:876-879

167. Lunt HA, Jacobson HGM (1944) The chemical composition of earthworm casts. Soil Sci 58:367-375

168. Maboeta MS, Van Rensburg L (2003) Vermicomposting of industrially produced wood chips and sewage sludge utilizing Eisenia foetida. Ecotoxicol Environ Saf 56:265-270

169. Madsen EL, Alexander M (1982) Transport of Rhizobium and Pseudomonas through soil. Soil Sci Soc Am J 46:557-560

170. Mahanil S, Attajarusit J, Stout MJ, Thipayong P (2008) Over expression of tomato phenol oxidase increases resistance to the common cutworm. Plant Sci 174:456-466

171. Maheswarappa HP, Nanjappa HV, Hegde MR (1999) Influence of organic manures on yield of arrowroot, soil physico-chemical and biological properties when grown as intercrop in coconut garden. Ann Agr Res 20:318-323

172. Mahmoud SA, Ramadan Z, Thabet EM, Khater T (1984) Production of plant growth promoting substance rhizosphere organisms. Zentrbl Mikrobiol 139:227-232

173. Makulec G (2002) The role of Lumbricus rubellus Hoffm. In determining biotic and abiotic properties of peat soils. Pol J Ecol 50:301-339

174. Marinari S, Masciandaro G, Ceccanti B, Grego S (2000) Influence of organic and mineral fertilisers on soil biological and physical properties. Bioresour Technol 72:9-17

175. Martin JP (1976) Darwin on earthworms: the formation of vegetable moulds. Bookworm Publishing, Ontario.

176. Masciandaro G, Ceccanti B, Gracia C (1997) Soil agro-ecological management: fertigation and vermicompost treatments. Bioresour Technol 59:199-206

177. Masciandaro G, Ceccanti B, Gracia C (2000) `In situ' vermicomposting of biological sludges and impacts on soil quality. Soil Biol Biochem 32:1015-1024

178. Meena RN, Singh Y, Singh SP, Singh JP, Singh K (2007) Effect of sources and level of organic manure on yield, quality and economics of garden pea (Pisum sativam L.) in eastern uttar pradesh. Vegetable Science 34:60-63

179. Mitchell A, Edwards CA (1997) The production of vermicompost using Eisenia fetida from cattle manure. Soil Biol Biochem 29:3-4

180. Mitchell A (1997) Production of Eisenia fetida and vermicompost from feed-lot cattle manure. Soil Biol Biochem 29:763-766

181. Mitchell MJ, Hornor SG, Abrams BI (1980) Decomposition of sewage sludge in drying beds and the potential role of the earthworm, Eisenia foetida. J Environ Qual 9:373-378

182. Mitchell MJ (1978) Role of invertebrates and microorganisms in sludge deomposition. In: Hartenstein R (ed) Utilisation of soil organisms in sludge management. natural technology information services, Springfield, Virginia. pp 35-50

183. Monroy F, Aira M, Domínguez J (2009) Reduction of total coliform numbers during vermicomposting is caused by short-term direct effects of earthworms on microorganisms and depends on the dose of application of pig slurry. Sci Tot Environ 407:5411-5416

184. Moody SA, Piearce TG, Dighton J (1996) Fate of some fungal spores associated with wheat straw decomposition on passage through the guts of Lumbricus terrestris and Aporrectodea longa. Soil Biol Biochem 28:533-537

185. Morra L, Palumbo AD, Bilotto M, Ovieno P, Ptcascia S (1998) Soil solarization: organic fertilization grafting contribute to build an integrated production system in a tomato-zucchini sequence. Colture-Protte 27:63-70

186. Munnoli PM, Da Silva JAT, Saroj B (2010) Dynamics of the soil-earthworm-plant relationship: a review. Dynamic soil, dynamic plant. 1-21

187. Muscolo A, Bovalo F, Gionfriddo F, Nardi S (1999) Earthworm humic matter produces auxin-like effect on Daucus carota cell growth and nitrate metabolism. Soil Biol Biochem 31:1303-1311

188. Muscolo A, Felici M, Concheri G, Nardi S (1993) Effect of earthworm humic substances on esterase and peroxidase activity during growth of leaf explants of Nicotiana plumbaginifolia. Biol Fertil Soils 15:127-131

189. Muscolo A, Panuccio MR, Abenavoli MR, Concheri G, Nardi S (1996) Effect of molecular complexity acidity of earthworm faeces humic fractions on glutamale dehydrogenase, glutamine synthetase, and phosphoenolpyruvate carboxylase in Daucus carota II cells. Biol Fertil Soils 22:83-88

190. Mylonas VA, Mccants CB (1980) Effects of humic and fulvic acids on growth of tobacco. I. Root initiation and elongation. Plant Soil 54:485-490

191. Nagavallemma KP, Wani SP, Stephane L, Padmaja VV, Vineela C, Babu Rao M, Sahrawat KL (2004) Vermicomposting: Recycling wastes into valuable organic fertilizer. Global Theme on Agrecosystems Report no.8. Patancheru 502324. International Crops Research Institute for the Semi-Arid Tropics, Andhra Pradesh. p 20

192. Nakamura Y (1996) Interactions between earthworms and microorganisms in biological control of plant root pathogens. Farming Jpn 30:37-43

193. Nakasone AK, Bettiol W, de Souza RM (1999) The effect of water extracts of organic matter on plant pathogens. Summa Phytopathology 25:330-335

194. Nardi S, Arnoldi G, Dell'Agnola G (1988) Release of hormone-like activities from Alloborophora rosea and Alloborophora caliginosa feces. J Soil Sci 68:563-657

195. Nardi S, Dell'Agnola G, Nuti PM (1983) Humus production from farmyard wastes by vermicomposting. Proc. Int. Symp. On Agricultural and Environmental Prospects in Earthworm Farming. Rome. pp 87-94

196. Ndegwa PM, Thompson SA, Das KC (2000) Effects of stocking density and feeding rate on vermicomposting of biosolids. Bioresour Technol 71:5-12

197. Nechitaylo TY, Yakimov MM, Godinho M, Timmis KN, Belogolova E, Byzov BA, et al. (2010) Effect of the earthworms Lumbricus terrestris and Aporrectodea caliginosa on bacterial diversity in soil. Microbial Ecol 59:574-587

198. Nethra NN, Jayaprasad KV, Kale RD (1999) China aster (Callistephus chinensis (L)) cultivation using vermicompost as organic amendment. Crop Research, Hisar 17:209-215

199. Nielson RL (1965) Presence of plant growth substances in earthworms demonstrated by paper chromatography and the Went pea test. Nature 208:1113-1114

200. Orozco FH, Cegarra J, Trujillo LM, Roig A (1996) Vermicomposting of coffee pulp using the earthworm Eisenia fetida: effects on C and N contents and the availability of nutrients. Biol Fertil Soils 22:162-166

201. Park SR, Cho EJ, Yu KH, Kim YS, Suh JJ, Chang CS (1996) Endogenous phenoloxidase from an earthworm Lumbricus rubellus. Tongmul Hakoehi 39:36-46

202. Parle JN (1963) A Microbiological Study of Earthworm Casts. J Gen Microbiol 31:13-22

203. Parthasarathi K, Ranganathan LS (1998) Pressmud vermicast are hot spots of fungi and bacteria. Ecol Environ Cons 4:81-86

204. Pathma J, Ayyadurai N, Sakthivel N (2010) Assessment of Genetic and Functional Relationship of Antagonistic Fluorescent Pseudomonads of Rice Rhizosphere by Repetitive Sequence, Protein Coding Sequence and Functional Gene Analyses. J Microbiol 48:715-727

205. Pathma J, Kamaraj Kennedy R, Sakthivel N (2011a) Mechanisms of fluorescent pseudomonads that mediate biological control of phytopathogens and plant growth promotion of crop plants. In: Maheswari DK (ed) Bacteria in Agrobiology: Plant Growth Responses, SpringerVerlag, Berlin. pp 77-105

206. Pathma J, Rahul GR, Kamaraj Kennedy R, Subashri R, Sakthivel N (2011) Secondary metabolite production by bacterial antagonists. Journal of Biological Control 25:165-181

207. Patil SL, Sheelavantar MN (2000) Effect of moisture conservation practices, organic sources and nitrogen levels on yield, water use and root development of rabi sorghum (Sorghum bicolor (L.)) in the vertisols of semiarid tropics. Ann Agric Res 21:32-36

208. Patriquin DG, Baines D, Abboud A (1995) Diseases, pests and soil fertility. In: Cook HF, Lee HC (eds) Soil Management in Sustainable Agriculture, Wye College Press, Wye. pp 161-174

209. Pedersen JC, Hendriksen NB (1993) Effect of passage through the intestinal tract of detritivore earthworms (Lumbricus spp.) on the number of selected gram-negative and total bacteria. Biol Fertil Soils 16:227-232

210. Petersen H, Luxton MA (1982) A comparative analysis of soil fauna populations and their role in decomposition process. Oikos 39:287-388

211. Phelan PL, Norris KH, Mason JF (1996) Soil management history and host preference by Ostrinia nubilatis: evidence for plant mineral balance mediating insect-plant interactions. Environ Entom 25:1329-1336

212. Phelan PL (2004) Connecting below-ground and above-ground food webs: the role of organic matter in biological buffering. In: Magadoff F, Well RR (eds) Soil Organic Matter in sustainable agriculture, CRC Press, Boca Raton. pp 199-226

213. Pinel N, Davidson SK, Stahl DA (2008) Verminephrobacter eiseniae gen. nov., sp. nov., a nephridial symbiont of the earthworm Eisenia foetida (Savigny). Int J Syst Evol Microbiol 58:2147-2157

214. Pitt D, Tilston EL, Groenhof AC, Szmidt RA (1998) Recycled organic materials (ROM) in the control of plant disease. Acta Hortic 469:391-403

215. Pizl V, Novokova A (1993) Interactions between microfungi and Eisenia andrei (Oligochaeta) during cattle manure vermicomposting. Pedobiologia 47:895-899

216. QiTian S (2004) Research on prevention and elimination of agricultural pests by ginkgo phenols phenolic acids. Chem Ind For Prod 24:83

217. Raguchander T, Rajappan K, Samiyappan R (1998) Influence of biocontrol agents and organic amendments on soybean root rot. Int J Trop Agri 16:247-252

218. Ramesh P (2000) Effects of vermicomposts and vermicornposting on damage by sucking pests to ground nut (Arachis hypogea). Indian J Agri Sci 70:334

219. Rao KR, Rao PA, Rao KT (2001) Influence of fertilizers and manures on the population of coccinellid beetles and spiders in groundnut ecosystem. Ann Plant Protect Sci 9:43-46

220. Rao KR (2002) Induce host plant resistance in the management sucking pests of groundnut. Ann Plant Protect Sci 10:45-50

221. Rao KR (2003) Influence of host plant nutrition on the incidence of Spodoptera litura and Helicoverpa armigera on groundnuts. Indian J Entomol 65:386-392

222. Ravindra NP, Raman G, Badri Narayanan K, Sakthivel N (2008) Assessment of genetic and functional diversity of phosphate solubilizing fluorescent pseudomonads isolated from rhizospheric soil. BMC Microbiol 8:230

223. Reeh U (1992) Influence of population densities on growth and reproduction of the earthworm Eisenia andrei on pig manure. Soil Biol Biochem 24:1327-1331

224. Ribeiro CF, Mizobutsi EH, Silva DG, Pereira JCR, Zambolim L (1998) Control of Meloidognye javanica on lettuce with organic amendments. Fitopatol Brasileira 23:42-44

225. Riffaldi R, Levi-Minzi R (1983) Osservazioni preliminari sul ruolo dell Eisenia foetida nell'umificazione del letame. Agrochimica 27:271-274

226. Rivera AMC, Wright ER, López MV, Fabrizio MC (2004) Temperature and dosage dependent suppression of damping-off caused by Rhizoctonia solani in vermicompost amended nurseries of white pumpkin. Phyton 53:131-136

227. Rodriguez JA, Zavaleta E, Sanchez P, Gonzalez H (2000) The effect of vermicomposts on plant nutrition, yield and incidence of root and crown rot of gerbera (Gerbera jamesonii H. Bolus). Fitopatol 35:66-79

228. Rodriguez-Kabana R (1986) Organic and inorganic amendments to soil as nematode suppressants. J Nematol 18:129-135

229. Rouelle J (1983) Introduction of an amoeba and Rhizobium Japonicum into the gut of Eisenia fetida (Sav.) and Lumbricus terrestris L. In: Satchell JE (ed) Earthworm Ecology: From Darwin to Vermiculture, Chapman and Hall, New York. pp 375-381

230. Sánchez-Monedero MA, Roig A, Paredes C, Bernal MP (2001) Nitrogen transformation during organic waste composting by the Rutgers system and its effects on ph, EC and maturity of the composting mixtures. Bioresour Technol 78:301-308

231. Sainz MJ, Taboada-Castro MT, Vilariño A (1998) Growth, mineral nutrition and mycorrhizal colonization of red clover and cucumber plants grown in a soil amended with composted urban wastes. Plant Soil 205:85-92

232. Saumaya G, Giraddi RS, Patil RH (2007) Utility of vermiwash for the management of thrips and mites on chilli (Capiscum annum) amended with soil organics. Karnataka J Agric Sci 20:657-659

233. Scheu S (1992) Automated measurement of the respiratory response of soil microcompartments: active microbial biomass in earthworm faeces. Soil Biol Biochem 24:1113-1118

234. Scheuerell SJ, Sullivan DM, Mahaffee WF (2005) Suppression of seedling damping-off caused by Pythium ultimum, and Rhizoctonia solani in container media amended with a diverse range of Pacific Northwest compost sources. Phytopathology 95:306-315

235. Schmidt O, Doubre BM, Ryder MH, Killman K (1997) Population dynamics of Pseudomonas corrugata 2140R LUX8 in earthworm food and in earthworm cast. Soil Biol Biochem 29:523-528

236. Sembdner G, Borgman E, Schneider G, Liebisch HW, Miersch O, Adam G, Lischewski M, Schieber K (1976) Biological activity of some conjugated gibberellins. Planta 132:249-257

237. Senesi N, Saiz-Jimenez C, Miano TM (1992) Spectroscopic characterization of metal-humic acid-like complexes of earthworm-composted organic wastes. Sci Total Environ 117–118:111-120

238. Sharma S, Pradhan K, Satya S, Vasudevan P (2005) Potentiality of earthworms for waste management and in other uses - A Review. The Journal of American Science 1:4-16

239. Sharpley AN, Syers JK (1976) Potential role of earthworm casts for the phosphorous enrichment of runoff waters. Soil Biol Biochem 8:341-346

240. Shi-wei Z, Fu-zhen H (1991) The nitrogen uptake efficiency from 15N labeled chemical fertilizer in the presence of earthworm manure (cast). In: Veeresh GK, Rajagopal D, Viraktamath CA (eds) Advances in Management and Conservation of Soil Fauna, Oxford and IBH publishing Co, New Delhi. pp 539-542

241. Siddiqui ZA, Mahmood I (1999) Role of bacteria in the management of plant parasitic nematodes: a review. Bioresour Technol 69:167-179

242. Sidhu J, Gibbs RA, Ho GE, Unkovich I (2001) The role of indigenous microorganisms in suppression of Salmonella regrowth in composted biosolids. Water Res 35:913-920

243. Simsek Ersahin Y, Haktanir K, Yanar Y (2009) Vermicompost suppresses Rhizoctonia solani Kühn in cucumber seedlings. J Plant Dis Protect 9:15-17

244. Singh R, Sharma RR, Kumar S, Gupta RK, Patil RT (2008) Vermicompost substitution influences growth, physiological disorders, fruit yield and quality of strawberry (Fragaria x ananassa Duch.). Bioresour Technol 99:8507-8511

245. Singh UP, Maurya S, Singh DP (2003) Antifungal activity and induced resistance in pea by aqueous extract of vermicompost and for control of powdery mildew of pea and balsam. J Plant Dis Protect 110:544-553

246. Singleton DR, Hendrixb PF, Colemanb DC, Whitmana WB (2003) Identification of uncultured bacteria tightly associated with the intestine of the earthworm Lumbricus rubellus (Lumbricidae; Oligochaeta). Soil Biol Biochem 35:1547-1555

247. Sinha RK, Agarwal S, Chauhan K, Valani D (2010) The wonders of earthworms and its vermicompost in farm production: Charles Darwin's 'friends of farmers', with potential to replace destructive chemical fertilizers from agriculture. Agricultural sciences 1:76-94

248. Sinha RK, Bharambe G, Chaudhari U (2008) Sewage treatment by vermifiltration with synchronous treatment of sludge by earthworms: a low-cost sustainable technology over conventional systems with potential for decentralization. The Environmentalist 28:409-420

249. Sinha RK, Heart S, Agarwal S, Asadi R, Carretero E (2002) Vermiculture technology for environmental management: study of the action of the earthworms Eisenia foetida, Eudrilus euginae and Perionyx excavatus on biodegradation of some community wastes in India and Australia. The Environmentalist 22:261-268

250. Sinha RK, Herat S, Valani D, Chauhan K (2009) Vermiculture and sustainable agriculture. Am-Euras J Agric and Environ Sci, IDOSI Publication 5:1-55

251. Sipes BS, Arakaki AS, Schmitt DP, Hamasaki RT (1999) Root-knot nematode management in tropical cropping systems with organic products. J Sustain Agr 15:69-76

252. Sreenivas C, Muralidhar S, Rao MS (2000) Vermicompost, a viable component of IPNSS in nitrogen nutrition of ridge gourd. Ann Agr Res 21:108-113

253. Steffen KL, Dan MS, Harper JK, Fleischer SJ, Mkhize SS, Grenoble DW, MacNab AA, Fager K (1995) Evaluation of the initial season for implementation of four tomato production systems. J Am Soc Hort Sci 120:148-156

254. Stephens PM, Davoren CW, Doube BM, Ryder MH (1993) Reduced superiority of Rhizoctonia solani disease on wheat seedlings associated with the presence of the earthworm Aporrectodea trapezoids. Soil Biol Biochem 11:1477-1484

255. Stephens PM, Davoren CW, Ryder MH, Doube BM, Correll RL (1994) Field evidence for reduced severity of Rhizoctonia bare-patch disease of wheat, due to the presence of the earthworms Aporrectodea rosea and Aporrectodea trapezoides. Soil Biol Biochem 26:1495-1500

256. Stephens PM, Davoren CW, Ryder MH, Doube BM (1994) Influence of the earthworm Aporrectodea trapezoides (Lumbricidae) on the colonization of alfalfa (Medicago sativa L.) roots by Rhizobium melilotti strain LS-30R and the survival of L5-30R in soil. Biol Fertil Soils 18:63-70

257. Stephens PM, Davoren CW (1997) Influence of the earthworms Aporrectodea trapezoides and A. rosea on the disease severity of Rhizoctonia solani on subterranean cloves and ryegrass. Soil Biol Biochem 29:511-516

258. Stone AG, Scheurell SJ, Darby HM (2004) Suppression of soilborne diseases in field agricultural systems: organic matter management, cover cropping and other cultural practices. In: Magdoff F, Weil (eds) Soil Organic Matter in Sustainable Agriculture, CRC Press LLC, Boca Raton. pp 131-177

259. Subler S, Edwards CA, Metzger PJ (1998) Comparing vermicomposts and composts. Biocycle 39:63-66

260. Sudhakar K, Punnaiah KC, Krishnayya PV (1998) Influence of organic and inorganic fertilizers and certain insecticides on the incidence of shoot and fruit borer, Leucinodes orbonalis Guen, infesting brinjal. J Entomol Res 22:283-286

261. Suhane RK (2007) Vermicompost. Publication of Rajendra Agriculture University, Pusa.

262. Summers G, Felton GW (1994) Prooxidation effects of phenolic acids on the generalist herbivore Helicoverpa zea: potential mode of action of phenolic compounds on plant anti-herbivory chemistry. Insect Biochem Mol Biol 24:943-953

263. Sunish KR, Ayyadurai N, Pandiaraja P, Reddy AV, Venkateshwarlu Y, Prakash O, Sakthivel N (2005) Characterization of antifungal metabolite produced by a new strain Pseudomonas aeruginosa PuPa3 that exhibits broad-spectrum antifungal activity and biofertility traits. J Appl Microbiol 98:145-154

264. Suthar S (2010) Evidence of plant hormone like sub-stances in vermiwash: An ecologically safe option of synthetic chemicals for sustainable farming. J Ecol Eng 36:1089-1092

265. Suthar S, Singh S (2008) Vermicomposting of domestic waste by using two epigeic earthworms (Perionyx excavatus and Perionyx sansibaricus). Int J Evniron Sci and Technol 5:99-106

266. Swathi P, Rao KT, Rao PA (1998) Studies on control of root-knot nematode Meloidogyne incognita in tobacco miniseries. Tobacco Res 1:26-30

267. Szech M, Rondomanski W, Brzeski MW, Smolinska U, Kotowski JF (1993) Suppressive effect of a commercial earthworm compost on some root infecting pathogens of cabbage and tomato. Biol Agric and Hortic 10:47-52

268. Szczech M, Smolinska U (2001) Comparison of suppressiveness of vermicomposts produced from animal manures and sewage sludge against Phytophthora nicotianae Breda de Haan var. nicotiannae. J Phytopathology 149:77-82

269. Szczech MM (1999) Suppressiveness of vermicomposts against fusarium wilt of tomato. J Phytopathology 147:155-161

270. Tajbakhsh J, Abdoli MA, Mohammadi Goltapeh E, Alahdadi I, Malakouti MJ (2008) Trend of physico-chemical properties change in recycling spent mushroom compost through vermicomposting by epigeic earthworms Eisenia foetida and E. andrei. J Agric Technol 4:185-198

271. Tan KH, Tantiwiramanond D (1983) Effect of humic acids on nodulation and dry matter production of soybean, peanut, and clover. Soil Sci Soc Am J 47:1121-1124

272. Thoden TC, Korthals GW, Termorshuizen (2011) Organic amendments and their influences on plant-parasitic and free living nematodes: a promosing method for nematode management. Nematology 13:133-153

273. Tiquia SM (2005) Microbiological parameters as indicators of compost maturity. J Appl Microbiol 99:816-828

274. Tiunov AV, Scheu S (2000) Microfungal communities in soil litter and casts of Lumbricus terrestris (Lumbricidae): a laboratory experiment. Appl Soil Ecol 14:17-26

275. Tiwari SC, Tiwari BK, Mishra RR (1989) Microbial populations, enzyme activities and nitrogen, phosphorous, potassium enrichment in earthworm casts and in the surrounding soil of pine apple plantation. Biol Fertil Soils 8:178-182

276. Tomati U, Grapppelli A, Galli E (1987) The presence of growth regulators in earthworm worked waste. In: Bonvicini Paglioi AM, Omodeo P (eds) On Earthworms.

Proceedings of International Symposium on Earthworms, Selected Symposia and Monographs, Union Zoologica Italian, 2, Modena, Mucchi. pp 423-435

277. Tomati U, Grapppelli A, Galli E (1988) The hormone-like effect of earthworm casts on plant growth. Biol Fertil Soils 5:288-294

278. Toyota K, Kimura M (2000) Microbial community indigenous to the earthworm Eisenia foetida. Biol Fertil Soils 31:187-190

279. Trevors JT (1984) Dehydrogenase activity in soil. A comparison between the INT and TTC assay. Soil Biol Biochem 16:673-674

280. Umesh B, Mathur LK, Verma JN, Srivastava (2006) Effects of vermicomposting on microbiological flora of infected biomedical waste. ISHWM Journal 5:28-33

281. Vadiraj BA, Siddagangaiah D, Potty SN (1998) Response of coriander (Coriandrum sativum L.) cultivars to graded levels of vermicompost. J Spices Aromatic Crops 7:141-143

282. Valdrighi MM, Pera A, Agnolucci M, Frassinetti S, Lunardi D, Vallini G (1996) Effects of compost-derived humic acids on vegetable biomass production and microbial growth within a plant (Cichorium intybus) soil system: a comparative study. Agric Ecosyst Environ 58:133-144

283. Valenzuela O, Gluadia Y, Gallardo S (1997) Use of vermicompost as a growing medium for tomato seedlings (cv. Pltense). Revista Cientifica Agropecuaria 1:15-21

284. Vaz-Moreira I, Maria E, Silva CM, Manaia Olga C, Nunes (2008) Diversity of Bacterial Isolates from Commercial and Homemade Composts. Microbial Ecol 55:714-722

285. Vinken R, Schaeffer A, Ji R (2005) Abiotic association of soil-borne monomeric phenols with humic acids. Org Geochem 36:583-593

286. Vivas A, Moreno B, Garcia-Rodriguez S, Benitez E (2009) Assessing the impact of composting and vermicomposting on bacterial community size and structure, and functional diversity of an olive-mill waste. Bioresour Technol 100:1319-1326

287. Webster KA (2005) Vermicompost increases yield of cherries for three years after a single application. EcoResearch, South Australia.

288. Weltzien HC (1989) Some effects of composted organic materials on plant health. Agric Ecosyst Environ 27:439-446

289. Wilson DP, Carlile WR (1989) Plant growth in potting media containing worm-worked duck waste. Acta Hortic 238:205-220

290. Yardim EN, Arancon NQ, Edwards CA, OliverTJ BRJ (2006) Suppression of tomato hornworm (Manduca quinquemaculata) and cucumber beetles (Acalymma vittatum and Diabotrica undecimpunctata) populations and damage by vermicomposts. Pedobiologia 50:23-29

291. Yardim EN, Edwards CA (2003) Effects of organic and synthetic fertilizer sources on pest and predatory insects associated with tomatoes. Phytoparasitica 31:324-329

292. Yasir M, Aslam Z, Kim SW, Lee SW, Jeon CO, Chung YR (2009) Bacterial community composition and chitinase gene diversity of vermicompost with antifungal activity. Bioresour Technol 100:4396-4403

293. Yasir M, Aslam Z, Song GC, Jeon CO, Chung YR (2009b) Eiseniicola composti gen. nov., sp. nov., with antifungal activity against plant pathogenic fungi. Int J Sys Evol Microbiol 60:268

294. Yeates GW (1981) Soil nematode populations depressed in the presence of earthworms. Pedobiologiaogia 22:191-202

295. Zhang BG, Li GT, Shen TS, Wang JK, Sun Z (2000) Changes in microbial biomass C, N, and P and enzyme activities in soil incubated with the earthworms Metaphire guillelmi or Eisenia foetida. Soil Biol Biochem 32:2055-2062

PART III

PLANT CULTIVARS IN ORGANIC AGRICULTURE

CHAPTER 10

CHARACTERISTICS IMPORTANT FOR ORGANIC BREEDING OF VEGETABLE CROPS

JASMINA ZDRAVKOVIC, NENAD PAVLOVIC, ZDENKA GIREK, MILAN ZDRAVKOVIC, AND DEJAN CVIKIC

10.1 INTRODUCTION

The concept of sustainable agriculture implies the application of the acquired knowledge in order to produce healthy food (LAZIC, 2008). The regulation of the World Health Organization, *Codex Alimentarius*, complies controlling the critical points of the production and the quality in the context of regulation, as well as using the natural resources. The ultimate aim to be achieved is the viable systems of production that are socially justified, economically payable and productive, and at the same time to protect health, improve community and animal welfare, and provide the safe environment.

This chapter was originally published as an open access article by the Sarbian Genetic Society. Zdravkovic J, Pavlovic N, Girek Z, Zdravkovic M, and Cvikic D. Characteristics Important for Organic Breeding of Vegetable Crops. Genetika *42,2 (2010). 223–233.*

Vegetable production based on sustainable agriculture is of interest to:

1. Small-scale, medium and large producers, through the value added and better conditions of sale at the market
2. Consumers, by better quality and healthy food that is produced in sustainable manners
3. Economy and industry, by greater profit from better products
4. Everybody, by higher quality of the environment

Tendencies of future plant selection will be on high production, better quality, optimization of production with low input of fertilizers and pesticides, tolerance to stresses and diseases (VAN WEAS, 2003).

Following the principles of the sustainable manners of agricultural production, the manners of work for each production system is defined, regarding the specific qualities of the agrosystem, including the conditions: soil, water, agricultural production, crops protection, breeding cattle, health of the cattle, cattle welfare, harvesting, processing and storing, welfare, health and safety of the population, animals and the landscape.

Therefore, the sphere of interest for the vegetable selection researchers is in the field of their contribution to the agricultural production within the scope of selecting the varieties, regarding the needs of the market, and in accordance with the environmental conditions, available resources; all these in order to preserve the fertility of the soil, prevent the development of the weeds, pests and diseases. Crops grown in organic system must have familiar seed origin, so called certified seed (BOLETIN OFICIAL 1999). Markets for this kind of vegetables are in expansion. However, there are no necessary quantities of seed for organic production (THOMPSON, 2000).

Agricultural Institutes in Serbia pay special attention to breeding of plant species and monitor the progress in this area. However, the finalization of these projects requires greater Government support. We are facing the lack of certified seed for organic production on the market today. Producers showed special interest for certain vegetable species such as peas, beans, green beans, onions, garlic, cucumbers, tomatoes and lettuce. Also, organic seed is crucial for scientific researches in selection, seed production and organic production of vegetables. Organic seed provides all necessary inputs for entire organic production (VALEMA, 2004).

The aim of this research was the characterization of organic vegetable breeding of certain species and seed production through quality, yield and resistance to plant pathogens in the agro-ecological conditions in Serbia.

10.2 SELECTION TECHNOLOGIES FOR CREATING VEGETABLE VARIETIES FOR SPECIFIC PURPOSES

Plant breeding is based on genetic variability, selection and recombination. It is multidisciplinary and creative work which is based on scientific disciplines such as genetics, physiology, molecular biology, growing technology and plant protection (VAN WEAS, 2003).

Compared to the methods available for conventional plant breeding, there are some limitations on the choice of the method for organic breeding. These methods can be classified as: permitted methods—intraspecific crossing, backcrossing, mass selection, individual selection, forbidden methods—interspecific crossing, protoplast fusion, genetic modification, induced mutations and conditionally permitted—use of hybrid varieties, somatic embryogenesis, meristem culture, and in vitro micropropagation anther culture.

1. The application of genetically modified organisms or their derivatives is banned in the organic production. Such a ban is included in our Organic Production and Organic Products Act ("Official Gazette of the RS", 62/06): "Genetically modified organisms and their derivatives cannot be used in organic production." (BERENJI, 2005).

2. Of the modern biotechnology methods, only the method of indirect selection via molecular markers is permitted because this method does not affect the change of the genetic plant construction.

3. The compromise for hybrid varieties that can be used in organic production is accepted, if they are fertile and if the sterility in the process of hybrid seed production is not chemically caused. Most

probably, in organic production will be developed the three-line (TC) and four-line (DC), and not the two-line hybrids that dominate in the conventional production.

4. The application of the cytoplasmic male sterile products is banned, except if the fertility is permanently restored and it continues in further generations of propagation.

5. Neither direct nor indirect application of the genetic material that contains induced mutations is permitted.

6. The application of silver nitrate, silver thiosulfate, synthetic hormones, antibiotics and colchicines is banned in organic plant breeding (BERENJI, 2008).

10.3 RESULTS ACHIEVED IN THE DEVELOPMENT OF VEGETABLE VARIETIES FOR SPECIFIC PURPOSES

10.3.1 CUCUMBER SELECTION

Cucumbers belong to leading species in the vegetable production in Yugoslavia. Continuous dissemination of its breeding range has been possible due to multiple usage and wide agro-ecological adaptability. Besides the fact that fruits are used in nutrition, both fresh and processed, it is also a raw material of the pharmaceutical industry.

However, breeding of this crop has been intensified only in recent 30 years in Serbia. Tendency of spreading production and increasing the use of cucumber in Serbian nutrition requires a wide range of varieties for pickling and for salad. Therefore, it is necessary to increase the number of cultivars for market and for consumers. Current trend of cucumber breeding in Serbia is heading towards the selection of genotypes with better quantitative characteristics. In the first place, these are varieties with non-bitter fruits, straight shape and with high yield. The exploiting of parthenocarpy of salad type of cucumber is also a current trend. New parthenocarp cucumber genotypes were intended for production in greenhouses.

Cucumber breeding to resistance to diseases is of great importance (PAVLOVIC et al., 2002). Downy (*Pseudoperonospora cubensis* (Berk and Curt.) Rostow) and Cucumber Mosaic Virus cause the greatest problems in cucumber growing in Serbia as well as in other parts of the world (METWALLY and WEHNER, 1990).

Finding and identifying the selection material for these, economically most important diseases is of the greatest priority. The use of chemicals in cucumber protection of downy has reached emergency proportions. This too can be overcome only by breeding resistant genotypes in organic systems of production. Today, the Institute for Vegetable Crops has cucumber lines that are highly tolerant to downy mildew in conditions of spontaneous infection.

The Institute for Vegetable Crops in Smederevska Palanka possesses a collection of cucumber germplasm consisted of over 100 divergent genotypes that are included in our breeding programs. Pickling cucumber hybrid characterized with high tolerance to this plant pathogen was created using experimental crossings. Based on the results from the trial at the Institute for Vegetable Crops, the pickling cucumber named Sirano F1 is marked as the pickling cucumber hybrid that is the most resistant to blight (PAVLOVIC et al., 2006).

10.3.2 BULB VEGETABLES SELECTION

Garlic draws more and more attention as an industrial plant and its production is perspective. Garlic is characterized by high contents of dry matter, proteins, fats, carbohydrates, vitamin C, thiamine, and B6. Furthermore, garlic is rich with minerals such as: Mg, Zn, Mn, Cu, Mo, and Se. It also possesses high energy values that can be paired with members of Fabaceae family. Through the new breeding programs, the chemical structure of garlic is emphasized. Based on detailed chemical analysis the best ecotypes are selected. The aim of the research was to create the ecotypes with the most favourable chemical composition, which would give valuable contribution to the process of utilization of this vegetable variety (PAVLOVIC et al., 2003a).

High biological value of onion is the result of its specific chemical composition dominated by sugars, vitamin C and characteristic ethereal oil. According to the average contents of vitamin C in bulbs (32.46 to 44.03 mg %), onion is a significant natural source of this vitamin. In our research higher genotypic variance and the phenotype variation coefficient is found, as compared to the ecological variance and the genotype variation coefficient. This suggests an important role of the genetic factors in the expression of this trait (PAVLOVIC et al., 2003b). It is confirmed with the broad-sense heritability values (0.75 and 0.76%). For dehydration in food industry only the onion genotypes with high percentage of dry matter contents are used. In many countries today, there are selection programs for the development of genotypes only for drying purposes. Variability of the chemical composition of onion in the field caused by climatic and soil conditions, as well as by the agricultural techniques is characteristic. Dry matter percentage in particular varieties suggests the part of the total bulb mass that can be used in the food production industry. In our research a high genotypic variability of the dry matter contents is found (PAVLOVIC et al., 2007).

10.3.3 TOMATO SELECTION

Tomato represents one of the most important sources of lycopene. It also contents a high level of other carotenoids (-carotene), vitamins (vitamin C), minerals, flavonoids and phenolic acid. Antioxidant effects of the substances affect on the reduction of the possibility for human to contract diseases; the substances affect on the proliferation of cancerous cells; and act as preventive for cardiovascular diseases. Nutritional estimation of the optimal antioxidant quantity through daily consummation of tomato is 9-18 mg of carotenoids, 175-400 mg of vitamin C, 3-4 mg of vitamin E, 50 mg of flavonoids, 0.4 mg of folate and 25-30 mg of lycopene. Tomato is consider as a rich source (in 100 g) of: vitamin C (20-29 mg), carotene (0.2-2.3 mg) and phenolic acid (1-2000 mg), regarding to the total antioxidant contents. There are also small portions of vitamin E (0.49 mg), flavonoids (0.5-5 mg) and traces of selenium (0.5-10 mg), copper (90 mg) and zinc (240 μg). It is recommended to consume 400 g per day of fresh tomato or other tomato products in five portions. Lycopene is one of the carotenoids that give naturally color to the tomato fruit and rank among the strongest antioxidants among all the other carotenoids. The intensive selection for tomato lycopene content has been performed. With adequate selection of the lines that will be used in hybrid development, it is expected that in process of the gene recombination we will get the most favorable nutritive contents ratio—especially antioxidants, with a particular emphasis on a lycopene (ZDRAVKOVIC et al., 2002a, 2003c, 2007).

Late blight, which is caused by the fungus *Phytophtora infestans*, emerges in tomato crops almost every year and causes considerable economic damages. Fungicide control of this parasite is not always effective and satisfactorily. The solution to this problem is in growing less sensitive or more resistant tomato varieties or hybrids. The research on tomato resistance to *Phytophtora infestans* is very complex due to high variability of the pathogen physiological races. Tomato genotypes that are the carriers of Ph-2 gene of resistance to this parasite were crossed with tomato genotypes with good production characteristics (yield and fruit quality) but more susceptible to this parasite. Successfully were selected tomato lines and hybrids that expressed a higher level of resistance than their parents (MIJATOVIC et al., 2007; ZDRAVKOVIC et al., 2004).

Special projects were launched aimed to the effect of partial drying part of a root (PRD-treatment) on the growth of tomato plant, photosynthesis, transpiration, water potential, peroxidase activation of the cell wall, yield, sugar contents, lycopene contents, mineral contents and dry matter content. This treatment causes the increase of peroxidase activity and sugar contents in mature tomato fruits (STIKIC et al., 2003).

Tomato fruit firmness can be achieved by entering a gene for delayed maturation (rin, nor, alc) in the selection of tomatoes. Therefore, the process stops the ripening on a certain level of maturity. As a result of interrupted maturation process, satisfactory fruit firmness occurs, but with slightly less sugar, lycopene, beta carotene, etc. (CVIKIC et al., 2000, ZDRAVKOVIC et al. 2008). Breeding for this trait in this manner are in opposite to the selection requirements for organic production. Fruits with greater firmness can be selected by accumulating firmness traits (ZDRAVKOVIC et al., 2007, 2008). Genotypes with "fruit firmness" gene cause long shelf life of mature tomato. (ZDRAVKOVIC et al., 2003, MARKOVIC et al., 2008).

Investigation of inheritance of yield and yield components in all plant species and in tomato is very important. Gene expression effects nutritional and quality characteristics and therefore the selection may lead to its increase and decrease (ZDRAVKOVIC et al., 2000, ATANASOVA and GEORGIEV, 2009).

The purpose of breeding crops with specific features designed for organic or other sustainable production requires researches in the field of seed production, so the results could be available to producers through new varieties. (VAN WEAS, 2003). Important features of research must be prices of seeds and final products. These aspects require a comparative analysis of conventional production and integrated crop management and organic methods (BRUMFIELD et al. 2000).

10.3.4 DRY BEANS SELECTION

The project of breeding dry beans resistant to stressful conditions of drought has been set out at the Institute for Vegetable Crops, and aim of the project was controlling the negative effects of a high temperature and low rainfall. In our research 62 genotypes were used. Pure lines suitable for further selection have been chosen. The number of nodes in the pure lines is not dependent on the water quantity for irrigation, which includes them in the next breeding phase for stressful conditions (ZDRAVKOVIC M. et al., 2004a; ZDRAVKOVIC M. et al., 2004b). The effects of two microbiological fertilizers (SOJ 1 and SOJ 2) have been investigated on dry beans, in the variation with and without additional nutrition with mineral fertilizers (KAN). The first pod height was recorded. Microbiological fertilizers do not affect on this trait, whereas they exhibited a significant effect on the bean mass per plant (JARAK et al., 2007).

10.3.5 LETTUCE SELECTION

Aimed at creating lettuce cultivars (*Lactuca sativa* L.) resistant to pathogens, the causal agents of plant diseases, and especially to virus diseases, research was carried out on the spontaneous flora in the locality of Pomoravlje and Sumadija where the genotypes of the species *Lactuca* sp. that are resistant to causal agents of virus diseases could be found. The interspecies hybrids *Lactuca virosa* L. x *Lactuca sativa* L., *L. saligna* L. x

L. sativa L., were investigated as possible sources of genetic variability. *L. saligna* L. and *L. virosa* L. represent only a part of the population related to *L. sativa* L. Wild varieties of this species belong to the weed flora. After crossing, viable achenes were obtained only in the crossing *L. sativa* L. x *L. saligna* L. At initial crossings two populations of *L. saligna* L. were used, one with and the other without anthocyanin. The seedlings of *L. saligna* L. without anthocyanin were lost after brought out on the field. In the process of the selection of F1 generation, 31 plants emerged. After transplantation on the field, only 19 plants survived. In 9 plants the fertility was provoked by colchicine, but the percentage of fertile achenes was low as compared to the number of achenes that were not viable. By collecting more genotypes of the species *Lactuca* sp. from spontaneous flora in the locality of Pomoravlje and Sumadija and investigating the possibilities of crossing with the cultivated lettuce (*Lactuca sativa* L.), the selection programs of this kind would be improved. Eventually, the final aim is to obtain the cultivar with the built-in genes of resistance to virus diseases and acceptable morphological characteristics (ZDRAVKOVIC et al., 2003b).

Our investigation was based on the problem of anthocyanin and vitamin C contents inheritance in F_1 progeny of lettuce. It was assumed that progenies with increased contents of these substances could be obtained. Diallel crossing of eight lettuce genotypes of different anthocyanin and vitamin C contents, classified into three varieties was performed. Parental and F_1 generations were investigated comparatively, and their mode of inheritance was determined. Concerning the inheritance of anthocyanin, dominant genes prevailed, and a higher content of this substance was succeeded in F_1 generation. Concerning the inheritance of vitamin C content, dominance mode of inheritance was recorded, when it was compared to the parents with the lower vitamin C content. Apart from the dominance mode of inheritance, significant additive gene effects in inheriting vitamin C content was also recorded (ZDRAVKOVIC et al. 2002a; ZDRAVKOVIC et al. 2002b).

10.4 CONCLUSION

Irreplaceability of plants in human diet as well as negative urban and industrial development impact on ecosystems imposes the need for planning the breeding programs for specific purposes. With the respect to the concept of viable agriculture and giving preference to biodiversity, intensifying and supporting such breeding programs at the Institutes becomes a necessity. The tendency of developing a new concept of agricultural production and greater consumption of healthy plants and their derivatives in human diet requires the existence of a broad range of vegetable varieties. It is of considerable importance to change the method of organizing the agricultural production, as well as passing new laws that would support and facilitate the introduction of the concept of viable development of the agro-system. The Institute for Vegetable Crops in Smederevska Palanka possesses a large collection of germplasm of vegetable varieties, which enables selecting the donors with favourable genes for some specific traits. This represents the basis for planning and modeling the ideotypes of vegetable varieties.

REFERENCES

1. Atanassova, B., H. Georgiev (2009): Part 4: Expression of heterosis by hybridization. Genetic improvement of Solanaceus Crops. Volume 2: Tomato. Science Publishers, Jersey, Plymouth. 113-151.
2. Berenji, J. (2004): Organic plant breeding. Book of abstracts. III Congress of Serbian Genetics, Subotica, Serbia and Montenegro, 87.
3. Berenji, J. (2005): Ethic aspects of GMO - transgenic plants. Journal of Scientific Agricurtural Research, 66 (5), 187-193.
4. Berenji, J. (2008): Organic plant breeding and organic seed production. Book of abstracts of V Scientific -research symposium on breeding and seed production, Vrnjacka Banja, Serbia, May 25-28th , 4.
5. Boletin Oficial de la Republica Argentina (1999): Ley 25127/1999 Production ecological, biological and organic. BO no. 2928,13/9/1999.
6. Brumfield, G.R, A. Rimal, S. Rieners (2000): Comparative Cost Analyses of Conventional Integrated Crop Management and Organic Methods. Horticultural Technology 10 (4): 785-793.
7. Cvikic, D. J. Zdravkovic, I. Ljubanovic-Ralevic., G. Surlan-Momirovic, Z. Susic (2000): b-Carotene content and its changing dynamics in the investigated nor and rin tomato genotypes. Acta horticulture, 579: 145-149
8. Jarak, M., M. Zdravkovic, S. Duric.and J. Zdravkovic (2007): Biofertilization in bean production. Proceedings I International Congress Food technology, quality and safety, Novi Sad, Serbia, 13-15. November, 135-141.

9. Lazic, B. (2008): Good agricultural practice – a part of sustainable agriculture. Magazine Agriculture, II Symposium "Healthy – Organic" Selenca, September 26 – 27th, 2 – 3.

10. Markovic Z., J. Zdravkovic, M. Damjanovic, M. Zdravkovic and R. Djordjevic (2008): New tomato hybrids with Long Shelf life. Acta horticulture 789, p.137-140.

11. Metwally M. I., T. C. Wehner (1990): Breeding cucumbers for fresh-market production in Egypt. CGC Rpt. No. 13, 12 – 14.

12. Mijatovic, M., J. Zdravkovic and Ž. Markovic (2007): Reaction of some tomato cultivars and hybrids to late blight (P. infestans Mont. De Bary). Acta Horticulture, 729, 463-466.

13. Pavlovic, N., L J. Stankovic, M. Mijatovic (2002a). The history of cucumber breeding in Yugoslavia. Book of proceedings. Cucurbitaceae, 8-12 December, Naples, USD 78-81.

14. Pavlovic, N., J. Zdravkovic and T. Sretenovic-Rajicic (2002b): Variability and heritability of the onion bulb fresh weight. Yearbook of the Faculty of Agriculture, University „Ss. Cyril and Methodius", Skopje, Macedonia, 47, 171-175.

15. Pavlovic, N., J. Zdravkovic and D. Cvikic (2003b): Content and variability of vitamin c in onion bulbs (Allium cepa L.). Book of proceedings 1st International Symposium "Food in the 21st century", Subotica, Serbia, November 2001, p: 638-642.

16. Pavlovic, N., J. Zdravkovic, D. Cvikic and D. Stevanovic (2003a): Genetic divergence of the cultivated ecotypes of spring garlic (Allium sativum L) on the territory of Yugoslavia. Acta Horticulture, 598, 187-192.

17. Pavlovic, N., D. Cvikic, J. Zdravkovic and B. Zecevic (2006): Cucumber (Cucumis sativus L.) breeding in the Centre for Vegetable Crops, Smederevska Palanka. In "Food production within European Law", pp. 92Teslic, BiH.

18. Pavlovic N., D. Cvikic, M. Zdravkovic, R. Dordevic and S. Prodanovic (2007): Variability and heritability coefficient of average dry matter content in onion (Allium cepa L.) bulbs. Genetika, 39 (1), 63 - 68.

19. Prodanovic, S. and G. Surlan-Momirovic (2006): Genetic resources for organic agriculture (monograph). Faculty of Agriculture, Belgrade. (ISBN86-7834-001-0 in Serbian)

20. Stikic, R., S. Savic, M. Srdic, D. Savic, J. Zovanovic, L J. Prokic and J. Zdravkovic (2003): Partial drying of tomato root-zone: physiological effects and implications for solute transport mechanisms. International Conference on Water-Saving Agriculture and Sustainable Use of Water and Land Resources (ICWSAWLR), J. Exp. Bot., 54, Supplement 1, 21, Yangling, Shaanxi, P.R. China.

21. Thompson ,G. (2000): International consumer demand for organic foods. Hort Technology 10 (4), 663- 674.

22. Valema, J. (2004): Challenges and opportunities in organic seed production. Processing of the First World Conference on organic seed "Challenges and opportunities for organic Agriculture and Seed Industry. Roma, Italy, July 2004, FAO, 4-6.

23. Van Weas, J. (2003): Impact of plant breeding on plant and food production - recent evolutions and perspectives of biotechnology. Book of proceedings of the 1st International Symposium "Food in the 21st Century", Subotica, Serbia, November 2001, 52-68.

24. Zdravkovic, J., Z. Markovic, M. Mijatovic, B. Zecevic, M. Zdravkovic (2000): Epystatic gene effects on the yield of the parents of F1, F2, BC1 and BC2 progeny. Acta Physiologiae Plantarum, Vol.22, No.3, 261-266.

25. Zdravkovic, J., Ž. Markovic, M. Zdravkovic and N. Pavlovic (2002): Tomato selection on increased value of lycopine. Proceedings EKO conference 2002, Novi Sad, book II, 281-287.

26. Zdravkovic, J., D. Stevanovic, T. Sretenovic-Rajicic and M. Zdravkovic (2002b): Genetic analysis of antocyanin and vitamin C contents in lettuce (Lactuca sativa L.). Acta Horticulturae, 579, 167-170.

27. Zdravkovic J., Ž. Markovic, M. Damjanovic, M. Zdravkovic, R. Dordevic (2003a): The expression of the rin gene in prolongated tomato fruit ripening (Lycopersicon esculentum Mill.). Genetika, Vol. 35, 2, 77-85.

28. Zdravkovic, J., L J. Stankovic and D. Stevanovic (2003b): Possibilities of using wild lettuce forms originating from the spontaneous Yugoslav flora in the selection for virus diseases of Lactuca sativa L. Acta Horticulture, 598, 243-246.

29. Zdravkovic, J., Ž. Markovic, and N. Pavlovic (2003c): Lycopene contents in tomato fruits (Lycopersicum esculentum Mill.) Book of proceedings 1st International symposium "Food in the 21st century", Subotica, Serbia, November 2001, p: 628-632.

30. Zdravkovic, J., M. Mijatovic, M. Ivanovic, Ž. Markovic and M. Zdravkovic (2004): Breeding of tomato hybrids for resistance to late blight casual agent (Phytophtora infestans Mont. de Barry). Proceedings of the XXXIV Annual ESNA Meeting, Novi Sad, Serbia and Montenegro, 305 – 308.

31. Zdravkovic J., Ž. Markovic, M. Zdravkovic, M. Damjanovic, N. Pavlovic (2007): Relation of mineral nutrition and content of lycopene and - carotene in tomato (Lycopersicon esculentum Mill.) fruits. Acta Horticulture 729, 177-181.

32. Zdravkovic J., Z. Markovic, D. Cvikic, M. Mijatovic, L J. Stankovic (2008): Firmness of tomato fruits depending on trait accumulation and incorporation of ripening inhibitor gene. Acta horticulture 789, 199-204.

33. Zdravkovic, J., (2008): Organic agriculture, biodiversity, breeding and seed production. Magazine agriculture, II Symposium „Healthy-Organic", Selenca 26-27th September, 6-10.

34. Zdravkovic, M., M. Damjanovic, J. Zdravkovic and Ž. Markovic (2004a): The reaction of some bean genotypes to drought. Book of proceedings VIII Symposium Biotechnology and agroindustry (vegetable, potato, decorative, aromatic and medical plants), November, Velika Plana, Serbia and Montenegro, p: 279-284.

35. Zdravkovic, M., M. Damjanovic, M. Jarak and J. ZdravkovIC (2004b): The effect of microbiological fertilizers to formation of the first pod and grain weight of the beans. Proceedings EKO Conference 2004, Novi Sad, book I,: 305-309.

CHAPTER 11

COLLABORATIVE PLANT BREEDING FOR ORGANIC AGRICULTURAL SYSTEMS IN DEVELOPED COUNTRIES

JULIE C. DAWSON, PIERRE RIVIÈRE,
JEAN-FRANÇOIS BERTHELLOT, FLORENT MERCIER ,
PATRICK DE KOCHKO, NATHALIE GALIC, SOPHIE PIN,
ESTELLE SERPOLAY, MATHIEU THOMAS, SIMON GIULIANO,
AND ISABELLE GOLDRINGER

11.1 INTRODUCTION

The regulatory system in many countries, particularly in the European Union, restricts the varieties available to farmers to those registered in an official catalogue (National or European). In the EU, to be commercialized, a variety has to be registered in the European catalogue and meet evaluation criteria including "distinctiveness, uniformity and stability" (DUS). Further regulations for variety registration vary by country but usually include "value for cultivation and use" standards that measure agronomic performance and technical end-use quality in conventional systems. This has resulted in a lack of suitable varieties available to organic farmers, since most modern varieties are developed for and tested in high-input conventional cropping systems. Thus, many farmers using organic

This chapter was originally published under the Creative Commons Attribution License. Dawson JC, Rivière P, Berthellot J-F, Mercier F, de Kochko P, Galic N, Pin S, Serpolay E, Thomas M, Giuliano S, and Goldringer I. Collaborative Plant Breeding for Organic Agricultural Systems in Developed Countries. Sustainability **2011,**3 (2011); 1206–1223. doi:10.3390/su3081206.

and low-input methods are unable to find a variety in the catalogue that is adapted to their agricultural environment. The vast majority (over 95%) of varieties used in organic agriculture were initially bred for conventional systems [1].

Due to the expense of registering a variety and the limited markets for organic seed, private-sector breeders must add organic breeding activities to existing conventional activities, and in general cannot spend more than 10% of their breeding investment on organic systems [2]. Because conventional systems can be buffered by inputs, the natural environmental variability encountered by modern varieties is generally limited, which makes it possible to breed varieties with apparent broad adaptation. This reduces the need for genetic diversity in commercial varieties [3] but these varieties are only superior in a narrow range of production environments where growing conditions are standardized and stress is minimized [4]. Many authors have pointed to a need for greater diversity in agriculture [1,3,5–12]. Due to greater heterogeneity of environmental conditions (both spatial and temporal) in organic systems, there is a particular need to increase genetic diversity on organic farms [13–18].

If the target environments are too different to select a single variety or population that has acceptable performance everywhere, breeders may target very focused regions for different varieties or heterogeneous populations that may then evolve specific adaptation. When varieties have different performance relative to each other when changing environments, variety-by-environment interactions, or more commonly, genotype by environment (G E) are present. In this case, an analysis of G E is a method of choosing the best variety or population for each target environment [4]. Usually, significant G E interactions imply that selection needs to be conducted in the target environment. Where this direct selection in the target environment is important, decentralization of the selection process is necessary and this often leads to participatory selection [19,20] because of the need to reduce differences between selection sites and the target environments. Decentralized selection on regional research stations may still not be representative of on-farm environments and management. Farmer participation brings more than just more representative testing locations, as farmers have in-depth knowledge of environmental conditions and plant traits that are adaptive under their conditions. The positive use of G E to develop locally-adapted varieties through decentralized selection and participatory breeding has been proposed as a solution to breeding for organic and low-input systems. The basis for many participatory plant breeding projects lies in the use of the genetic diversity present in farmer varieties and in farmers' specific knowledge and evaluation of varietal traits of importance in their agricultural environment. The need for specific traits (both agronomic and quality) for different environments and management practices is currently not considered in the official varietal registration process. In developed countries, interest in on-farm breeding and conservation has been primarily from organic and low-input farmers because they face many challenges not present in conventional systems, including more heterogeneous environmental conditions, the absence of adapted varieties, and a lack of interest by the commercial seed sector in producing varieties for these systems [15,18].

Farmers interested in on-farm breeding often look for more diverse varieties because they want these varieties to be able to evolve specific adaptation to their conditions, and because genetic heterogeneity may buffer crop responses to unpredictable environmental conditions. The agronomic benefits of diversity include improved durability of disease resistance and reduction of disease severity [16,21] and greater buffering capacity of heterogeneous populations [22,23]. In heterogeneous populations, phenotypic stability may arise from genetic diversity that allows the flexible expression of component traits that lead to higher stability for complex traits such as yield and quality. In addition to contributing to the development of well-adapted varieties, the conservation of genetic diversity within varieties is also important to maintain the adaptive potential of these varieties. Contrasted selection pressures over multiple farms will preserve the greatest level of diversity at the meta-population level (i.e., considering all the farms together), even if each population loses a portion of its initial diversity in the process of selection for local adaptation [24–28]. Selection and conservation objectives benefit from the ability to evaluate and use the widest range of genetic diversity available.

Because of the benefits drawn from diversity under organic conditions, organic farmers in developed countries have been key in the development of seed exchange and seed saving organizations [29]. The work presented in this paper was possible because of the engagement and activities of the Réseau Semences Paysannes (RSP, Farmers' seed network), a farmers' association dedicated to the conservation, cultivation and exchange of diverse varieties. Farmers involved in this network grow historic varieties and landraces because they are more suitable for their management practices and because of superior quality characteristics for on-farm processing and direct sales to consumers. They also have a strong commitment to preserving biodiversity, with some farmers growing over 200 different varieties in small (5 m^2) plots and continuously experimenting with new varieties from different genebanks and collections. Demeulenaere et al. [5] showed that a small group of farmers growing these types of populations can make a significant contribution to the conservation of genetic diversity when compared to the diversity present in specific genebank accessions. On-farm cultivation of these diverse varieties is an effective method of in situ conservation which also conserves the evolutionary process [30–34].

In this paper, we first present results from an experiment on the evolution of wheat landraces cultivated on-farm over 3 years, with farmer mass selection in the final year. This researcher-led project was designed to address the need for more information on how on-farm selection and management affects the diversity of cultivated species and how these practices can be optimized for both the conservation of important genetic resources and the development of well-adapted varieties. Second, we present results from a farmer-led plant breeding project that was initiated by farmers in the RSP to create locally adapted varieties and increase farmer autonomy in terms of varietal choice and seed production. In the first case farmers participated in a researcher-led project and in the second, researchers participated in a farmer-led project. Finally, the process of developing a collaborative plant breeding program for organic agriculture is presented in the last two sections of the discussion, including lessons learned from our current work and their implications for its future direction. It is challenging to develop sustainable collaborations with research strategies that work for scientists and farmers, and the role of the RSP farmers' association was critical in creating a successful model. We consider collaborative plant breeding to be a method where interested parties participate equally in the development, implementation and evaluation of a plant breeding program. In our case, this included farmers who also mill their grain and have on-farm bakeries, farmers' associations and public-sector researchers. We hope that our experience will prove useful to other project teams working to do applied on-farm research and plant breeding.

11.2 MATERIALS AND METHODS

Two on-farm studies of farmer varieties of bread wheat in organic systems were conducted from 2006 to 2010. The first project studied variety evolution and farmer mass selection in wheat populations within the framework of the European project Farm Seed Opportunities (FSO), designed principally to study the evolution, diversity and adaptation of non-conventional varieties, such as landraces, historic varieties and varietal mixtures, referred to as farmer varieties hereafter. The second is an ongoing project between the RSP and researchers from the Institut National de la Recherche Agronomique (INRA). This project is creating new populations by crossing wheat landraces and other public varieties and is referred to as the *Croisements du Roc* after the farm where the crosses were made.

11.2.1 FARM SEED OPPORTUNITIES

Eight farmer varieties of bread wheat and two modern varieties were studied from 2006 to 2009 as part of the FSO project. These varieties were grown for 2 years on organic farms, 4 in France, 2 in Italy and 2 in the Netherlands. In the 3rd year, 2008–2009, replicated trials on each farm compared a sample of the variety from the originating farmer (original version) to the version grown for the two previous years on their farm (3rd generation version)(see Figure 1). An additional step was added in 2009 for three farmers who chose to participate to measure the effects of farmer mass selection. These farmers first gave their feedback on the varieties studied in order to choose five of these varieties for the selection protocol. The varieties chosen represent different genetic structures and included Piave (PI), a landrace from NE Italy; Rouge de Bordeaux (RB), a historic variety from SW France; Redon (RD), a mixture of landraces of the same name from western France; Renan (RN), a modern variety considered to be the reference for organic agriculture; and Zonnehoeve (ZH), a mixture of two modern varieties from the 1990s, grown as a mixture for 15 years on an organic farm in the Netherlands (see Serpolay et al. [35] for more details on the varieties).

Before harvest, each of the farmers were asked to select 20 spikes (from different plants) in the 2 replicates of the 3rd generation version of each varieties. Phenotypic traits were measured at maturity for these selected individuals and for 20 other individual plants chosen at random from each replicate of the origin and 3rd generation version. These traits included plant height (PH), last leaf-to-spike distance (LLSD), spike length (SL), grain weight per spike (GW/spike), and thousand kernel weight (TKW). LLSD was considered important by farmers based on their observations, because a longer LLSD distances the spike from the canopy and potential foliar disease spores, may reduce humidity around the spike making conditions less favorable to disease development, and may improve grain filling under conditions where leaves no longer photosynthesize due to drought, heat or disease stress. The other individual plant traits were chosen based on the descriptors used for variety registration (which evaluate the distinctiveness, uniformity and stability of varieties), in order to characterize farmer varieties with respect to modern varieties.

The versions were then compared (origin vs. 3rd generation without selection, and selections within the 3rd generation vs 3rd generation without selection) for these phenotypic traits. The analysis of on-farm evolution and selection differentials were done using an ANOVA model: Y = u + farm + rep(farm) + variety + version(variety) + farm*variety + farm*version(variety) + error. Tests of differences between versions were done using Tukey's multiple comparison procedure for two cases: (i) between the original and 3rd generation (without conscious selection) versions within each farm and variety; and between selections and the 3rd generation (without conscious selection) version within each farm and variety. In 2009–2010, progeny head-rows of selected and non-selected spikes of all four farmers varieties and Renan were grown at the research station of INRA le Moulon. Plots of selected and non-selected progeny of Rouge de Bordeaux (RB), Blés de Redon (RD) and Renan (RN) were grown on one participating farm to analyze the response to selection. Results from the on-farm trial are presented here.

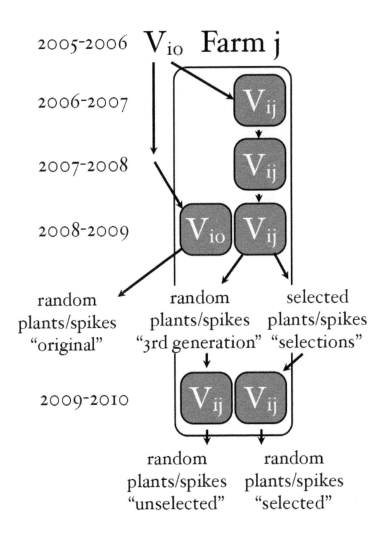

FIGURE 1: FSO project design for each variety i farm j. Vio indicates seed harvested from the variety i in its region of origin. All varieties were cultivated on each farm from 2006–2009.

11.2.2 CROISEMENTS DU ROC

On the initiative of a farmer active in the RSP, a participatory plant breeding project was started in 2005 with researchers from INRA le Moulon. Crosses were made on his farm between different wheat landraces and more recent varieties of interest. The first (F1) and second (F2) generations of progeny of 90 different families (one family is derived from each cross) were also evaluated on his farm in 2006–2007 and 2007–2008. Selections were made of individual spikes in several of the F2 families. Seed of these spikes was bulked for each family and evaluated side-by-side in the F3 generation with the corresponding unselected bulk at Moulon in 2008–2009. The protocol for plant and spike measurements was the same as that of the FSO project. Analysis was done using an ANOVA model: Y = u + rep + family + version(family) + error for the trial at le Moulon; tests between the selected and non-selected versions within each family were made with Tukeys multiple comparison procedure.

F2 families were harvested in bulk in 2008 on-farm and seed was distributed to 14 additional farmers in the autumn of 2008 for the evaluation of the F3 generation on their farms. Management of on-farm trials was the responsibility of the farmers, and researchers made phenotypic measurements on 25 plants and spikes per F3 family on each farm at maturity. This was the first year of diffusion of progeny from these crosses within the farmer seed network, and was an experimental year to see if farmers were interested and how the network could enhance the use of results from these types of trials. In the 2009–2010 year, all but two farmers continued their populations for the F4 generation using seed they harvested from F3 plants in 2009, and an additional 12 farmers joined the project, primarily cultivating F3 populations from the remaining seed of the 2008 harvest. In addition to on-farm evaluation, crosses were made in the spring of 2010 and 2011 to start new populations. These crosses were made between parents chosen by farmers for their particular environments, and a workshop on creating new populations through crossing different varieties was held at le Moulon in February 2010.

In 2010–2011, the on-farm trials have been substantially modified to better meet the needs of scientists and farmers. Trials are now structured into regional platforms with surrounding satellite trials where each farmer tests the most promising populations for their farm and a larger number of populations are tested on a central farm for each region. This network of on-farm trials will permit an analysis over multiple locations and years for each family and exchanges among participating farmers in order to provide quantitative and qualitative information on the different families to aid in making selection decisions. This new strategy is the result of farmers and researchers learning during the first years of the *Croisements du Roc* project and during the FSO project and the lessons taken from these two experiences and new methods for collaborative plant breeding are described at the end of the discussion.

11.3 RESULTS

11.3.1 FARM SEED OPPORTUNITIES

There were often similar levels of intra-varietal variability between farmer and modern varieties, indicating that the strong selection for genetic homogeneity to meet regulatory criteria had little impact on the phenotypic variability of certain traits when assessed on organic farms [35]. This could be due to high levels of micro-environmental variation within plots or residual genetic variation that is not expressed in conventional trials. Several farmer varieties had high values of traits related to productivity outside their region of origin, which underlines the need for experimentation and exchange across regions with diverse material in order to find and develop appropriate varieties for organic systems [35].

In general there were fewer and more moderate significant differences due to on-farm evolution without selection than there were significant selection differentials (see Figure 2). Natural selection increased plant height (PH) twice and decreased it 3 times, with non-significant differences 10 times, while selection differentials for PH were significant for all varieties and all farms except for ZH at farm HF and was always in the direction of increasing plant height. Natural selection decreased the LLSD twice and had no other significant differences. LLSD was more variable, with fewer significant selection differentials, and significant changes in both directions. Natural selection increased GW/spike twice and decreased it

6 times, with 7 non-significant differences. GW/spike always had a significant positive selection differential except for RD at farm HF. KN/spike (not shown) presented similar results to GW/spike. Natural selection increased TKW once, decreased it 4 times, with 10 non-significant differences. TKW also had significant positive selection differentials except for RB at farms GC and HF and RN at farm HF where it was negative, and for ZH at farm GC where it was not significant.

Figure 3 shows selection differentials for PH and GW/Spike in 2009 for RB, RD, and RN selected by FM and the selection response for the same populations grown in 2010 at FM. KN/Spike showed the same pattern as GW/Spike. Selection differentials were positive for all three varieties for PH, KN/Spike and GW/Spike, but only RB showed significant positive responses to selection for these traits. There were no significant responses for LLSD or TKW.

11.3.2 CROISEMENTS DU ROC

Significant quantitative responses to selection were found in many F3 families at Le Moulon. Table 1 shows significant differences between F3 progeny of selected and non-selected F2 spikes. There were significant responses to selection for many families, a few are shown as examples. There were also 9 of 36 families with no significant changes for any trait measured and several which only showed a significant response to selection for a few traits. There were more significant differences for PH and LLSD, whileGWand TKW had fewer families that showed a significant response to selection. However, PH and LLSD changed in both directions while the change in GW and TKW was always positive.

For the F3 and F4 populations, it was possible to qualitatively differentiate among families grown at the same farm, and there were usually one or two families that stood out to the farmer as being particularly interesting in the first year. Farmers chose to either continue with all families for the second year or to discard certain families and add new ones for the 2009–2010 growing season. Evaluations from 2010 were used to decide on which populations to keep on each farm for 2010–2011. In 2010–2011 a new methodology was adopted by the selection network, described in the discussion section.

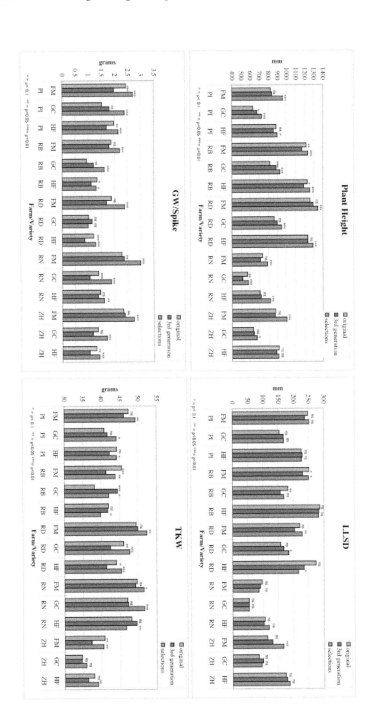

FIGURE 2: Tests of significant differences by variety and farm due to on-farm evolution over 3 years without human selection and tests of significance for selection differentials.

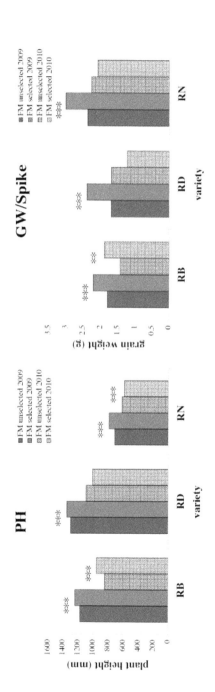

FIGURE 3: Selection differentials (2009) and response to selection (2010) for Rouge de Bordeaux (RB), Bles de Redon (RD) and Renan (RN) grown on the farm of FM. Significance levels are: ** p < 0:01, *** p < 0:001.

TABLE 1: Examples of families with significant changes among F3 progeny from selected and non-selected spikes in the F2 generation evaluated at the research station of Le Moulon.

Family	PH P-value	PH Change	LLSD P-value	LLSD Change	GW/Spike P-value	GW/Spike Change	TKW P-value	TKW Change
24	0.0187	–	0.4038	ns	0.6568	ns	0.0021	+
34	0.0001	+	<.0001	+	0.723	ns	0.0051	+
42	0.0786	+	0.0016	+	0.5407	ns	0.602	ns
60	0.0956	+	0.0442	–	0.6457	ns	0.752	ns
64	<.0001	+	0.0001	+	0.0366	+	0.3011	ns
80	<.0001	+	0.0114	+	0.0158	+	0.0884	+
90	0.5706	ns	0.2792	ns	0.0014	+	0.0005	+

PH = Plant height from the ground to the top of the spike not including the awns; LLSD = distance between the base of the spike and the flag leaf; SL = length of the spike from the base to the tip not including the awns; GW/Spike = weight of grain per spike; TKW = thousand kernel weight; P = values for the difference between selected and non-selected progeny within each F3 family using ANOVA and Tukeys mcp. The direction of change from non-selected to selected progeny is given for $p < 0.1$; –non-selected>selected, +non-selected<selected, ns = $p > 0.1$.

11.4 DISCUSSION

11.4.1 SELECTION WITHIN FARMER VARIETIES

Traits such as PH are very easy to see in the field and every farmer chose plants that were taller than the average for the plot, but that had not lodged. The majority of organic farmers involved in this network state that they would like varieties that are vigorous and tolerant of weed pressure early in the growing season but strong enough not to lodge. In contrast, long LLSD is a characteristic sought by farmers but did not have a significant selection differential in most cases. This trait may be more difficult to evaluate in the field or may not be as important as traits such as GW/spike and PH when farmers are making their selections.

GW/spike seems to be fairly easy to evaluate in the field, as there were almost always significant selection differentials. This trait was not always favored by natural selection, even if we may reasonably hypothesize that it is connected to plant fitness. There may be a trade-off with plant height, where tall plants are favored during the growing season due to competition for light, but those that have heavy spikes have a greater tendency to lodge and be lost at harvest. TKW had a significant selection differential in all but one location for one variety, and was positive in all but 3 cases when it was significant. This is somewhat surprising due to the difficulty of visually assessing the TKW of spikes without looking at the grain.

In the on-farm evaluation of the response to selection, one historic variety, RB, responded positively to selection for PH and GW/spike while a landrace population, RD, and the modern variety RN did not, despite significant selection differentials in 2009. This difference in varietal capacity to respond to selective pressure is in agreement with the farmers observation that RB was one of the varieties that performed very well outside its region of origin, indicating greater adaptability than landraces such as RD. It also corresponds to molecular genetics findings of high levels of internal diversity of farmer-managed populations of RB [5]. In contrast, the particular sample of the landrace RD used for this study was composed by a farmer starting from a small number of spikes selected in accessions from the national genebank, possibly inducing a genetic bottleneck and a reduction of adaptive potential. Ex situ conservation in gene banks may lead to reduced genetic diversity due to genetic drift in small population sizes [9,36]. The lack of response to selection in Renan is not surprising as we expected the observed selection differential to be due primarily to micro-environmental factors that led to phenotypic differences among plants within a single plot in the selection generation. However, these results must be interpreted with caution as they are only from one location. A common garden experiment with all selections from all varieties and farms at the research station le Moulon is currently being analyzed for more generalizable results on response to selection.

Results from the overall FSO project showed that farmer varieties can be heterogeneous for some traits but homogeneous for others, and can be more or less adaptable depending on the trait and range of environmental conditions. The response to natural selection over three years was variable, with traits changing in both directions or not at all, showing that for certain traits, farmer selection is necessary to maintain or improve certain traits. A strong selection differential was expected for some traits that are observed easily in the field, for other traits such as TKW the presence of a strong selection differential was more surprising. The results for some traits are more complex to interpret, and it is not always clear if the lack of a selection differential or response was due to a lack of genetic diversity for those traits or because the traits used for varietal characterization may not be the most appropriate to measure the effects of on-farm selection when this selection is much more holistic in nature. Molecular genetic studies are in progress on the base populations to determine the levels of genetic variation and the on-farm evolution of both neutral variation and variation linked to hypothesized adaptive traits such as earliness.

Significant responses were found after only one cycle of selection in farmer varieties, but this is dependent on the variety and sufficient genetic diversity needs to be present for selection to have an effect. Since farmer selection is based much more on a global evaluation of the plants and populations, and since this often produces desirable agronomic changes in the field based on the experience of farmers in the RSP over several years, quantitative measures and analysis may be most useful when focused on a small number of traits that are more difficult to evaluate by farmers, such as protein content, or that respond unsatisfactorily to their selection, such as LLSD. Quantitative measures may also be used after selection to monitor these changes from one generation to the next, especially for traits that have negative correlations, or to document and characterize these populations as they evolve.

While the EU directive on conservation varieties is an improvement over existing standards which prohibit the exchange of any variety not registered in the official catalogue, it is fairly restrictive in the types of varieties that can be registered in the special catalogue for conservation varieties. It still includes the distinct, uniform and stable criteria for registration, with slightly relaxed criteria of evaluation. In addition, they are limited to a restricted zone of origin and percentage of the area cultivated to that species and to limited quantities for commercial use [37]. The new directive therefore does not include the range of farmer varieties currently of interest in organic systems, and may not be adequate for some types of on-farm conservation and plant breeding activities, especially those conducted by established farmers seed groups, which rely on a broad network seed exchanges to maintain and enhance diversity and agronomic performance.

11.4.2 RESPONSE TO SELECTION IN EARLY GENERATIONS

The response to selection in early generations that was measured at Le Moulon was of interest to researchers and farmers because it allowed us to evaluate different selection strategies, or example, deciding the generation at which individual spike selection within segregating population might be efficiently introduced, or the number of years to observe a bulk population before selecting among bulks. There were many significant changes when comparing F3 progeny from selected and unselected F2 generation spikes. Some families had more significant changes than others, which may be a function of the heterogeneity of the parental varieties or the degree of difference between the two parents. Within families that responded to selection for multiple traits, some traits responded more frequently to selection in early generations. This includes traits such as PH that are easy to assess visually when selecting spikes in the field, and those that are not, such as TKW. Other traits may not change in early generations, or may change in the opposite direction of what is desired, such as LLSD. This may mean that for certain traits it is better to wait to conduct selection till later generations, and a potential function of the collaboration with researchers could be to identify which traits respond to selection and which do not, or to identify easy techniques in the field to avoid undesirable changes in the populations. Further analyses will investigate the relation between the parents of the crosses and the response to selection in the F2 progenies.

Results in terms of means and variances for each trait measured for all the *Croisements du Roc* populations evaluated on their farms were distributed to each farmer, as well as the results from the side-by-side trials of selected and unselected F3 families at Moulon. While there were differences among families tested at the same farm, often with one or two families being identified as very promising by the farmer, farmers and researchers always use more than one year to evaluate the potential of new population varieties and the early-generation cross populations are no exception.

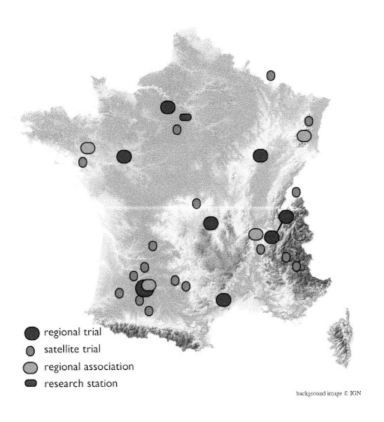

FIGURE 4: Map of on-farm collaborative wheat breeding trials in 2010–2011.

The on-farm trials from 2008-2010, in the F3-F4 generations presented certain difficulties in drawing conclusions across all field sites, because the specific populations grown on each farm were not the same from one farm to the next. This was due to both the limited availability of seed of each family and because families were targeted to the environment where they were most likely to be adapted based on their parentage and the phenotypic evaluation done on one farm in 2008. In addition to the statistical challenges with this trial design, the farmers needed more flexibility in choosing populations to keep and discard. The ideal trial setup from a statistical point of view, with the same populations replicated within and among farms, was not feasible from a practical standpoint.

11.4.3 LESSONS LEARNED AND METHODS DEVELOPED FOR THE FUTURE OF THE PROJECT

One of the main themes that emerged in discussions among farmers and scientists is the different approaches of farmers and scientists to the selection process and to the evaluation of plant and variety performance. Scientists have a more analytical, quantitative approach, while farmers have an appreciation of the interactions of the plant with its environment and the overall performance of plant populations in their fields. In this case of participatory selection, the goal is to bring together the strengths of both these methods to improve the on-farm selection of populations that respond to the needs of organic farmers. While, in general, scientists have been criticized for being overly reductionist, and farmers have been criticized for being less rigorous in their experiments, in this case participants have an appreciation and respect for the expertise and experience that each person brings to the collective project.

The FSO project was seen by many farmers as too top-down, primarily directed by researchers. In particular, the development of a collaborative project was difficult due to the EU requirements that all project documents and reports be submitted in English, which facilitates international work among scientists but creates barriers to farmer participation in project meetings and written publications. Researchers agreed with this assessment in general, as the project was designed to respond to a specific need for information in order to shape the regulatory framework for conservation varieties and was not intended to be a program of participatory plant breeding. The information gained and lessons learned from the FSO project can be used to in collaborative projects, as it helped researchers learn more about farmers evaluation of their varieties and their motivations in cultivating these non-conventional varieties. It is hoped that the results submitted to the European Commission will result in a regulatory framework that is more conducive to on-farm conservation and selection.

The project *Croisements du Roc* began as an initiative of farmers within the RSP and has been co-constructed by farmers, the RSP and the research team. The majority of farmers involved in the *Croisements du Roc* project also cultivate historic varieties or landraces within the larger RSP association. There is an ongoing discussion on the advantages and disadvantages of making crosses rather than selecting plants within landraces or making mixtures of landraces of interest. This includes questions of which methods of plant breeding are ethically acceptable in organic systems, whether planned crosses have a place in the creation of farmer varieties, and whether this affects the integrity of the plant or variety. Crossing landraces provides more diversity for selection for new conditions, especially when landraces have lost most of their initial diversity through reduction in the land area cultivated or through conservation ex situ in a genebank. This presents an alternative when there is not diversity present in a landrace or when varieties are not working in a mixture. However, the in situ conservation of landraces with on-farm mass selection to improve local adaptation and performance, and the approach of mixing landraces and allowing the mixed population to evolve in situ are also very valuable as methods of introducing diversity and maintaining evolutionary potential. Current studies in parallel to the *Croisements du Roc* project are evaluating the longer-term evolution and adaptation of these populations in the farmer network.

11.4.4 STRUCTURE OF COLLABORATIVE BREAD-WHEAT BREEDING PROJECT

The project *Croisements du Roc* has evolved into a collaborative plant breeding project with shared responsibility among farmers, farmers' associations and researchers in the context of a new European project that has much more flexibility in terms of the structure of research projects. From the preliminary year of on-farm trials, there appears to be much interest on the part of farmers, but there are also many questions on methodology, ranging from practical to theoretical and ethical. Discussions over methods and strategies have led to a better understanding of the needs and constraints of each partner, and the goal is for the project to remain flexible, within the limits of some minimal standards that all participants agree to implement, with participants able to choose their level of involvement over time. In terms of methods of experimental design, there is always a tension between researchers need for designs that give greater precision and allow for appropriate statistical analysis and farmers need for simplicity in terms of management. This has led researchers to look into alternatives to traditional experimental designs such as randomized complete block designs which may be difficult for farmers without small scale equipment to implement.

A broader reflection on the methods of organization of the on-farm trials to meet both researcher and farmer needs led to the new structure of the trials. To provide greater flexibility for the farmers and accommodate researchers need for quantitative data to analyze population adaptation and evolution, a regional platform satellite trial model was implemented in 2010–2011, based on methods developed by Snapp [38]. Figure 4 shows the division of on-farm trials in the French network and Figure 5 is a schematic of the overall organization of the collaborative wheat breeding program.

One farm in each region, preferably with access to small plot equipment, was designated to host the regional platform trial. This trial consisted of about 30 different populations with two checks chosen by the farmers and replicated twice each. Plant phenotypic data is taken to characterize the populations, comparable to data collected in previous years as described in the material and methods section, and agronomic traits such as plot yield and protein are measured. Satellite trials have no restrictions on the number of populations or their consistency from one year to the next, which gives farmers much more flexibility in deciding which populations to try, keep and discard. These satellite trials do not collect quantitative data, but farmers record visual scores at key stages during plant development (winter survival and early spring vigor, heading time, and maturity). Farmers also recorded these observations at the regional platform trials during summer field visits in 2011 and discuss the merits of different populations and breeding objectives. Phenotypic observations include traits of greatest interest to the farmers. Regional associations and the RSP are responsible for the collection of farmer observations on satellite and regional farms and for the coordination and facilitation of project activities.

Collaborations with farmer-bakers will be used to assess quality for artisanal breadmaking as soon as sufficient grain is available. It is hoped that this design will enable the network to use data collected on the research platforms to assess the performance of populations on their farms and to choose new populations or parents for crosses that will perform well under each farmers' specific environmental conditions. The role of regional associations and the national RSP must not be underestimated in making this network of trials possible and in organizing field days and farm visits to encourage the exchange of ideas and selected populations among participants.

11.5 CONCLUSIONS

While farmers may always base their selection, appropriately, on their knowledge of the workings of specific organic agricultural systems and their intuitive assessment of plant global performance on their farms, scientific analysis may be useful for documenting these populations and showing the effectiveness of farmer selection for the conservation and improvement of cultivated populations. A combined approach of evolution under natural selection and directed farmer selection within evolving populations may be useful in developing varieties adapted to organic systems [10]. In fact, certain traits may not be favored by natural selection and may require conscious selection to maintain landrace qualities which is not provided if these varieties are conserved in genebanks.

The value of landraces lies in their history of farmer selection and adaptation to diverse environmental conditions worldwide, their capacity to continually evolve and their resilience to heterogeneous environmental conditions. This creative process of variety development through on-farm selection has largely been lost in industrialized agricultural systems. Collaborative plant breeding in developed countries, starting either with existing landraces or with new populations developed from crosses and farmer-selected variety mixtures can renew this process, reclaiming the knowledge of selection and autonomy of seed production for farming communities. This process of continual crop evolution is key to adapting crops to changing climatic conditions and complex environmental stresses which are present in organic and low-input environmental systems, both in developed and developing countries.

On-farm selection by farmers who are also millers and bakers or who work closely with them is also key to developing varieties that have the right characteristics for high quality artisanal breads and regional products.

On-farm selection is thus complementary to on-farm conservation, using the diversity within an existing farmer network to maintain and enhance local adaptation and crop performance. While on-farm selection requires a significant commitment from farmers and researchers to be successful, a relatively small group of dedicated farmers and researchers can serve a broader public interest in conserving genetic diversity for important agricultural species through selecting diverse populations across a wide range of environments. These populations, while started by one farmer or the research group, can then be tested and selected in many on-farm environments. The exchange of observations and ideas among participants in the collaborative selection project within the RSP is very rich for both farmers and scientists, highlighting the value of a network approach to plant breeding, rather than an approach were all communication is between individual farmers and the research institution. Because the project was started and driven by farmers, it has far greater durability than a project tied to a particular grant or funding source. By working in this network we hope to increase farmer autonomy in variety creation and to develop a sustainable

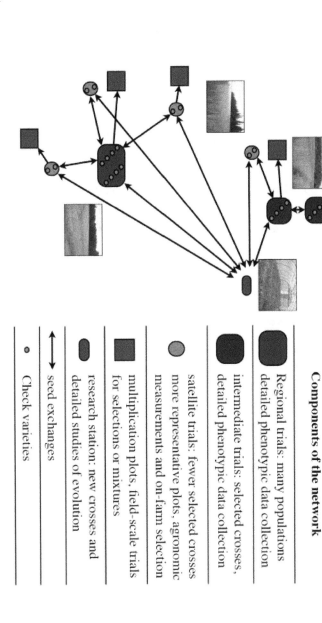

Components of the network

Regional trials: many populations
detailed phenotypic data collection

intermediate trials: selected crosses,
detailed phenotypic data collection

satellite trials: fewer selected crosses
more representative plots, agronomic
measurements and on-farm selection

multiplication plots, field-scale trials
for selections or mixtures

research station: new crosses and
detailed studies of evolution

seed exchanges

Check varieties

FIGURE 5: Schematic of the organization and components of the French collaborative wheat breeding project.

process of varietal innovation, leading not necessarily to new fixed varieties, but rather to well-adapted heterogeneous populations for each farm which continue to evolve.

REFERENCES

1. Lammerts van Bueren, E.T.; Backes, G.; Vriend, H.; Ostergaard, H. The role of molecular markers and marker assisted selection in breeding for organic agriculture. Euphytica 2010, 175, 51–64.
2. Löschenberger, F.; Fleck, A.; Grausgruber, H.; Hetzendorfer, H.; Hof, G.; Lafferty, J.; Marn, M.; Neumayer, A.; Pfaffinger, G.; Birschitzky, J. Breeding for organic agriculture: The example of winter wheat in Austria. Euphytica 2008, 163, 469–480.
3. Phillips, S.L.; Wolfe, M.S. Evolutionary plant breeding for low input systems. J. Agric. Sci. 2005, 143, 245–254.
4. Ceccarelli, S. Positive interpretation of genotype by environment interactions in relation to sustainability and biodiversity. In Plant Adaptation and Crop Improvement; Cooper, M., Hammer, G.L., Eds.; CAB International: Wallingford, Oxon, UK, 1996; pp. 467–486.
5. Demeulenaere, E.; Bonneuil, C.; Balfourier, F.; Basson, A.; Berthellot, J.F.; Chesneau, V.; Ferté, H.; Galic, N.; Kastler, G.; Koenig, J.; et al. Étude des complémentarités entre gestion dynamique à la ferme et gestion statique en collections. em Les Actes du BRG 2008, 7, 117–138.
6. Finckh, M.R.; Gacek, E.S.; Goyeau, H.; Lannou, C.; Merz, U.; Mundt, C.C.; Munk, L.; Nadziak, J.; Newton, A.C.; de Vallavieille-Pope, C.; et al. Cereal variety and species mixtures in practice, with emphasis on disease resistance. Agronomie 2000, 20, 813–837.
7. Hajjar, R.; Jarvis, D.I.; Gemmill-Herren, B. The utility of crop genetic diversity in maintaining ecosystem services. Agric. Ecosyst. Environ. 2008, 123, 261–270.
8. Heal, G.; Walker, B.; Levin, S.; Arrow, K.; Dasgupta, P.; Daily, G.; Ehrlich, P.; Maler, K.G.; Kautsky, N.; Lubchenco, J.; et al. Genetic diversity and interdependent crop choices in agriculture. Resour. Energy Econ. 2004, 26, 175–184.
9. Horneburg, B.; Becker, H.C. Crop adaptation in on-farm management by natural and conscious selection: A case study with lentil. Crop Sci. 2008, 48, 203–212.
10. Murphy, K.M.; Lammer, D.; Lyon, S.R.; Carter, B.; Jones, S.S. Breeding for organic and low-input farming systems: An evolutionary-participatory breeding method for inbred cereal grains. Renewable Agric. Food Syst. 2005, 20, 48–55.
11. Newton, A.C.; Begg, G.S.; Swanston, J.S. Deployment of diversity for enhanced crop function. Ann. Appl. Biol. 2009, 154, 309–322.
12. Zhu, y.; Chen, H.; Fan, J.; Wang, Y.; Li, Y.; Chen, J.; Fan, J.X.; Yang, S.; Hu, L.; Leung, H.; et al. Genetic diversity and disease control in rice. Nature 2000, 406, 718–722.
13. Altieri, M.A. The ecological role of biodiversity in agroecosystems. Agric. Ecosyst. Environ. 1999, 74, 19–31.

14. Ceccarelli, S.; Grando, S. Decentralized participatory plant breeding: An example of demand driven research. Euphytica 2007, 155, 349–360.

15. Desclaux, D. Participatory Plant Breeding Methods for Organic Cereals. In Proceedings of the COST SUSVAR/ECO-PB Workshop on Organic Plant Breeding Strategies and the Use of Molecular Markers, Driebergen, The Netherlands, 17–19 January 2005; Lammerts Van Bueren, E.T., Goldringer, I., Ostergard, H., Eds.; Louis Bolk Institute, Driebergen, The Netherlands, 2005; pp. 17–23.

16. Finckh, M.R. Integration of breeding and technology into diversification strategies for disease control in modern agriculture. Eur. J. Plant Pathol. 2008, 121, 399–409.

17. Østergård, H.; Finckh, M.R.; Fontaine, L.; Goldringer, I.; Hoad, S.P.; Kristensen, K.; Lammerts van Bueren, E.T.; Mascher, F.; Munk, L.; Wolfe, M.S. Time for a shift in crop production: Embracing complexity through diversity at all levels. J. Sci. Food Agric. 2009, 89, 1439–1445.

18. Wolfe, M.S.; Baresel, J.P.; Desclaux, D.; Goldringer, I.; Hoad, S.; Kovacs, G.; Löschenberger, F.; Miedaner, T.; Østergard, H.; Lammerts van Bueren, E.T. Developments in breeding cereals for organic agriculture. Euphytica 2008, 163, 323–346.

19. Sperling, L.; Ashby, J.A.; Smith, M.E.; Weltzien, E.; McGuire, S. A framework for analyzing participatory plant breeding approaches and results. Euphytica 2001, 122, 439–450.

20. Ceccarelli, S.; Grando, S.; Bailey, E.; Amri, A.; El-Felah, M.; Nassif, F.; Rezgui, S.; Yahyaoui, A. Farmer participation in barley breeding in Syria, Morocco and Tunisia. Euphytica 2001, 122, 521–536.

21. Wolfe, M.S. Crop strength through diversity. Nature 2000, 406, 681–682.

22. Ceccarelli, S. Specific adaptation and breeding for marginal conditions. Euphytica 1994, 77, 205–219.

23. Finckh, M.R.;Wolfe, M.S. Diversification strategies. In The Epidemiology of Plant Disease, 2 ed.; Cooke, B.M., Gareth Jones, D., Kaye, B., Eds.; Springer: Dordrecht, The Netherlands, 2006.

24. Enjalbert, J.; Dawson, J.C.; Paillard, S.; Rhoné, B.; Rousselle, Y.; Thomas, M.; Goldringer, I. Dynamic management of crop diversity: From an experimental approach to on-farm conservation. C. R. Biol. 2011, 334, 458–468.

25. Enjalbert, J.; Goldringer, I.; Paillard, S.; Brabant, P. Molecular markers to study genetic drift and selection in wheat populations. J. Exp. Bot. 1999, 50, 283–290.

26. Goldringer, I.; Enjalbert, J.; David, J.; Paillard, S.; Pham, J.L.; Brabant, P. Dynamic management of genetic resources: A 13 year experiment on wheat. In Broadening the Genetic Base of Crop Production; Cooper, H.D., Spillane, C., Hodgkin, T., Eds.; IPGRI/FAO: Rome, Italy, 2001; pp. 245–260.

27. Lavigne, C.; Reboud, X.; Lefranc, M.; Porcher, E.; Roux, F.; others. Evolution of genetic diversity in metapopulations: Arabidopsis thaliana as an experimental model. Genet. Sel. Evol. 2001, 33, S399–S423.

28. Porcher, E.; Giraud, T.; Goldringer, I.; Lavigne, C. Experimental demonstration of a causal relationship between heterogeneity of selection and genetic differentiation in quantitative traits. Evolution 2004, 58, 1434–1445.

29. Osman, A.; Chable, V. Inventory of initiatives on seeds of landraces in Europe. J. Agric. Environ. Int. Dev. 2009, 103, 95–130.

30. Almekinders, C.J.M.; de Boef, W.; Engels, J. Synthesis between crop conservation and development. In Encouraging Diversity: The Conservation and Development of Plant Genetic Resources; Almekinders, C., de Boef, W., Eds.; Intermediate Technology Publications: London, UK, 2000; pp. 330–338.

31. Berthaud, J.; Clément, J.C.; Empearire, L.; Louette, D.; Pinton, F.; Sanou, J.; Second, G. The role of local level gene flow in enhancing and maintaining genetic diversity. In Broadening the Genetic Base of Crop Production; Cooper, H.D., Spillane, C., Hodgkin, T., Eds.; IPGRI/FAO: Rome, Italy, 2001; pp. 81–103.

32. Elias, M.; McKey, D.; Panaud, O.; Anstett, M.C.; Robert, T. Traditional management of cassava morphological and genetic diversity by the Makushi Amerindians (Guyana, South America): Perspectives for on-farm conservation of crop genetic resources. Euphytica 2001, 120, 143–157.

33. Louette, D.; Charrier, A.; Berthaud, J. In situ conservation of maize in Mexico: Genetic diversity and maize seed management in a traditional community. Econ. Bot. 1997, 51, 20–38.

34. Smith, M.E.; Castillo G., F.; Gomez, F. Participatory plant breeding with maize in Mexico and Honduras. Euphytica 2001, 122, 551–565.

35. Serpolay, E.; Dawson, J.C.; Chable, V.; Lammerts Van Bueren, E.; Osman, A.; Pino, S.; Silveri, D.; Goldringer, I. Phenotypic responses of wheat landraces, varietal associations and modern varieties when assessed in contrasting organic farming conditions in Western Europe. Org. Agric. 2011, doi 10.1007/s13165-011-00116.

36. Parzies, H.K.; Spoor, W.; Ennos, R.A. Genetic diversity of barley landrace accessions (Hordeum vulgare ssp. vulgare) conserved for different lengths of time in ex situ gene banks. Heredity 2000, 84, 476–486.

37. Commission, E. Commission Directive 2008/62/EC of 20 June 2008 providing for certain derogations for acceptance of agricultural landraces and varieties which are naturally adapted to the local and regional conditions and threatened by genetic erosion and for marketing of seed and seed potatoes of those landraces and varieties, 2008. Available online: http://eur-lex.europa.eu (accessed on 21 April 2011).

38. Snapp, S.S. Mother and baby trials: A novel trial design being tried out in Malawi. Target Newsletter of the Southern Africa Soil Fertility Network 1999, 17, 8.

CHAPTER 12

PHENOTYPIC CHANGES IN DIFFERENT SPINACH VARIETIES GROWN AND SELECTED UNDER ORGANIC CONDITIONS

ESTELLE SERPOLAY, NICOLAS SCHERMANN, JULIE DAWSON, EDITH T. LAMMERTS VAN BUEREN, ISABELLE GOLDRINGER, AND VÉRONIQUE CHABLE

12.1 INTRODUCTION

Seed is an input of importance in agriculture. For each crop, farmers have to choose, according to their farming system, from a range of different types of varieties, based on genetic, phenotypic, commercial, social or cultural characteristics. In Europe, the seed market is regulated: commercialized seeds have to meet a number of standards set by law. The main one is the registration of the variety in an official catalogue, which requires it to meet Distinctiveness, Uniformity and Stability (DUS) criteria [1]. These criteria are evaluated by visual assessments in trials performed by specialized institutions. Today, most marketed varieties (created by seed companies) are F1 hybrids or pure lines, which are visually very uniform and stable since they are genetically homogeneous (all the plants of the variety have the same genotype). Other types of varieties, like populations or open pollinated varieties are less homogeneous (more diverse) from a genetic

This chapter was originally published under the Creative Commons Attribution License. Serpolay E, Schermann N, Dawson J, van Bueren ETL, Goldringer I, and Chable V. *Phenotypic Changes in Different Spinach Varieties Grown and Selected under Organic Conditions.* Sustainability **2011**,3 (2011); 1616–1636. doi:10.3390/su3091616.

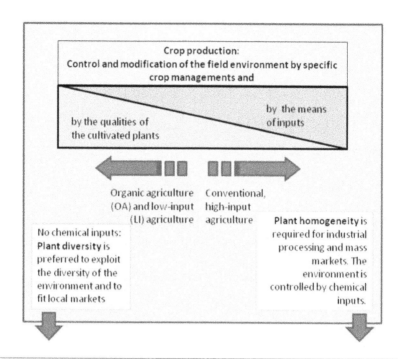

Characteristics of varieties: traits	
* adaptation to greater environmental heterogeneity for stability of performance * adaptation to the absence or low use of inputs and water tolerance of weed competition, diseases, pests and drought * qualities for local marketing and artisanal processing	* High yield potential in the most favorable conditions * Homogeneity and stability of crops, suitability for industrial processing

Characteristics of varieties: genetic types	
Intravarietal genetic heterogeneity Populations, mixtures of pure lines or mixtures of populations (autogamous). Populations (allogamous species)	Intravarietal genetic homogeneity Pure lines (autogamous) F1 hybrids (allogamous species and some autogamous species)

Characteristics of varieties: germplasm	
Historic and landrace varieties, crosses among historic, landrace and other farmer varieties	Improved varieties for conventional agriculture, backcrosses with historic and landrace varieties, and sometimes wild or cultivated related species

Characteristics of varieties: breeding techniques	
Methods which respect the naturalness of the species and avoid biotechnologies	Biotechnologies and conventional breeding methods

Fulfillment of needs in varieties	
Very little investment of private sector. PPB programs involving farmers and public sector since the end of the 20th century	Strong private and public sector investment. Strong consolidation of breeding companies.

Legislation issues	
In the EU, well known landraces can be registered under a catalogue of "conservation varieties", but seed multiplication, cultivation and commercialization is limited to a small region of origin. A new approach is needed to meet organic seed demand for heterogeneous varieties	To be marketed, varieties need to answer to DUS norms (Distinctness, Uniformity and Stability) to be registered into an official catalogue.

FIGURE 1: Strategies of organic and conventional farming systems, focused on seeds and variety use.

and phenotypic point of view. Thus they do not fulfill the legislative DUS requirements and cannot be easily commercialized in the European seed market. They are gaining interest in Europe, especially for low-input (LI) and organic agriculture (OA) farming [2]. In the following, we will refer to organic agriculture (OA) for simplicity, but what is said of OA also applies to low-input agriculture in which limited quantities of chemical inputs are used. To achieve coherence with IFOAM principles, many organic farmers are interested in varieties that can be re-sown, that also have intrinsic genetic diversity because (i) diversity may buffer the variability of cultivation conditions [3], (ii) genetic diversity enables the farmer to select within the variety to adapt it to the specific requirements of its farm (specific markets for example) and (iii) such varieties allow farmers to achieve seed self-sufficiency, especially for allogamous species.

Figure 1 illustrates the strategies used to optimize organic and conventional systems: the first one is based mainly on the choice of plants and the other is mainly based on the proper use of inputs. All intermediate situations are possible in which a balance is established between plant adaptation and inputs. Because professional breeding has primarily targeted conventional agriculture, available varieties on the market present several shortcomings in regards to the aims of organic agriculture. The vast majority of commercial varieties are genetically homogeneous: pure lines for autogamous species, F1 hybrids for allogamous species and F1 hybrids for many autogamous vegetable species. Resowing seeds resulting from the crop of these commercial varieties either will not give a proper crop in the case of F1 hybrids, or is subjected to a fee for pure lines in some European Union (EU) countries, such as in France. Furthermore, there are very few varieties specifically bred for organic and low-input conditions. This is the case for spinach, the crop under discussion in this article [4]. Finally, using little or no chemicals asks for more resilience from the varieties in order to best exploit environmental conditions. Therefore, due to the specific conditions of OA (great variability of the environment, agronomic practices, end-uses, markets), farmers would like to tailor the varieties to their needs through exchanges with other farmers and selection in their fields. In the case of OA, genetically diverse varieties can benefit from on-farm participatory plant breeding [5-9] in addition to classical breeding, even when classical breeding is conducted under organic conditions.

Given the growing need for historical and landrace (farmer) varieties bred on-farm [10], the EU needed more scientific information on how such varieties evolve and the associated legislative issues (among others DUS criteria), since seed legislation was developed for homogeneous varieties in a conventional agriculture framework [1]. In the literature, results can be found on the evolution of farmer varieties based on genetic markers, for instance [11-13], but there are few results based on phenotypic traits that are of direct relevance to the farmers [12,14]. Thus, in the EU, research programme Farm Seed Opportunities [15], experiments over three years studied on-farm cultivation and selection in contrasting environments of farmer varieties of wheat, maize, bean and spinach. In this paper, we report on the experiments with spinach (*Spinacia oleracea* L.), an allogamous vegetable species. In addition, a similar trial on bread wheat, an autogamous crop species, is reported [10]. Results to be published later on maize, an allogamous crop species, will enable us to have more insights on the short-term evolution of farmer varieties, and their potential adaptation to farmers' conditions.

Specifically, our study tries to address the following questions: (1) Does farm cultivation and selection over two years lead to statistically significant phenotypic changes in spinach farmer varieties? To what extent have the tested varieties diverged in a two-year timeframe? Are varieties still distinct after on-farm cultivation? (2) If changes are observed, how can we relate them to the cultivation and type of selection experienced by the different varieties? (3) Are there differences of within-variety diversity levels between a F1 hybrid control and the farmer varieties under organic conditions?

12.2 MATERIALS AND METHODS

In our study, phenotypic changes of different spinach (*Spinacia oleracea* L.) population varieties were explored during an experiment over three years, from 2007 to 2009. The plants were cultivated and selected by organic farmers in contrasted environments (two farmers in Western France and one in The Netherlands). The changes of each variety were assessed by planting all varieties together in the same environment and comparing the original seed sample with the variety cultivated and selected on-farm for two cycles. Figure 2 illustrates the procedure of the trials and Table 1 gives some information about the environments of the different trials. In the rest of the paper, we have called a variety cultivated by a farmer in a given environment over two years a "version" of this variety. The "original version" is the original seed sample. For example, the variety 'Eté de Rueil' was cultivated by two farmers (FD and MC). In our trial, we compared three versions of this variety: the MC version (the variety cultivated for two years by MC), the FD version (the variety cultivated for two years by FD), and the original version (initial seed lot).

The varieties tested were European populations (open pollinated varieties) of spinach which were historically registered in the European variety catalogue but which have now entered the public domain. Seeds came from national gene banks (CGN—The Netherlands—and GEVES—France) and small-scale seed companies (La Semeuse and Germinance, France). Twenty-one varieties were chosen to represent a diverse range of traits that could be of interest for European conditions. All were indicated to be spring varieties. The different varieties tested are presented in Table 2, with those chosen by the farmers for the experiment indicated.

12.2.1 TRIAL DESIGN AND OBSERVATIONS

The experiment was performed in three steps: (i) 2007, selection among varieties: cultivation by all the farmers of all twenty-one varieties, with farmers choosing at least one variety to be grown and selected on their farms for the following years; (ii) 2007 and 2008, selection within varieties: on-farm mass selection within the chosen variety(ies) and on-farm multiplication; (iii) 2009: evaluation of all the varieties chosen and selected by the farmers in a common trial (one location, Le Rheu, Brittany, France). As a first step, the farmers chose the variety(ies) to experiment during cultivation and non-suitable varieties were eliminated before flowering. Each farmer chose among the varieties according to their ideotype for fresh-market spinach (see Table 3).

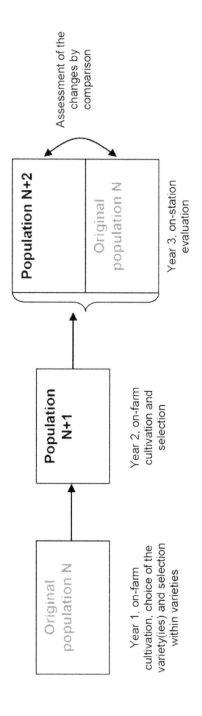

FIGURE 2: Scheme of the "Farm Seed Opportunity" EU project experiment.

TABLE 1: Information about the trials.

Name of the trial	AVO	FD	MC	Le Rheu (only 2009)
Location	Duiven, The Netherlands, 51.95 N, 5.99 E	St Martin D'Arcé, France, 47.58 N, 0.07 W	Morlaix, France, 48.62 N, 3.84 W	Le Rheu, France, 48.09 N, 1.80 W
Soil	2007: sandy, very poor fertility; 2008: heavy clay, poor for spinach	2007: sandy, quite rich; 2008: sandy, quite rich	2007: deep silt soil; 2008: new soil (new greenhouse built in 2005)	silty clay
Pre-crop	2007: faba beans; 2008: carrots	2007: carrots and cucumbers for seed production and grass production during winter; 2008: two years black radish for seed production	2007: broccoli; 2008: cucumbers	potatoes
Preparation of soil (work and fertilization)	2007: ploughing, cow manure application at a low rate; 2008: ploughing, 35–40 tonnes/ha of cow manure	2007: superficial work only, compost applied at a low rate; 2008: superficial work only, low rate of compost applied in autumn	2007: ploughing, manure; 2008: superficial work twice, compost applied before cucumber crop	ploughing, no fertilization
Sowing	2007: 15th March, in-line crop covered for 2 weeks after sowing; 2008: 1st of April, direct seeded	2007: 13th March, in-line; 2008: 1st October 2007 (winter trial), direct seeded	2007: 10th of April, sowing in cubes of soil, 23rd of April, transplanting in one meter-wide beds covered with plastic mulch; 2008: sowing in cubes of soil in mid-March, planting in one meter-wide beds covered with plastic mulch; 2009: sowing in cubes of soil in Nov. 2007, transplanting mid-December 2007 in one meter-wide beds covered with plastic mulch in a greenhouse	25th February, sowing in cubes of soil in mid-March, planting in one meter-wide beds covered with plastic mulch
Intervention during cultivation	2007: Regular irrigation; 2008: Regular irrigation	2007: nothing; 2008: nothing	2007: nothing; 2008: irrigation	nothing

TABLE 2: Origin of the varieties tested by each farmer the first year of the experiment.

Variety name	Seed bank or origin	Accession number	Year of registration if registered	Name of the farmer if chosen
Hollandia	CGN	9421	1943	
Verbeterde Hollandia	CGN	9420	1943	MC
Vroeg Reuzenblad	CGN	9644	1943	
Breedblad Scherpzaad Zomer	CGN	9400	1943	
Proloog	CGN	9440	1961	
Resistoflay	CGN	9442	1963	
Duetta	CGN	14170	1962	
Pre Vital	CGN	9468	1962	
Spinoza	CGN	9451	1963	
Virtuosa	CGN	9466	1963	
Amsterdams Reuzenblad	CGN	14179	1943	
Advance	CGN	14181	1955	
Viking	CGN	9463	1943	MC
Nobel	CGN	14173	1943	
D'été de Rueil	La semeuse			MC and FD
Viking-Matador	Germinance			MC
Matador foncé	GEVES	10595		
Monstrueux de Viroflay	GEVES			
Alwaro	GEVES	13054		MC
Supergreen	GEVES	14601		MC
Monarch Long Standing	GEVES	14751		AVO

When the farmer had chosen several varieties, they were multiplied individually under cages in order to avoid intercrossing. On-farm, the plot size was at least 100 plants and 5 m² per variety. Selection was carried out differently according to the farmer, and could include the elimination of unsuitable plants, the selection of plants corresponding to a given type, or only through natural selection in their environmental conditions. Three farmers participated in the experiment (MC, FD and AVO). All of them sowed the spinach in spring the first year and two of them (MC and FD) decided to sow the varieties in winter for the second year because, according to their experience, they identified the varieties as winter varieties (although the gene bank identified them as spring varieties). MC cultivated the varieties under plastic tunnels and in one-meter-wide beds covered with plastic mulch, FD and AVO cultivated the varieties outside and without plastic mulch.

The common trial of the third year was cultivated in spring, outside in beds with plastic mulch. It was a split-plot design with three replicates. Sub-blocks were composed of 32-plant plots of the different versions of one population variety, i.e., population N from the gene bank and population(s) N + 2 from farmer(s). We added in this common trial a commercial F1 hybrid variety (variety 'Lazio', Voltz seeds) in order to compare levels of intra-varietal variability with the population varieties. Phenotypic traits were observed on 15 plants per replicate for each version of each variety.

Measurements were made on one well-developed and representative leaf of each plant (at the harvestable stage). The traits measured are traits used in the DUS evaluation, however, for DUS tests, traits are assessed at the variety level, considering that the plants are homogeneous. Homogeneity (uniformity) is evaluated at a global level by the number of "off-types" (plant which clearly do not look the same than others). Stability is evaluated by comparison of different seed lots of the variety (year N and year N − 5 for example) and distinctness is validated by visual comparison with other varieties and by comparison with the breeder's description of different traits. In our study, we observed development traits at plot level (bolting and flowering indexes) and the other phenotypic traits at the plant level. The traits observed are presented in Table 4.

12.2.2 DATA ANALYSIS

Each variety was analysed separately. ANOVA tests were performed for quantitative traits Y (i.e., petiole length, leaf blade length and width and leaf color parameters), according to the model: $Y_{ijk} = \mu + replicate_i + version_j + \varepsilon_{ijk}$ where "version" effect refers to original/farmer versions of a variety. Chi square tests of the version effect were performed on the distributions of the semi-quantitative traits (i.e., anthocyanin, petiole and

TABLE 3: Selection criteria of each farmer for each variety and varieties' changes.

Farmer	Criteria of selection	Variety selected	Significant changes compared to original version
AVO	Ideotype: plants late to bolt and flower Selection method: transplanting only the latest plants	"Monarch Long Standing"	Shorter petioles Less bolting and flowering on May 29
FD	Ideotype: nothing, wish of increasing diversity Selection method: no selection in the field but elimination of the spiny seeds after harvest ('Eté de Rueil' is supposed to be a variety with non-spiny seeds)	"Eté de Rueil"	More anthocyanin More acute leaves More bolting and flowering on May 22 Less bolting and flowering on May 29
MC	Ideotype: plants with big, dark green, narrow, spear shaped and smooth leaves Selection method: negative selection by elimination of the plants not sufficiently of the type wished	"Eté de Rueil"	More « yellow » leaves (chromameter) More anthocynin on the petiole More acute leaves Thicker leaves More bolting and flowering on May 22 and 29
		"Supergreen"	Narrower leaves More erect leaves More bolting on June 4 More flowering on May 29
		"Verbeterde Hollandia"	More « yellow » and « green » leaves (chormameter) More flowering on May 29
		"Viking"	More « yellow », « greener » and lighter leaves (chromameter) Less erect leaves More blistered leaves
		"Viking Matador"	Longer leaves More bolting on May 22

TABLE 3: *Cont.*

Farmer	Criteria of selection	Variety selected	Significant changes compared to original version
		"Alwaro"	Longer petioles
			Longer leaves
			More « yellow » and « greener » leaves (chromameter)
			More erected leaves
			More flowering on May 22 and 29
			More flowering for all the dates of observation

leaf blade attitude, leaf blade blistering and intensity of green, shape of apex and thickness) after pooling the data of the three replicates. For both ANOVA and Chi square tests, the tests were performed with functions "aov" and "fisher.test" using R software [16] with a significance threshold of 5%. For bolting and flowering index, we transformed the index in number of plants bolted or flowered at a given date. For this purpose, we multiplied the percentage in the middle of the range of the index by the basic number of plants per plot (32)—we applied 12.5% for index 1, 50% for index 2 and 87.5% for index 3. We applied chi square tests on the new count data. However, in Table 5, for these traits, means of the values in the 3 replicates are given in the 0 to 3 scale.

We performed an ascending hierarchical classification on the variety x version means for all traits except bolting and flowering indexes in order to assess multi-trait changes of varieties. Data were standardized to a mean of zero and variance of one prior to the analysis, and the Euclidian distance and Ward agglomeration criterion (Ward's minimum variance) were used (function « hclust » in the R freeware was used). Ascending hierarchical classification is a multivariate analysis that leads to groupings of the most similar variety x version combinations. However, one must keep in mind for this multivariate analysis that results are dependent on the choice of distance among individuals and agglomeration method. For the latter, simulation studies have shown that there is not one best choice

TABLE 4: Traits observed.

Trait	Details, scale or unit
Petiole length	mm
Leaf blade length	mm
Leaf blade width	mm
Leaf blade intensity of green	Visual assessment: 1-very light, 2-light, 3-medium, 4-dark, 5-very dark
Leaf color	Measured with a chromameter—model Minolta CR-200—(average of 3 measurements per leaf*, each measurement gives 3 parameters: L (light saturation), A (yellow-blue axis, negative values yellow, positive values blue, and B (green-red axis, negative values green, positive values red) (Hunter Labs, 1996))
Anthocyanin on the stem	1-presence, 2-absence
Petiole attitude	1-horizontal, 2-semi-erect, 3-erect
Leaf blad attitude	1-semi-pendulous, 2-horizontal, 3-semi-erect, 4-erect
Leaf blade shape of apex	1-acute, 2-obtuse, 3-rounded
Thickness	1-very fine, 2-fine, 3 medium, 4-thick, 5-very thick
Leaf blade blistering	1-absent or very weak, 2-weak, 3-medium, 4-strong, 5-very strong
	Bolting index
	0-not bolted (0%), 1-very little bolted (1 to 25%), 2-moderately bolted (25 to 75%), 3-bolted (75 to 100%)
Bolting index	3 dates of observation: 22 May, 29 May, 4 June, 2009
Petiole length	mm
Leaf blade length	mm
Leaf blade width	mm
Flowering index	0-not flowered (0%), 1-little flowered (1 to 25%), 2-medium flowered (25 to 75%), 3-flowered (75 to 100%)
	3 dates of observation: 22 May, 29 May, 4 June 2009

The 3 measurements were taken in the three zones of the leaf indicated in Figure 3.

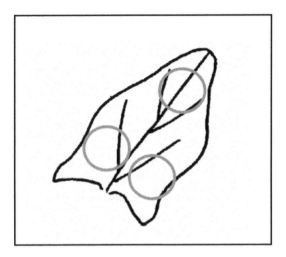

FIGURE 3: Zones of measurements of leaf blade color with the chromameter.

for all cases, even though Ward's minimum variance criterion was among those with the best overall performance [17], this is why we have chosen this criterion.

To compare the level of intra-varietal diversity between populations (N versions) and the commercial variety, we calculated the Simpson's diversity index for qualitative criteria ($\Sigma_i\, p_i$, where p_i are the frequencies of the different modalities) and the difference between the 90% and the 10% quantiles for the quantitative criteria.

12.3 RESULTS AND DISCUSSION

12.3.1 CHANGES WITHIN THE VARIETIES

Table 5 presents the mean values of the traits observed for original and farmers' versions for each variety and the significance of the observed differences.

Significant differences between the original version and the version cultivated by farmers were detected on at least one trait for all the varieties and for all traits, except for intensity of green color.

Most of the varieties showed significant changes for the A and B parameters of color, bolting and flowering indexes (three to five). Morphological traits (leaf attitude, thickness or blistering) were the traits that evolved the least. Intensity of green did not show significant changes for any of the varieties. We studied (data not shown) the correlation between the visual intensity of green score (from one to five) and each parameter of the chromameter (L, A and B) separately. Visual observation and chromameter parameters were always well correlated, for example, decreasing intensity of green (most of the cases observed in the table above) corresponded with increasing of L value ("whiter"), decreasing of A value ("greener") and increasing of B value ("more yellow"). Thus, we can conclude that there were changes of color in some cases even if the visual intensity of green score did not show significant changes. The chromameter measurements completed the visual observations because they decomposed the different dimensions of color, which the eye cannot do.

All varieties showed statistically significant, $p < 0.05$ changes within different measured traits for each variety. Some varieties changed for only one or a few traits ('Monarch Long Standing', 'Verbeterde Hollandia' and 'Viking Matador'), whereas others showed changes for a greater number of traits ('Alwaro', 'Viking' or 'Eté de Rueil'). Figure 4 illustrates the different amplitudes of changes with two varieties: 'Verbeterde Hollandia' and 'Alwaro'.

'Alwaro' is the variety that showed the most differences between the original and farmer's (MC) versions. MC version has significantly longer petioles and leaves (+1.3 cm for both relatively to original version). Color showed significant changes for parameter A (decrease, "greener") and parameter B (increase, "more yellow"). MC version had more erected petioles and is earlier to bolt and flower than the original (five development indexes are different from the original version).

'Eté de Rueil' was cultivated by two different farmers (MC and FD). Changes were observed on both farms but were not identical. Indeed, the MC version changed significantly for the B color parameter and leaf thickness (the MC version was thicker than original version).

Both the MC and FD versions had more plants with anthocyanin, more rounded leaves and both were earlier to bolt and flower than the original version. It is interesting to note that when both MC and FD versions showed differences, it was always in the same direction.

'Monarch Long Standing' changed for the petiole length and two bolting indexes (the AVO version was later than the original version). The MC version of 'Supergreen' had narrower and more erect leaves than the original version. It was also earlier for two development indexes. 'Verbeterde Hollandia' showed modifications for two color parameters (A and B) and for one flowering index (the MC version was earlier than the original version).

'Verbeterde Hollandia'—year N + 2,   'Verbeterde Hollandia'—year N
farmer MC version   original version

'Alwaro'—year N + 2,   'Alwaro'—year N
farmer MC version   original version

FIGURE 4: The photos show the two versions of variety 'Verbeterde Hollandia', which showed very few phenotypic changes and the two versions of variety 'Alwaro' which showed several significant phenotypic changes over the two years of on-farm cultivation.

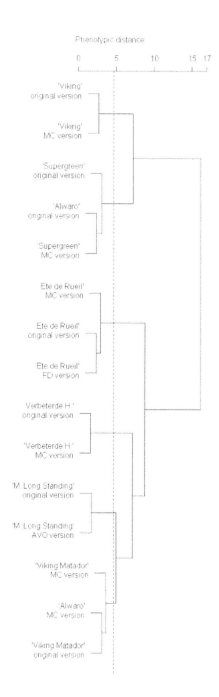

FIGURE 5: Ascending Hierarchical Classification (AHC) of the different varieties for all the measured morphological traits. The dotted line is drawn to split the tree into six groups.

MC version of 'Viking' showed changes for all the color parameters (L, A and B) and had also less erect petioles and more blistered leaves than the 'Viking' original version. 'Viking' differs from the others varieties by the fact that it did not change for any phenological trait while all other varieties showed changes for at least one of those traits. 'Viking Matador' showed changes for two traits: MC version had longer leaves and was earlier than original version for one bolting index.

Based on the ascending hierarchical classification (AHC), we classified all the versions of all the varieties considering all the traits observed (except bolting and flowering indexes) (Figure 5). We can draw two observations from Figure 5. If we separate the tree into six groups, each group comprises the different versions of a single variety except two groups that contain one version of 'Alwaro' each (and thus there is no 'Alwaro' group). If we separate the tree into two groups, the two versions of 'Alwaro' are still in different groups, in contrast to the different versions of all the other varieties, which are always in the same group. The varieties appear to have changed but conserved their phenotypic identity, except 'Alwaro,' which seems to have significantly diverged from its original version. AHC confirms the field observations and the results of individual trait analysis: 'Alwaro' is the variety that shows the most phenotypic changes. According to the AHC, the 'Alwaro' original version is close to 'Supergreen' and 'Alwaro' MC is close to 'Viking Matador'.

From these two first analyses (Table 5 and Figure 5), we observed that all the varieties tested showed changes for at least one trait, but they conserved their variety identity (the different versions of each variety were always grouped in the AHC). 'Alwaro' is a particular case as it has changed significantly for morphological and development aspects and it may be appropriate to consider it two different varieties at this point in time. For such a short period of time (two years), we were not expecting many phenotypic changes, and expected phenological criteria to change more often than morphological traits. However, the traits used as varietal descriptors often showed significant changes and this surprised us. We next discuss different explanatory factors that could contribute to the observed changes.

The choice of the experimental method was based on the method of DUS testing, but we also wanted to detect fine changes, for which the DUS method is not adapted. This is why we chose to observe one representative leaf per plant on a representative sample of individual plants instead of taking one measurement on the population in its entirety (the DUS method). Spinach plants are difficult to characterize because the harvested organ is the leaf, during the vegetative growth period of the plant. Furthermore, as each plant has multiple leaves, one observation does not necessarily represent the plant in its entirety, as could be considered for the maize ear for example. However, we tried to balance this potential source of error by sampling a large number of individuals per plot and rigorously adhering to a set protocol in leaf selection and measurement.

Other than the phenological traits, all the traits we observed are also included in DUS tests. These are traits that display genetic differences between varieties and that are not too much influenced by the environment. This is also likely the case for phenologicaly traits, which are of adaptive significance, as is known in other species [14]. Most of these traits are also likely quantitatively inherited. Therefore, we expected they would react to natural and/or human selection. The changes observed can thus be linked with natural adaptation (pressure of the environment), possibly farmers' selection, and also a certain degree of chance (genetic drift or accidental pollination during the multiplication phase, even though multiplication was controlled very carefully). The changes could also be due to different multiplication methods. Varieties coming from gene banks are maintained with a strict scheme, with the aim of conserving a specific phenotype of the variety. The method applied by farmers differed from this scheme and this could explain part of the observed changes.

At the least, we can say that the varieties tested in this study have changed in only two years of cultivation. We interviewed farmers on their selection criteria throughout the project to see if human selection could have had a directional influence on the phenotypic changes observed.

12.3.2 EFFECT OF FARMER SELECTION ON CHANGES IN THE VARIETIES

Selection criteria applied by the three farmers of this study are described in Table 3. AVO selected plants only on precocity and 'Monarch Long

Standing' has evolved on this aspect in the direction wished by the farmer (earlier bolting and flowering on 29 May).

'Eté de Rueil' showed changes for six criteria after cultivation at FD although no human selection was applied. It is interesting to note that the variety evolved in the same way when selected by MC.

MC selected all the varieties chosen according to the same criteria. All the varieties showed changes but not always in the direction he wished. For example, although MC selected dark plants later to bolt and flower, we noted that, on the whole, most of MC versions had a shorter cycle and whiter, "more yellow" and "greener" leaves (changes that we can interpret as "lighter green" according to the correlations observed between intensity of green and the L, A and B color parameters, data not shown). MC also selected plants with spear shaped leaves. Three varieties showed narrower or longer leaves ('Supergreen', 'Viking Matador' and 'Alwaro'), which tended towards a spear shape. From those results, it does not seem possible to distinguish which changes were due to the environment, human selection or others factors: human selection did not always produce changes in the direction wished by the farmer, and FD did not perform any selection and this variety did show phenotypic changes.

We measured changes as a difference between the phenotypic expressions of two versions of the same variety. However, even if we recorded the selection criteria of the farmers, we cannot say that we only measured the adaptive response to farmers' selection. The phenotypic expression of the plants depends on the genotype of the plant, and also on the environment (especially for spinach, which is very sensitive to nitrogen and water for example). Our study in the last year of the project took place in a different environment than the on-farm environments of selection/adaptation of the varieties. So, for example, although our observations underline that MC versions are earlier than the original versions, we cannot claim that MC failed in his selection of plants late to bolt and flower. MC's versions are earlier in our trial conditions, but perhaps not in its farm conditions. We know that he cultivated the spinach under plastic tunnels, and the second year the trial was conducted in winter instead of spring. Thus, when the seeds produced in these conditions were cultivated in our trial conditions (in spring in the field), we cannot conclude that the changes observed would have been the same at MC's farm.

In order to evaluate the efficiency of farmers' selection (or adaptive response to selection pressure), farmers planted their versions of the spinach varieties alongside the original versions in the last year of the project. The results of these on-farm trials were not statistically exploitable because farmers did not sow replicates, but this allowed the farmers to visually compare the varieties that they had selected to the original versions. We can report some observations made by the farmers without quantitative evaluations. For example, at MC, selected versions of the varieties were globally less blistered (a desired change) and at AVO, the original version of 'Monarch Long Standing' was earlier to bolt even though AVO had selected only the plants late to bolt. Unfortunately, the last year of the trial failed at FD due to weather conditions. The on-station and on-farm trials were complementary, as the on-station trial could evaluate the changes of varieties at a global level (all varieties coming from different farms) while the on-farm trials allowed farmers to evaluate the efficiency of their selection.

In this study, MC and AVO applied selection pressure and FD let the variety respond to natural selection. However, the types of changes observed were similar whether in response to natural or farmers' selection. So in our case, the type of selection (natural or human) does not seem to have an influence on the intensity of changes (the degree of significance and number of traits with significant differences). Changes in the FD version of 'Eté de Rueil' are comparable to changes in the MC version and even to the changes in other varieties. This shows the importance of so-called "natural evolution" or importance of the environmental impacts for cultivated plants. We can wonder if this capacity to respond to selection pressure could be linked with the allogamous reproductive strategy of spinach. However, in an article by Dawson [10] on a similar experimentation of the Farm Seed Opportunities project on wheat, an autogamous species, the same kind of results are shown for population varieties subjected to natural selection on-farm. It would be of interest and importance for breeding to compare the intensity and efficiency of human and natural selection in terms of the adaptive potential of the varieties. With this view, a more thorough experimentation would be needed where all varieties are selected by farmers in each environment, alongside the variety cultivated with only natural selection pressure.

12.3.3 HOMOGENEITY OF THE VARIETIES

For quantitative criteria, levels of heterogeneity varied according to the variety, but the F1 hybrid was not specifically less heterogeneous than the other varieties. Except for the FD version of 'Eté de Rueil', which always had a higher level of heterogeneity, differences of heterogeneity levels between other farmers' versions and their respective original version were observed, but without a systematic pattern. For qualitative criteria, levels of homogeneity of the farmers' version were almost always comparable to those of the respective original versions or to the F1 hybrid. No significant differences were observed. Figure 6 gives an example of the results observed for qualitative criteria.

There was no correlation between changes in mean values observed in the varieties and their heterogeneity, or between changes in mean values and changes in the level of heterogeneity within versions of a variety. Simpson's diversity index (calculated for qualitative traits) and the interquantile range (calculated for quantitative traits) do not distinguish the F1 hybrid from the other varieties tested. Furthermore, the farmers' versions of varieties appeared to have conserved their level of heterogeneity whether they were selected or not.

When there was divergence in the mean values of different versions of the same variety, the intra-varietal diversity may have increased even if within each version, the same amount of phenotypic diversity exists compared with the original version. Each version may have the same internal level of diversity, but the global diversity represented by all the versions of the variety may have increased. This has been demonstrated by Goldringer et al. for phenotypic changes and diversity of agromorphological traits in wheat populations subjected to selection pressure in contrasting environments [18]. This new diversity and the rapidity of the changes observed indicate that population varieties may be good starting points for on-farm selection and adaptation.

Farmer varieties are often said to be too diverse to meet the uniformity criterion of registration in the European Union, however, our study showed that the level of heterogeneity of the populations was often no different from that of the hybrid for traits used in the DUS evaluation.

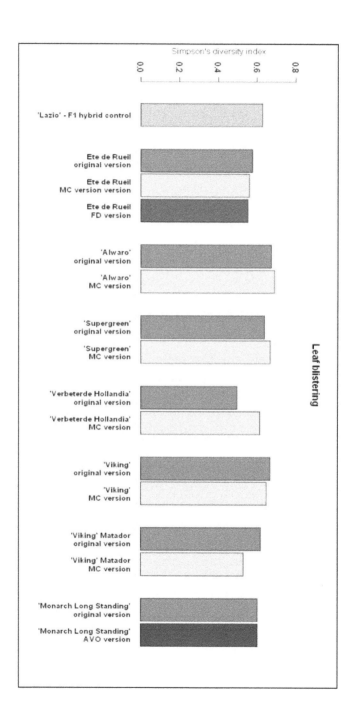

FIGURE 6: An example of intra-varietal heterogeneity of the different spinach varieties for the qualitative trait leaf blistering.

As the populations chosen for the experimentation were historically in the official catalog, they had to meet the uniformity criteria when they were initially registered. Even after they were removed from the catalog, they were conserved ex situ with very few plants and conservative selection during the regeneration process. However, our results question the assumption that F1 hybrids are always more uniform than population varieties; this appears not to be the case for many traits under organic conditions.

12.4 GENERAL DISCUSSION AND PERSPECTIVES

We found phenotypic changes in population varieties of spinach after two generations of on-farm breeding. But have the varieties changed so much that they are no longer distinct? The results of the ascending hierarchical classification clearly showed that one variety has evolved to such a point that its two versions were not in the same group. This leads to certain questions about what constitutes a new variety in this context, where limit is between one variety and another, and how can we define and measure variety distinctness for population varieties. From a scientific point of view, we would like to understand the reasons for the phenotypic changes (and maybe evolution) of a variety, and how the adaptive and evolutionary potential of varieties can be conserved. The changes observed within tested populations after two generations would cause this type of variety to be rejected from catalogue registration based on the criterion of stability of the European catalogue varieties. Seed companies or gene banks that maintain open pollinated populations practice conservative selection based on a certain phenotypic description of the variety, which may eliminate genetic diversity that was initially present.

In our case, farmers chose varieties corresponding to their own criteria, and then selected the variety based on their objectives. The two processes are different and complementary in terms of maintaining useful genetic diversity in agricultural species, and our results are specific to on-farm conservation and breeding where farmers are seeking varieties with more variability, as discussed in the introduction.

The possibilities for adaptation of the varieties to specific environmental conditions are enhanced by their intrinsic variability. This is promising for farmers who would like to adapt populations to their own conditions (environments and markets) and it is of great interest for organic and low input agriculture in which chemical inputs are prohibited. Farmers' use of their own saved seed is also related to the sustainability of farms from an economic perspective. If farmers can make the varieties evolve according to their own objectives and markets, they would not need to buy seeds for each crop and each year and their autonomy would be enhanced. It is one of the conclusions of the FSO Project that the regulation should be re-valuated in order to create a specific framework for population varieties and seed saving activities where varieties are not stable as defined by the current catalogue regulation [1]. However, registration is not necessarily an issue in on-farm breeding activity. Farmers involved in PPB have organized seed associations to manage breeding and seed exchanges [19], and those activities constitute a separate seed system from the dominant commercial seed system.

12.5 CONCLUSIONS

This study showed that the populations tested have phenotypically changed over two years of on-farm cultivation and selection. The changes detected in such a short period of time are a sign that the observed population varieties could be adapted to diverse conditions. It is of great interest for organic and low input farmers who are interested in developing their own varieties and adapting them to their particular conditions. There is a need for further research on the evolution of farmer varieties, specifically related to biological questions of adaptability and evolutionary capacities of allogamous crops and how the legislative framework take into account the biological reality of those varieties, recognizing the farmers' role in creating and maintaining genetic diversity in cultivated species.

REFERENCES

1.   Chable, V.; Louwaars, N.; Hubbard, K.; Baker, B.; Bocci, R. Plant breeding, variety release and seed commercialisation: Laws and policies of concern to the organic sector. In Organic Crop Breeding; Lammerts van Bueren, E.T., Myers, J.R., Eds.; Wiley-Blackwell: Hoboken, NJ, USA, in press.

2. Wood, D.; Lenne, J.M. The conservation of agrobiodiversity on-farm: Questioning the emerging paradigm. Biodivers. Conserv. 1997, 6, 109-129.
3. Lammerts van Bueren, E.T. Ecological concepts in organic farming and their consequences for an organic crop ideotype. Neth. J. Agric. Sci. 2002, 50, 1-26.
4. Colley, M.R.; Navazio, J.P. New F1 and open-pollinated spinach varieties for organic systems: A farmer, seed company, and non-profit model of participatory breeding. In Abstract Book of Eucarpia Symposium Plant Breeding for Organic and Sustainable, Low-Input Agriculture: Dealing with Genotype-Environment Interactions, 7–9 November 2007, Wageningen, The Netherlands; Lammerts van Bueren, E.T., Østergård, H., Goldringer, I., Scholten, O., Eds.; Wageningen University: Wageningen, The Netherlands; p. 69.
5. Ceccarelli, S.; Grando, S. Decentralized participatory plant breeding: An example of demand driven research. Euphytica 2007, 155, 349-360.
6. Chable, V.; Conseil, M.; Serpolay, E.; Le Lagadec, F. Organic varieties for cauliflowers and cabbages in Brittany: From genetic resources to participatory plant breeding. Euphytica 2008, 164, 521-529.
7. Murphy, K.; Lammer, D.; Lyon, S.; Carter, B.; Jones, S.S. Breeding for organic agriculture and low-input farming systems: An evolutionary-participative breeding method for inbred cereal grains. Renewable Agric. Food Syst. 2004, 20, 48-55.
8. Newton, A.C.; Akar, T.; Baresel, J.P.; Bebeli, P.J.; Bettencourt, E.; Bladenopoulos, K.V.; Czembor, J.H.; Fasoula, D.A.; Katsiotis, A.; Koutis, K.; Koutsika-Sotiriou, M.; et al. Cereal landraces for sustainable agriculture: A review. Agron. Sustain. Dev. 2010, 30, 237-269.
9. Philipps, S.L.; Wolfe, M.S. Evolutionary plant breeding for low-input systems. J. Agric. Sci. 2005, 143, 245-254.
10. Dawson, J.C.; Serpolay, E.; Giuliano, S.; Schermann, N.; Galic, N.; Osman, A.; Pino, S.; Goldringer, I. Phenotypic diversity and evolution of farmer varieties of bread wheat in organic agricultural systems in Europe. Sustainability, in press.
11. Kumar, S.; Bisht, I.S.; Bhat, K.V. Population structure of rice (Oryza sativa) landraces under farmer management. Ann. Appl. Biol. 2010, 156, 137-146.
12. Pressoir, G.; Berthaud, J. Population structure and strong divergent selection shape phenotypic diversification in maize landraces. Heredity 2004, 92, 95-101.
13. Vaz Patto, M.C.; Moreira, P.M.; Almeida, N.; Satovic, N.; Pego, S. Genetic diversity evolution through participatory maize breeding in Portugal. Euphytica 2008, 161, 283-291.
14. Goldringer, I.; Prouin, C.; Rousset, M.; Galic, N.; Bonnin, I. Rapid differentiation of experimental populations of wheat for heading time in response to local climatic conditions. Ann. Bot. 2006, 98, 805-817.
15. Chable, V.; Goldringer, I.; Dawson, J.; Bocci, R.; Lammerts van Bueren, E.T.; Serpolay, E.; Gonzalez, J.M.; Valero, T.; Levillain, T.; Van der Burg, J.W.; et al. Farm seed opportunities: A project to promote landrace use and renew biodiversity. In Bioversity Technical Bulletin, No. 15; Negri, V., Maxted, N., Vetelainen, M., Eds.; Bioversity International: Rome, Italy, 2009, pp. 266-274.
16. R Development Core Team. R: A language and environment for statistical computing. R Foundation for Statistical Computing: Vienna, Austria; Available online: http://www.R-project.org/; 2011 (accessed on 2 September 2011).

17. Milligan, G.W. A review of Monte Carlo Tests of cluster analysis. Multivariate Behav. Res. 2008, 16, 379-407.
18. Goldringer, I.; Paillard, S.; Enjalbert, J.; David, J.L.; Brabant, P. Divergent evolution of wheat populations conducted under recurrent selection and dynamic management. Agronomie 1998, 18, 413-425.
19. Chable, V.; Le Lagadec, F.; Supiot, N.; Léa, R. Participatory research for plant breeding in Brittany, the Western region of France: Its actors and organisation. In Proceedings of the 1st IFOAM Conference on C7—Organic Animal and Plant Breeding, Breeding Diversity, Santa Fe, NM, USA, 25–28 August 2009.

Table 5 is not available in this version of the article. To view the missing table, please see the original version of the article, as cited on the first page of this chapter.

PART IV

ENVIRONMENTAL EFFECTS OF ORGANIC AGRICULTURE

CHAPTER 13

SOIL ENZYME ACTIVITIES, MICROBIAL COMMUNITIES, AND CARBON AND NITROGEN AVAILABILITY IN ORGANIC AGROECOSYSTEMS ACROSS AN INTENSIVELY-MANAGED AGRICULTURAL LANDSCAPE

TIMOTHY M. BOWLES, VERONICA ACOSTA-MARTHNEZ, FRANCISCO CALDERÓN, AND LOUISE E. JACKSON

13.1 INTRODUCTION

Agricultural landscapes exhibit a high degree of spatial variability, including variation in soil physicochemical characteristics and agroecosystem management (Drinkwater et al., 1995 and Vasseur et al., 2013), which can affect the activity and composition of the soil biota (Acosta-Martínez et al., 2008 and Schipanski and Drinkwater, 2012). Soil microbes mediate the biochemical transformations of organic matter that underpin essential ecosystem functions, including decomposition, mineralization of plant available nutrients, and nutrient retention. Organic production relies on these microbially-derived ecosystem functions and thus may be a model system for ecological intensification of agriculture (Jackson et al., 2012). By focusing on building and utilizing soil organic matter (SOM) as opposed to using synthetic fertilizers, organic production systems differ greatly from conventional systems; organic management in many research station trials has been shown to improve soil fertility (Burger and Jackson, 2003 and Gattinger et al., 2012), reduce nutrient losses (Drinkwater and Wagoner, 1998, Kramer et al., 2006 and Syswerda et al., 2012), and reduce global warming potential (Burger et al., 2005 and Cavigelli et al., 2013) while supporting similar crop yields in certain contexts (Seufert et al., 2012).

Yet, such research station-based experiments may belie the challenge of evaluating multiple ecosystem services on working organic farms across actual landscapes that vary in topography, soil type, commodities, and motivations of farmers for making the organic transition (Darnhofer et al., 2005 and Williams and Hedlund, 2013). Organic farms also use many different nutrient management strategies (Guthman, 2000 and Darnhofer et al., 2010) even when growing the same crop in the same region (e.g. Drinkwater et al., 1995 and García-Ruiz et al., 2008). While this heterogeneity could help explain some of the ambiguous results of landscape-scale comparisons of organic and conventional farms relative to site-specific experiments (e.g. Williams and Hedlund, 2013), we lack basic understanding of how heterogeneity affects soil microbial activity and community composition and the implications for soil ecosystem functions and agroecosystem management.

The quantity and quality of SOM and carbon (C) and nitrogen (N) inputs are the overriding controls on soil microbial biomass and activity (Fierer et al., 2009 and Kallenbach and Grandy, 2011). Thus, distinct organic amendments (e.g. manure, leguminous cover crops, and composted materials) can stimulate microbial biomass differently through increases in labile organic matter (Marriott and Wander, 2006, Smukler et al., 2008 and Kallenbach and Grandy, 2011) and/or total soil C on time frames from months to decades (Drinkwater and Wagoner, 1998 and Kong et al., 2005). However, little is known about how the quantity and composition of SOM and nutrient inputs (e.g. C:N ratio) affect microbial communities and their enzyme activities, and in turn, transformations of C, N, phosphorus (P), and sulfur (S) on organic farms. The total enzymatic activity of soil, derived from active microorganisms and the stabilized pool in clay–humus complexes (Tabatabai, 1994 and Burns et al., 2013), plays a major role in the depolymerization of structurally diverse polymeric macromolecules, which is considered the rate-limiting step in decomposition and nutrient mineralization potential of soil (Schimel and Bennett, 2004).

Organic management increases overall enzyme activity (Mäder et al., 2002, García-Ruiz et al., 2008 and Moeskops et al., 2010), but activities of specific enzymes may change depending on the composition of the amendments and the relative availability of nutrients, as well as other factors, such as soil type and its unique characteristics, e.g. pH and texture (Acosta-Martínez et al., 2007, Sinsabaugh et al., 2008 and Štursová and Baldrian, 2010). Given the relatively constrained C:N:P ratios of microbial biomass (Cleveland and Liptzin, 2007), enzymatic activity might be expected to enhance the availability of the most limiting nutrients in order to meet microbial metabolic demands (Sinsabaugh et al., 2008 and Allison et al., 2011). For instance, in grassland and forest soils, long-term N fertilization increased the activity of soil enzymes involved in labile C breakdown (Ajwa et al., 1999, Saiya-Cork et al., 2002 and Tiemann and Billings, 2010) with similar trends in conventionally-managed agricultural soils (Bandick and Dick, 1999 and Piotrowska and Wilczewski, 2012).

Properties of SOM and organic amendments may also influence microbial community composition and in turn, microbial activity and associated ecosystem processes (Fraterrigo et al., 2006 and Reed and Martiny, 2013). Increases in the fungal:bacterial ratio have been linked to increases in soil C and the C:N ratio across landscapes (Fierer et al., 2009 and de Vries et al., 2012) and in response to organic management (Bossio et al., 1998) as well as various organic amendments, such as conifer-based compost (Bernard et al., 2012) and vetch cover-cropping (Carrera et al., 2007). Other studies have shown increases in phospholipid fatty acid biomarkers for arbuscular mycorrhizal fungi (AMF) in response to composted green waste as well as long-term organic management (Bossio et al., 1998, Moeskops et al., 2010 and Moeskops et al., 2012). While management that supports fungal communities has been suggested as a means of increasing agroecosystem N retention and other functions (de Vries and Bardgett, 2012 and Jackson et al., 2012), changes in microbial community composition may be relatively constrained in agricultural landscapes with a legacy of intensive agricultural management (Fraterrigo et al., 2006 and Culman et al., 2010), even in response to organic management (Williams and Hedlund, 2013). Indeed, in agricultural soils that are intensively managed, microbial activity tends to change more quickly in response to organic management than community composition (Burger and Jackson, 2003).

The overall objective of this study is to examine how soil physico-chemical characteristics and nutrient management practices affect soil microbial activity and microbial community composition in organic agricultural systems, using an on-farm approach with several participating farmers. This study is part of a larger project examining plant–soil–microbial interactions and multiple ecosystem functions across a set of organic farms selected to be representative of the local landscape using geographic information system (GIS) techniques (Bowles et al., ms. in preparation). Thirteen organically-managed fields growing Roma-type tomatoes (*Solanum lycopersicum* L.) were selected in Yolo County, part of the Sacramento Valley of California, an agricultural landscape dominated by high-input conventional agriculture with a diverse array of crops. The focus is on the period of maximal tomato nutrient demand when microbial activity is most important for crop productivity. There were two main hypotheses. First, farm fields would differ in soil microbial biomass and enzyme activities, and these differences would depend on the quantity and composition of SOM as well as other factors related to the type of organic amendments. Second, microbial community composition would be influenced by nutrient management practices but with fewer differences across the fields relative to enzyme activities given the overall lack of diversity in the soil biota in this landscape, which appears to be related to high disturbance and low complexity (Culman et al., 2010).

The specific objectives of this study are to: 1) characterize the variability of soil properties and organic management practices across a number of organically-managed Roma-type tomato fields; 2) determine patterns of soil enzyme activities and fatty acid methyl esters (FAMEs) to indicate microbial community composition and relate them with soil properties and management practices; and 3) consider the implications for microbially-derived ecosystem functions for management of different types of organic farms across this landscape. On 13 organic fields differing in nutrient management practices, soil physicochemical characteristics; microbial biomass C and N; activities of soil enzymes involved in C, N, P, and S cycling; and FAMEs were measured and analyzed with multivariate techniques to model the relationships among these factors. The on-farm approach provided a wide range of farming practices and soil characteristics to reveal how microbially-derived ecosystem functions can be effectively manipulated to enhance nutrient cycling capacity.

13.2 MATERIALS AND METHODS

13.2.1 AGROECOSYSTEM CHARACTERISTICS

The organically-managed fields in this study were on similar parent material (mixed alluvium) in a 1579 km^2 landscape including all of the arable land in Yolo County, California, which is situated along the western side of the Sacramento Valley. Yolo County has a Mediterranean-type climate with cool, wet winters and hot, dry summers. Annual precipitation in 2011 was 403 mm and the mean maximum and mean minimum temperatures were 21.7 and 7.3 °C, respectively, compared to 462 mm, 23.1 °C, and 8.4 °C for the previous 20 years (CIMIS, 2013). Organic farming has a long history in this area, with roots over 30 years ago (Guthman, 2004), and is relatively widespread and continuing to grow (Jackson et al., 2011). Different land use histories (e.g. history of cultivation, time in organic agriculture) and natural edaphic variability provide a range of soil characteristics.

Roma-type tomatoes are widely grown in this region for conventional and organic markets, and for both processing and direct-marketing to local consumers. The California Certified Organic Farmers (CCOF) directory was used to identify certified organic farming operations growing tomatoes in the study area (CCOF, 2011). CCOF is the primary organic certifier in this region of California (Guthman, 2004). All growers identified from this directory were contacted during winter 2010–11 to assess plans for growing Roma-type tomatoes and to gauge interest in the project. Eight growers expressed interest and we identified a total of 13 fields in which they expected to transplant tomatoes in early April 2011. All fields were transplanted within two weeks of one another. Nutrient inputs varied across farms (Table 1) with two general groups based on primary organic matter amendment (manure or composted green waste). Several farms also used a vetch winter cover crop alone or in conjunction with other amendments and some applied other nutrient sources (e.g. seabird guano, Chilean nitrate, fish emulsion) as a sidedressing or through drip irrigation. Tillage was used on all fields and was of similar intensity. Soil series identified from the SSURGO database are all considered highly productive (Table 1; Soil Survey Staff, Natural Resources Conservation Service, 2011) and had similar mineralogy (Schafer and Singer, 1976).

13.2.2 SOIL SAMPLING AND ANALYSES

Surface soil samples were collected in June 2011. Sampling was timed to coincide with tomato anthesis and early fruit development, a critical phenological and agronomic period in which tomato nutrient demand is high and growers often add supplemental nutrients. Fields were all sampled within two weeks of one another, an average of 68 days after transplanting. In each field, six plots were established at random locations within a 0.25 ha area to monitor soil and plants over the course of the season. An intact soil core (15 cm in diameter, 0–15 cm deep) was removed from each plot in between two tomato plants, situated 15 cm from the centerline of the planting row.

Soil samples were kept on ice until processing within 4 h for different analyses. After thoroughly mixing the soil sample, field-moist soil was used in determination of microbial biomass C (MBC) and N (MBN) within 24 h of sampling (see below). Inorganic N was extracted from moist soils with 2 M KCl and analyzed colorimetrically for ammonium (NH_4^+) and nitrate (NO_3^-; Foster, 1995 and Miranda et al., 2001). Olsen P was determined using the methods outlined by Olsen and Sommers (1982) at the University of California Agriculture and Natural Resources (ANR) Analytical Laboratory. Soil pH was determined on air-dried samples using a 1:2.5 soil/water ratio. Gravimetric water content (GWC) was determined by drying at 105 °C for 48 h. Air dried soil samples were sieved to 2 mm, ground, and analyzed for total C and N at the UC Davis Stable Isotope Facility. Particle size was determined by the laser diffraction method according to Eshel et al. (2004). Additionally, a ~50 g subsample was immediately frozen at −80 °C for

TABLE 1: Field management and soil types at the 13 organic tomato fields in Yolo County, California, USA.

Field	Years in organic[a]	Primary organic inputs[b]	Secondary nutrient inputs[c]	Soil type[d]
1	4	Manure	None	Tehama loam
2	8	Manure	None	Tehama loam
3	NA	Manure	None	Capay silty clay
4	3	Vetch	Guano, soluble	Tehama loam
5	8	Manure, vetch	None	Capay silty clay
6	11	Manure, vetch	Guano	Brentwood silty clay loam
7	3	Compost, vetch	Pellets, soluble	Yolo silt loam
8	2	Compost	Pellets, soluble	Yolo silt loam
9	11	Manure, vetch	Guano	Yolo silt loam
10	8	Compost	Chilean nitrate	Yolo silt loam
11	5	Compost	Chilean nitrate	Yolo silt loam
12	26	Compost	Soluble	Yolo silt loam
13	8	Compost	Chilean nitrate	Yolo silt loam

[a] *Years since certification (does not include transition years). This information was not available for field 3, but it is certified organic.*
[b] *Compost and manure were applied in fall 2010, with the exception of field 5, in which manure was applied in early spring prior to tomato transplanting. Winter vetch cover crops were incorporated prior to transplanting. Compost was composted green waste with a C:N ranging from 15 to 18. Manure was poultry manure or poultry litter with a C:N ranging from 9.8 to 15.*
[c] *Guano refers to seabird guano (12-12-2.5). Pellets were pelletized poultry manure (6-3-2). Chilean nitrate (16-0-0) is NaNO3, a mined mineral product. Soluble refers to solubilized organic fertilizers, especially fish emulsions, which have a range of nutrient concentrations. Guano, pellets, and Chilean nitrate were all applied as a sidedressing close to tomato transplanting. Small amounts (less than 7 kg-N ha−1) of soluble fertilizers are applied through the drip line periodically throughout the growing season.*
[d] *Tehama loam: fine-silty, mixed, superactive, thermic Typic Haploxeralfs; Capay silty clay: fine, smectitic, thermic Typic Haploxererts; Brentwood silty clay loam: fine, smectitic, thermic Typic Haploxerepts; Yolo silt loam: fine-silty, mixed, superactive, nonacid, thermic Mollic Xerofluvents*

subsequent analysis of FAME profiles and potential soil enzyme activities (see below).

13.2.3 MICROBIAL COMMUNITY ANALYSES

MBC and MBN were determined by the chloroform fumigation extraction method (Vance et al., 1987 and Wu et al., 1990). Organic C was quantified using a Dohrmann Phoenix 8000 UV-persulfate oxidation analyzer (Tekmar-Dohrmann, Cincinnati, OH) and organic N was quantified using alkaline persulfate oxidation (Cabrera and Beare, 1993). No correction factors were applied. K_2SO_4 extractable organic carbon (EOC) and nitrogen (EON) were calculated as organic C or N, respectively, quantified in non-fumigated samples (Ros et al., 2009).

Soil microbial community composition was characterized using FAME profiles. FAME analysis was performed on a 3-g field-moist equivalent sample using the ester-linked FAME procedure of Schutter and Dick (2000). FAME analysis was conducted using an Agilent 6890 N gas chromatograph with a 25 m × 0.32 mm × 0.25 μm (5% phenyl)-methylpolysiloxane Agilent HP-5 fused silica capillary column (Agilent, Santa Clara, CA) and flame ionization detector (Hewlett Packard, Palo Alto, CA) with ultra-high purity hydrogen as the carrier gas. Absolute amounts of FAMEs (nmol g^{-1} soil) were calculated according to Zelles (1996) using the 19:0 internal standard and these values were subsequently used to calculate mol percent by dividing each individual FAME by the total sum of all FAMEs. Selected FAMEs were used as microbial markers according to previous research (Zelles, 1999), and included Gram-positive (Gram+) bacteria (i15:0, a15:0, i17:0, a17:0), Gram-negative (Gram−) bacteria (cy17:0, cy19:0), and actinomycetes (10Me16:0, 10Me17:0, 10Me18:0). Fungal markers included saprophytic fungi (18:1ω9c, 18:2ω6c, and 18:3ω6c) and AMF (16:1ω5c). FAME 20:4ω6c was used as a marker for soil microfauna (protozoa and nematodes) and mesofauna (e.g. Collembola) (Stromberger et al., 2012 and references therein). Bacterial sums were calculated using the Gram+, Gram−, and actinomycete markers; fungal sums were calculated using both saprophytic and AMF fungal markers, and the fungal/bacteria ratio was calculated by dividing the fungal sum by the bacterial sum.

Activities of nine soil enzymes indicative of C-cycling (α-galactosidase, β-glucosidase), C/N-cycling (β-glucosaminidase), N-cycling (aspartase, l-asparaginase, urease), P-cycling (acid phosphatase, alkaline phosphomonoesterase) and S-cycling (arylsulfatase) were evaluated. These enzyme activities were assayed using 1 g of air-dried soil with their appropriate substrate and incubated for 1 h (37 °C) at their optimal pH as described by Tabatabai (1994) and Parham and Deng (2000). Enzyme activities were assayed in duplicate with one control, to which substrate was added after incubation and subtracted from the sample value.

13.2.4 STATISTICAL ANALYSES

Box plots were graphed using the default settings from the *ggplot2* package in R (Wickham, 2009 and R Development Core Team, 2012). The horizontal line is the mean, and upper and lower "hinges" are the first and third quartiles, respectively. Upper and lower "whiskers" extend to the highest or lowest value, respectively, within 1.5 times the inter-quartile range (the distance between the first and third quartiles). Data beyond this range are plotted as points.

Linear relationships between selected soil physicochemical characteristics were tested with the *lm()* function in R using field averages (i.e. n = 13) in order to examine broad relationships across the fields. Field was considered an explanatory factor (13 levels) in analysis of variance (ANOVA) models with separate analyses for each response variable, including 9 enzyme activities and 9 taxonomic groups compiled from indicator FAMEs. All models were statistically significant at the $p < 0.001$ level. F-statistics, i.e. the ratio of variance among fields to variance within fields, derived from these analyses were used to compare the relative magnitude of the field effect for each variable. Hence, greater between field variability for a given enzyme (or taxon) relative to other enzymes (or taxa) is reflected by a larger F-statistic.

Principal components analysis (PCA) was performed in the vegan package in R (Oksanen et al., 2012) using a correlation matrix. Component scores for each of the six plots within a field were used to generate 95% confidence ellipses around each field using the *ordiellipse()* function within *vegan*. PCA of FAMEs used data expressed as mol percent.

TABLE 2: Soil properties measured at the 13 organic tomato fields in Yolo County, California, USA in the 0–15 cm surface layer (se = standard error).

Field	Total C (g kg⁻¹)		Total N (g kg⁻¹)		C:N		pH		Clay (%)		Silt (%)		Sand (%)		Soil texture
	Mean	se	Mean	se	Mean	se	Mean	se	Mean	se	Mean	se	Mean	se	
1	6.7	0.17	0.8	0.03	8.2	0.16	6.7	0.11	13.9	0.80	47.1	2.77	39.1	3.57	Loam
2	9.6	0.23	1.2	0.02	8.3	0.08	6.8	0.10	21.4	0.27	59.3	0.36	19.3	0.56	Silt loam
3	10.7	0.17	1.3	0.03	8.2	0.07	6.7	0.08	19.4	0.12	55.6	0.45	25.1	0.40	Silt loam
4	11.1	0.22	1.4	0.03	8.1	0.09	6.6	0.06	17.6	0.36	61.4	1.30	21.0	1.35	Silt loam
5	11.2	0.22	1.4	0.03	8.1	0.06	6.3	0.06	18.1	0.45	60.1	0.67	21.9	1.04	Silt loam
6	12.5	0.53	1.4	0.07	9.2	0.14	6.3	0.04	16.2	0.19	63.5	0.45	20.3	0.52	Silt loam
7	12.8	0.52	1.4	0.05	9.0	0.10	7.2	0.04	14.6	0.29	52.1	0.59	33.3	0.86	Silt loam
8	13.2	0.38	1.5	0.04	9.0	0.07	6.8	0.02	9.7	0.21	48.4	0.43	41.9	0.49	Loam
9	13.9	0.24	1.6	0.03	9.0	0.15	6.4	0.04	14.0	0.06	58.1	0.26	27.8	0.31	Silt loam
10	16.5	0.33	1.7	0.04	9.7	0.24	6.6	0.06	15.7	0.16	63.8	0.48	20.6	0.62	Silt loam
11	17.1	0.48	1.8	0.04	9.8	0.12	6.9	0.04	17.5	0.16	67.3	0.34	15.3	0.45	Silt loam
12	18.1	0.76	2.0	0.08	9.3	0.05	6.8	0.04	11.2	0.21	47.8	0.40	41.0	0.52	Loam
13	20.0	0.56	2.1	0.05	9.7	0.10	6.5	0.04	16.4	0.38	66.4	1.15	17.3	1.41	Silt loam
Mean	13.3		1.5		8.9		6.7		15.8		57.7		26.4		
CV	28.0		22.1		7.2		3.7		20.2		12.2		35.3		

Since the PCA showed patterns among fields for the FAME and soil enzyme data, a constrained ordination technique was then used to evaluate relationships between these data and soil physicochemical factors through redundancy analysis (RDA). RDA combines regression and PCA and allows the direct analysis of how a set of response variables is structured by a set of explanatory variables (Borcard et al., 2011). RDA constrains ordination axes to be linear combinations of explanatory variables. Soil NH_4^+, NO_3^-, and Olsen P were $\ln(x + 1)$ transformed to help correct positive skewing and all variables were standardized prior to analysis. Forward selection of soil physicochemical factors (i.e. 15 variables: soil C and N, soil C:N, clay, silt, sand, pH, Olsen P, MBC, MBN, EOC, EON, $^+_4$NH-N, $^-_3$NO-N, and GWC) was performed independently for each set of response variables (either enzymes or FAMEs) to derive a parsimonious set of explanatory variables based on a double stopping rule of both alpha level (p < 0.05) and adjusted R^2 (Blanchet et al., 2008). RDA was performed with the *rda()* function in the *vegan* package.

Canonical variation partitioning (Borcard et al., 1992) was used to determine the relative importance of soil physicochemical properties and FAMEs in explaining variation in soil enzyme activities using adjusted R^2 values to obtain unbiased estimates (Peres-Neto et al., 2006). Soil factors were the same as used in RDA, as identified in the forward selection procedure. The same selection procedure was used to derive a parsimonious set of indicator FAMEs (mol percent) to explain enzyme activities. The analysis was performed in R using the *varpart()* function in the *vegan* package. Significance of the fractions (i.e. explained fractions of variation accounted for by the sum of the canonical axes) was tested by partial redundancy analyses and permutational significance tests (1000 permutations).

13.3 RESULTS

13.3.1 SOIL PROPERTIES AND C AND N POOLS

The 13 organically-managed Roma-type tomato fields had similar soil texture; measurements classified three fields as loams and ten as silt loams (Table 2). Clay content ranged from 9.7 to 21.4% and had a coefficient of variation (CV) of only 20.2 (Table 2). A three-fold range of soil organic C (6.7–20.0 g C kg^{-1} soil) and N (0.8–2.1 g N kg^{-1} soil) occurred across the set of fields, and soil C:N ranged from 8.1 to 9.8 (Table 2). Soil C and N were highly correlated with one another ($p < 0.0001$, $r^2 = 0.95$). Soil pH was near neutral (6.3–7.2) and varied the least of all measured soil properties with a CV of 3.7. Soil NH_4^+, NO_3^-, and Olsen P had the most variation of all measured variables with CVs of 70.8, 127.5, and 76.7, respectively, and were positively skewed, especially NO_3^- (Table 3). Field 4 had the highest level of NO_3^- (44.9 μg-N g^{-1} soil) while field 1 had the lowest (0.2 μg-N g^{-1} soil).

MBC ranged from 67.7 μg-C g^{-1} soil in field 1, to 165.8 μg-C g^{-1} in field 13, but it did not consistently increase across the C gradient (Table 3). Overall, MBC had a positive relationship with soil C ($p = 0.03$, $R^2 = 0.36$), soil N ($p = 0.02$, $R^2 = 0.41$), and silt ($p = 0.05$, $R^2 = 0.31$), although the low R^2 values indicate that none of these variables was highly associated with the variation in MBC. MBN was not significantly related to any soil variable other than MBC ($p = 0.003$, $R^2 = 0.56$) and showed more variability than MBC with a CV of 39.8 vs. 21.9 for MBC. Total FAMEs, which can be considered an alternative measure of microbial population size, had a positive but weak relationship with MBC ($p = 0.05$, $R^2 = 0.31$) and similar associations with soil C and N as MBC. Total FAMEs also had positive relationships with EOC ($p = 0.02$, $R^2 = 0.41$) and EON ($p = 0.02$, $R^2 = 0.40$).

13.3.2 PATTERNS OF SOIL ENZYME ACTIVITIES

The activities of nine soil enzymes showed different trends across the 13 fields (Fig. 1). Phosphodiesterase activity showed the strongest field effect (F-statistic $= 27.3$); higher F-statistic values indicate the relative magnitude of the "field effect", i.e., a higher ratio of variance among fields to variance within a field (see methods section). Six out of nine enzyme activities had an F-statistic greater than 20. β-glucosaminidase activity was the least variable of the measured enzymes (F-statistic $= 8.2$), followed by the activities of l-asparaginase (F-statistic $= 14.1$) and urease (F-statistic $= 19.1$). The geometric mean of enzyme activities (Table 2), an indicator of the overall metabolic potential, showed a strong positive relationship with

TABLE 3: Soil nutrient and biological properties measured at the 13 organic tomato fields in Yolo County, California, USA in the 0–15 cm surface layer (se = standard error).

Field	MBC (µg g⁻¹)		MBN (µg g⁻¹)		Total enzyme activity[a]		Total FAME (nmol g⁻¹ soil)		EOC[b] (µg g⁻¹)		EON[b] (µg g⁻¹)		NH₄-N (µg g⁻¹)		NO₃-N (µg g⁻¹)		Olsen P (µg g⁻¹)	
	Mean	se	Mean	se	Mean	se	Mean	se	Mean	se	Mean	se	Mean	se	Mean	se	Mean	se
1	67.7	4.95	7.2	0.67	28.5	1.26	80.5	3.63	22.9	1.96	2.9	0.45	0.28	0.02	0.19	0.13	20.0	1.25
2	76.1	3.03	4.6	0.79	38.2	1.02	80.6	7.38	32.9	0.94	5.0	0.20	0.38	0.03	0.54	0.08	30.3	2.65
3	142.7	4.91	14.0	1.73	67.3	2.73	114.5	6.48	48.9	4.23	9.5	0.72	0.37	0.03	4.07	1.25	56.5	5.57
4	126.9	11.25	19.2	2.77	61.3	1.83	105.9	5.39	36.6	2.50	4.8	0.37	1.16	0.26	44.91	12.76	29.2	1.11
5	119.5	5.67	9.8	1.22	77.5	3.53	137.6	4.67	44.3	2.16	7.2	0.34	1.19	0.07	10.54	1.56	28.3	3.36
6	122.3	9.41	15.3	0.94	66.5	3.95	118.6	16.09	40.3	4.14	5.8	1.10	0.51	0.03	18.45	8.12	32.9	4.15
7	108.5	4.59	11.7	2.24	49.2	2.87	86.5	5.55	41.4	2.18	8.2	0.65	1.43	0.18	13.12	3.19	29.2	1.78
8	121.6	4.44	23.0	1.43	73.1	3.81	92.7	9.73	41.6	1.75	4.1	0.18	0.34	0.06	4.27	1.81	4.9	0.21
9	122.9	6.53	16.1	0.70	62.6	2.02	100.6	5.37	43.2	2.03	5.8	1.06	0.54	0.12	18.13	7.00	18.3	0.86
10	132.2	7.26	13.5	1.27	62.5	1.23	159.4	8.53	57.8	1.77	10.9	0.29	0.44	0.03	1.48	0.26	16.1	1.80
11	120.4	6.84	12.5	1.38	59.8	2.50	118.5	7.53	57.0	3.37	12.5	0.43	0.24	0.05	2.89	0.55	12.8	1.41
12	100.1	8.88	7.9	1.60	66.8	3.75	132.0	10.76	91.9	4.07	18.0	0.83	0.18	0.03	2.94	0.36	93.1	2.70
13	165.8	11.92	20.0	1.14	83.8	3.39	126.3	10.45	73.0	2.40	12.7	0.52	0.42	0.04	4.21	0.57	15.8	1.03
Mean	117.5		13.4		61.3		111.8		48.6		8.2		0.57		9.67		29.8	
CV	21.9		39.8		24.7		21.2		37.1		52.3		70.8		127.5		76.6	

[a] Total enzyme activity is calculated as the geometric mean of activities of all nine enzymes tested here (García-Ruiz et al., 2008).
[b] EOC: K₂SO₄ extractable organic C; EON: K₂SO₄ extractable organic N.

MBC ($p < 0.001$, $R^2 = 0.72$) but weaker relationships with MBN ($p = 0.02$, $R^2 = 0.40$) and soil C ($p = 0.02$, $R^2 = 0.40$).

When soil physicochemical factors were used to constrain the ordination of all nine enzyme activities with RDA (Fig. 2), the model accounted for 65% of the total variation ($p < 0.001$). Based on permutation tests, all explanatory variables retained in the model were significant ($p < 0.05$) in constraining enzyme activities.

Increases in all enzyme activities along axis 1 (42% of total variation, 60% of fitted variation) were positively correlated with MBC, which was more highly associated with this axis than any other variable (Fig. 2a), as indicated by the length and direction of its vector. Axis 2 (14% of the total variation and 20% of the fitted variation) largely differentiated C- vs. N-cycling enzymes, with P- and S-cycling enzymes falling intermediate. Activities of C-cycling enzymes β-glucosidase and α-galactosidase and C/N-cycling enzyme β-glucosaminidase loaded positively on axis 2 and were associated with soil inorganic N, especially NO_3^-, and MBN. Activities of N-cycling enzymes aspartase and l-asparaginase loaded negatively along axis 2 and were associated with soil C and N, soil C:N, and EOC and EON.

Non-overlapping confidence ellipses for most of the fields indicated a unique suite of enzyme activities within each field (Fig. 2b). Clusters of fields reflected the primary organic amendment used. Fields 1 and 2, with the lowest soil C and N and receiving manure, had the lowest values along axis 1 while field 13, with the highest soil C and N and receiving composted green waste, had the highest values. Other fields had similar, positive values along axis 1 (except field 7) but were strongly differentiated by axis 2. Fields 10, 11, 12, and 13, in which composted green waste was applied, had negative values along axis 2, such that increases in N-cycling enzyme activities corresponded with indicators of higher SOM pools. The remaining fields, in which manure or vetch was used (with the exception of field 8), had positive values along axis 2, such that increases in C-cycling enzyme activities corresponded with higher inorganic N and MBN. The pattern of sites and enzyme activities in the RDA resembled that of an ordination by PCA (Supplemental Fig. 1).

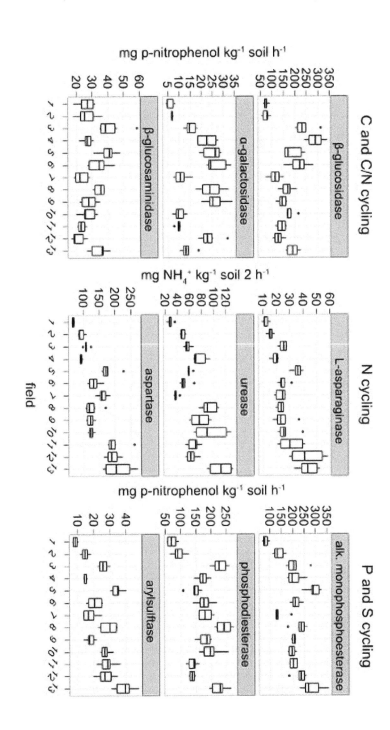

FIGURE 1: Box plots of soil potential enzyme activities at the 13 organic tomato fields in Yolo County, California, USA in the 0–15 cm surface layer.

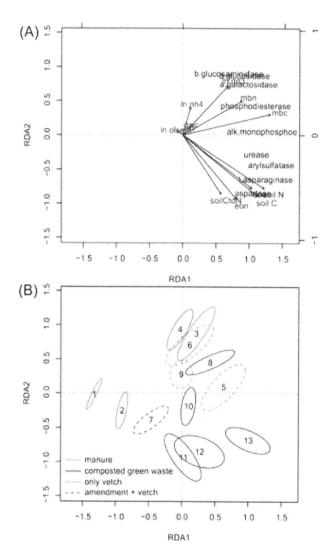

FIGURE 2: Redundancy analysis of soil potential enzyme activities constrained by soil physicochemical properties (0–15 cm) at the 13 organic tomato fields in Yolo County, California, USA. A parsimonious set of explanatory factors was generated using a forward selection procedure (see methods; Blanchet et al., 2008). (a) Vectors represent selected soil physicochemical properties. Axes 1 and 2 represent 42 and 14% of the total variation, respectively, with the whole model accounting for 65% of the variation. (b) 95% confidence ellipses (see methods) for RDA field scores (6 per field), numbered by field and colored by primary organic amendment. Dotted lines indicate fields where a vetch cover crop was grown in addition to the organic amendment (either manure or composted green waste) used.

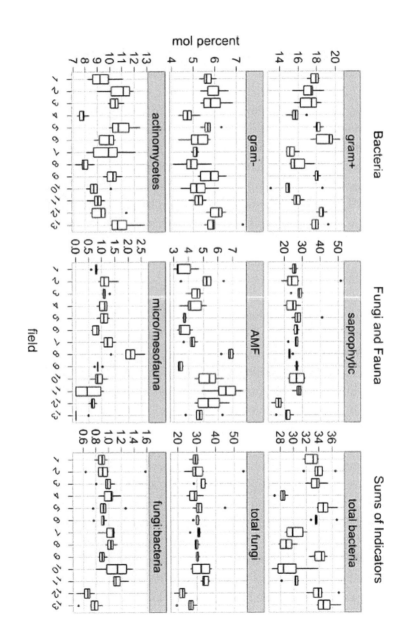

FIGURE 3: Box plots of relative abundance of indicator FAMEs in soil from the 0–15 cm surface layer at the 13 organic tomato fields in Yolo County, California, USA. Each group is composed of one or more FAMEs shown to be associated with the particular taxon (see methods).

13.3.3 PATTERNS OF FAMES

Overall, the relative abundance of indicator FAMEs (Fig. 3) was less variable across fields than soil enzyme activities, as reflected by smaller F-statistics. Based on relative abundance, markers for Gram+ bacteria showed more variation across fields (F-statistic = 11.5) than that of Gram− bacteria (F-statistic = 5.4) or actinomycetes (F-statistic = 8.2). Markers of saprophytic fungi were generally similar across the fields (F-statistic = 3.7) while that of AMF showed more variation (F-statistic = 15.2) with the highest relative abundance in fields 8, 10, 11, 12, and 13. A marker for soil micro/mesofauna showed the most variation of the FAME indicators (F-statistic = 19.4). The sum of all biomarkers for fungi (i.e. total fungi) was relatively consistent across the fields (F-statistic = 3.5), while the sum of all markers for bacteria (i.e. total bacteria) was more variable (F-statistic = 16.9). Biomarkers for total fungi and total bacteria accounted for a mean of 30.1 and 32.2% of the total FAMEs, respectively, which reflected a fungi:bacteria FAME ratio close to one (mean = 0.94) for most fields, except in fields 12 and 13, where it was appreciably lower.

When soil physicochemical factors were used to constrain the ordination of the relative abundance of 14 indicator FAMEs with RDA (Fig. 4), the full RDA model accounted for only 29.3% of the variation ($p < 0.001$). The first two RDA axes accounted for 13.1 and 11.5% of the overall variation and 36.7 and 32.1% of the fitted model (Fig. 4a). All explanatory variables retained in the model were significant ($p < 0.05$) in constraining indicator FAMEs based on permutation tests. Olsen P, GWC, and pH played a role in the dispersion of fields along the first axis, while soil texture variables (clay and silt) and EOC varied mainly along the second axis (Fig. 4b). While it was retained in the selection procedure and significant, MBC did not appear to play a strong role in the ordination of indicator FAMEs, based on the length of its vector. Higher Olsen P was associated with bacterial markers while higher GWC and pH were associated with fungal and micro/mesofaunal markers. EOC was correlated with the gram-positive bacterial marker i17:0.

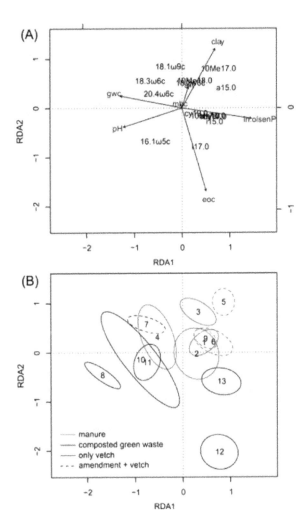

FIGURE 4: RDA of indicator FAMEs in soil from the 0–15 cm surface layer at the 13 organic tomato fields in Yolo County, California, USA, constrained by soil physicochemical properties. A parsimonious set of explanatory factors was generated using a forward selection procedure (see methods; Blanchet et al., 2008). (a) Vectors represent selected soil physicochemical properties. Axes 1 and 2 represent 13.1 and 11.5% of the total variation, respectively, with the whole model accounting for 29.3% of the variation. (b) 95% confidence ellipses (see methods) for RDA field scores (6 per field), numbered by field and colored by primary organic amendment. Dotted lines indicate fields where a vetch cover crop was grown in addition to the organic amendment (either manure or composted green waste) used.

TABLE 4: Partitioning variation in soil potential enzyme activities as a function of soil physicochemical factors and indicator FAMEs at the 13 organic tomato fields in Yolo County, California, USA in the 0–15 cm surface layer.

	Percent variation explained[a]	Explanatory variables[b]	Variables retained
Overall model	64.9***	16	
Soil physicochemical factors	37.7***	11	MBC, total N, $^-_3$NO-N, GWC, total C, MBN, $^+_4$NH-N, C:N, Olsen P, DOC, DON
FAME indicators	6.1***	10	20.4ω6c, 16.1ω5c, 10Me18.0, i15.0, a15.0, 10Me16.0, a17.0, cy17.0, cy19.0, 10Me17.0

[a] *Variation explained by the full model (i.e. soil factors and FAMEs) consists of both the respective contributions of each set of explanatory variables as well as their covariation. Significance was assessed using 1000 permutations of either the full RDA model or the appropriate partial redundancy analysis model, with ***P < 0.001.*
[b] *A parsimonious set of explanatory variables was derived using forward selection based on a double stopping rule of both an alpha criterion and an adjusted R^2 within each set of explanatory variables individually (Blanchet et al., 2008 and Borcard et al., 2011).*

The first axis of a PCA of relative abundances of these indicator FAMEs showed a similar pattern as the RDA, with three out of four fungal markers as well as the micro/mesofaunal marker grouping away from all other markers, while axis 2 did not yield any clear pattern (Supplemental Fig. 2). In general, the fields were more dispersed in the RDA ordination relative to the PCA, reinforcing a role of soil physicochemical factors in explaining variation among fields in the microbial community.

The pattern of FAME biomarkers was also somewhat related to the type of primary organic amendment that was applied in the past year in both the RDA and PCA (Fig. 4b and Supplemental Fig. 2). Fields using manure as the primary organic amendment and several fields using composted green waste formed separate clusters (except field 12), with both groups containing fields that also used a vetch cover crop. The only field using a vetch cover crop without compost or manure clustered with compost.

13.3.4 RELATIVE INFLUENCE OF SOIL PHYSICOCHEMICAL FACTORS AND FAMES ON ENZYME ACTIVITIES

Canonical variation partitioning showed the relative influence of soil phys-icochemical factors and microbial community composition on potential enzyme activities by quantifying both the unique and shared proportion of variability accounted for by each set of explanatory factors (Table 4). Soil physicochemical factors and the relative abundance of indicator FAMEs together explained 64.9% of the total variation in soil potential enzyme activities, compared to 37.7% for soil factors alone ($p < 0.001$). Indicator FAMEs uniquely explained only 6.1% of the variation ($p < 0.001$). The remaining variation, which cannot be attributed uniquely to either explanatory dataset, totaled 27.2%.

13.4 DISCUSSION

This research approach provides insight into how microbial community function and composition respond to the variation that exists across organic farm fields where farmers are growing the same crop in a single landscape. The two hypotheses were supported in the following ways. First, distinct profiles of soil potential enzyme activities indicated unique potential metabolic capacities across the fields, such that C-cycling enzyme activity increased with inorganic N availability while N-cycling enzyme activity increased with C availability. Second, although FAMEs suggested that microbial community composition was less variable across fields than potential enzyme activities, there were slight community differences that were related to the use of compost vs. manure as the primary organic amendment. Overall, however, the general similarity among fields for particular taxonomic indicators, such as saprophytic fungi, is consistent with another nearby study in this intensively managed landscape (Young-Mathews et al., 2010). These patterns suggest that differences in organic agroecosystem management have strongly influenced soil nutrients and potential enzyme activity, but without a major effect on soil microbial communities in this landscape. Development of better indicators of microbial functions in organic systems may help farmers evaluate and discover management options that continue to improve the nutrient cycling capacity of the soil.

13.4.1 PATTERNS AND DETERMINANTS OF SOIL POTENTIAL ENZYME ACTIVITY

In this landscape, the majority of the variation in potential enzyme activities could be explained by soil characteristics related to nutrient availability and microbial biomass, which are well-known to be strongly influenced by management on relatively short times scales, as well as soil properties well-known to be influenced by both management and soil type at longer time scales (e.g. soil C and N). Soil variables determined by soil type (e.g. texture) did not contribute to explaining variation in enzyme activities, which may be partly a result of the similar soil types and relatively narrow range of soil textures sampled. MBC was positively correlated with increases in the potential activity of most enzymes, as reflected in the first axis of the RDA, and strongly related to the geometric mean of enzyme activity, an indicator of overall microbial metabolic capacity (García-Ruiz et al., 2008). Most fields showed fairly similar values along this axis, except for fields at opposite ends of the SOM gradient. At the low extreme are fields 1 and 2 with soil C below 10 g kg^{-1} and much lower MBC and enzyme activities than other fields. Below a certain level of MBC, microbial functioning may be reduced. At the high extreme is field 13, with high soil C, MBC, and potential enzyme activities, especially for those involved in nutrient release. The other fields, with mid-range values for soil C and MBC, suggest that a diverse set of soil conditions and nutrient management strategies result in similar overall soil metabolic capacity, albeit with differences in the activity of certain enzymes related to C vs. nutrient cycling processes. Interestingly, this may also be reflected in tomato yields. Nine of the 13 fields had similar yields ($104.0 \pm 3.6 \text{ tons ha}^{-1}$, mean $\pm$ SE) that were above the Yolo County average in 2011 (86 tons ha^{-1}), which included both conventional and organic Roma-type tomatoes (Bowles et al. ms. in preparation).

More subtle patterns in the potential activity of C vs. N cycling enzymes were apparent along the second RDA axis. Microbes regulate extracellular enzyme production to acquire limiting nutrients, so changes in enzyme activities may reflect patterns of microbial nutrient limitations and hence nutrient availability (Allison et al., 2007, Allison et al., 2011, Sinsabaugh et al., 2008 and Burns et al., 2013). The strong association among soil inorganic N, particularly NO_3^-, and the activities of C-cycling enzymes (β-glucosidase, α-galactosidase) and a C/N-cycling enzyme (β-glucosaminidase) suggest a shift toward increased C acquisition as N becomes readily available. Other studies have shown increased activity of cellulases (i.e. enzymes that catalyze degradation of cellulose, including β-glucosidase and α-galactosidase in this study) in response to N fertilization (Bandick and Dick, 1999, Sinsabaugh et al., 2005 and Piotrowska and Wilczewski, 2012). Reduced activity of enzymes involved directly in N mineralization (e.g. urease and amidase) with higher inorganic N availability has also been shown in agricultural systems (Dick et al., 1988 and Bandick and Dick, 1999) and agree with our results of reduced potential activity of l-asparaginase and aspartase in several fields with higher NO_3^- (e.g. field 4). Higher levels of soil NO_3^- were typically found in fields with intermediate levels of soil C and N in conjunction with application of a labile N source (e.g. seabird guano), which was likely rapidly mineralized and nitrified.

In contrast, greater potential activity of two N-cycling enzymes (l-asparaginase and aspartase) but lower activity of C-cycling enzymes occurred in fields with higher soil C and N where composted green waste was applied as a primary organic matter source. In such situations, an abundant supply of diverse C sources may have resulted in N limitation for the microbial community and hence, greater production of enzymes to mineralize N. The high concentrations of EOC and EON and the low concentrations of soil NH_4^+ and NO_3^- in fields 10, 11, 12, and 13 support this hypothesis. EOC and EON are comprised of a diverse array of organic molecules, including free amino acids (Yu et al., 2002 and Paul and Williams, 2005) that would include substrates for l-asparaginase and aspartase (Frankenberger and Tabatabai, 1991 and Senwo and Tabatabai, 1996). Furthermore, we hypothesize that rapid microbial and plant uptake of mineralized N likely kept soil NH_4^+ and NO_3^- concentrations low, even while the supply rate may have been high, given the activity of these enzymes. High rates of both gross mineralization and microbial immobilization have been observed in an organic tomato system on similar soil in the same landscape (Burger and Jackson, 2003). An alternative hypothesis is that increased C availability increased denitrification and subsequently lowered soil NO_3^-; however, N_2O emissions over two growing seasons were negligible in a separate case study of an organic Roma tomato field managed by one of the growers involved in this research (Smukler et al., 2010) as well as in other organic tomato systems in this area (Burger et al., 2005). Thus, denitrification was probably low.

The lack of association between P availability, as indicated by Olsen P, and the potential activity of P-cycling enzymes, phosphodiesterase and alkaline phosphomonoesterase, is in contrast with previous work that demonstrates a negative relationship between phosphatase activity and P availability in non-agricultural systems (Olander and Vitousek, 2000 and Allison et al., 2007). Across the 13 fields, these enzyme activities appear related to microbial biomass; for instance, phosphodiesterase activity had the strongest positive relationship with MBC of any enzyme ($p < 0.001$ R^2 = 0.727), suggesting that soil microbial biomass was more important than P availability in regulating investments in phosphatases across these fields.

The relative importance of microbial community composition vs. environmental factors in regulating enzyme expression remains unclear (Sinsabaugh et al., 2005, Allison et al., 2007, Frossard et al., 2012 and Reed and Martiny, 2013). In this study, microbial community composition explained little unique variation in potential enzyme activities relative to soil physicochemical properties. The plasticity of the resident microbial community to respond to environmental conditions may be high, as suggested by the relatively large fraction of variation in potential enzyme activities explained by soil physicochemical characteristics (37.7%). Moreover, a large fraction of variation was also explained jointly by soil factors and FAMEs (~27% of the total variation in the canonical variance partitioning analysis), indicating that microbial communities did influence activity under specific environmental conditions, despite the low variation in community composition across this landscape (see below).

13.4.2 PATTERNS AND DETERMINANTS OF MICROBIAL COMMUNITY BIOMASS AND COMPOSITION

Soil microbial community composition, as measured by FAMEs, was not as strongly differentiated among individual fields as soil potential enzyme activities, based on F-statistics that were predominately lower for FAMEs. Rather, FAMEs formed weak clusters in both the PCA (Supplemental Fig. 2) and RDA (Fig. 4) that were associated at least in part with the primary organic amendment used (manure, composted green material, and/ or vetch cover crop). The clusters in the PCA and the RDA, as well as the relatively low proportion of FAME variation accounted for in the RDA with the measured soil physicochemical factors (<30%), suggests that unmeasured attributes of the organic amendments may exert strong effects on the microbial community, or that past management is still having effects. Microbial communities unique to the type of organic amendment may also have an inoculating effect and contribute to the differentiation of the microbial community composition (Marschner et al., 2003 and Lazcano et al., 2008).

Microbial community composition was associated with factors determined by parent soil type (e.g. clay and silt) as well as those influenced by a combination of soil characteristics and management (e.g. Olsen P and pH). Similarly, Bossio et al. (1998) showed that soil type followed by specific management operations (e.g. cover crop incorporation or manure application) were the primary factors in governing the composition of microbial communities in a cropping system experiment comparing organic and conventional management on similar soils in the same landscape. Of these 13 fields, those that used manure as a primary organic amendment clustered together in the RDA and were associated with increased Olsen P and increases in Gram+ and Gram− bacteria and with decreases in fungal and micro/mesofaunal markers. Increased P availability has been shown to negatively affect fungi in other agricultural landscapes (Lauber et al., 2008 and Williams and Hedlund, 2013) and can result from manure application over time (Clark et al., 1998). Another example of possible management effects is field 12, in which the microbial community was strongly differentiated from other fields. It had been in organic management more than twice as long (26 years) as any other field in the study. High EOC and EON in this field may be indicative of a diversity of organic moieties built up over time with organic management (Aranda et al., 2011) and supportive of a more unique microbial community composition (Giacometti et al., 2013).

Despite differences in management, the relative abundance of FAMEs indicative of saprophytic fungi and the fungi:bacteria ratio was much more consistent across fields than bacterial FAMEs. Disturbance intensity and frequency appear to be lower in landscapes where studies have shown positive relationships between the fungi: bacteria ratio and soil C and soil C:N (Fierer et al., 2009 and de Vries et al., 2012). Saprophytic fungi are well-known to be particularly sensitive to certain management practices, especially tillage and fertilization (Minoshima et al., 2007). Tillage intensity was similar across the 13 fields, since these organically-managed farms rely on cultivation as a means of weed control and for incorporating organic amendments. While organic management may increase fungi relative to conventional management (Bossio et al., 1998), routine soil disturbance may represent a strong filter for fungi, such that only a resistant subset persist in arable soils in this area (Calderón et al., 2000 and Young-Mathews et al., 2010). In contrast to saprophytic fungi, the relative abundance of a FAME biomarker for AMF, $16.1\omega5c$, was distinctly different across fields. A significant negative relationship ($p = 0.017$, $R^2 =$ 0.45) was found between the relative abundance of this marker and Olsen P when field 12, which has high Olsen P, is excluded from the analysis. This is in line with other studies in agricultural landscapes (Williams and Hedlund, 2013) and reflects the sensitivity of AMF to P availability.

13.4.3 ON-FARM APPROACH TO MICROBIAL COMMUNITY FUNCTIONING

The on-farm approach used in this study provided a range of SOM characteristics (e.g. a three-fold range of total soil C and N) and organic nutrient management practices to investigate how these factors influence soil microbes and ecosystem functions while controlling for other factors to the extent possible in a real landscape. Narrowing a landscape's extent to a smaller geographic area allowed for sampling generally similar soils, in terms of texture, mineralogy, and parent material, while still encompassing a range of farming strategies. Focusing on the same crop controlled for the effect of plant species and plant functional traits, which also strongly influence the microbial community (Gardner et al., 2011). Since timing (relative to seasonal and agronomic events) can exert a strong influence on enzyme activities and microbial community composition, we carefully sampled during a defined crop phenological period when nutrient demand was maximal (i.e. anthesis to early green fruit stage), which provided insight into soil functions at a crucial time for crop productivity. Furthermore, the fields in this study were planted within a two-week period, and this minimized the differential effects of temperature and rainfall on soil biology and plant growth across fields. Since farmers in this area use irrigation and summers are reliably hot and dry, inter-annual variability may be reduced compared to locations with less predictable summer weather.

Differing nutrient management practices and SOM characteristics across these fields reinforces the need for robust indicators of microbially-derived ecosystem functions to support management decisions, since one-size-fits-all recommendations are not viable in such heterogeneous systems. While MBC would have differentiated fields with apparently compromised soil quality (i.e. fields 1 and 2) from others, it would not have differentiated more subtle variation related to potential activities of C- and N-cycling enzymes, which may have important implications for ecosystem functioning, such as sufficient N availability with low potential for N loss. Differences in enzyme activities in concert with specific C and N pools may eventually be useful to farmers for improving site-specific management to balance these types of tradeoffs. Such research complements nearby research station-based experiments (e.g. Bossio et al., 1998 and Kong et al., 2011), which were by design and necessity limited to a relatively narrow set of practices at a single field. In turn, such experiments disentangle the relative effects of individual management practices and examine long-term trends, which can be challenging in a landscape approach. A dynamic interplay between site-specific experimental research and landscape-scale surveys of working farms may be the most promising route to improving understanding and management of microbial processes for ecological intensification of agriculture.

REFERENCES

1. Acosta-Martínez et al., 2007 V. Acosta-Martínez, L. Cruz, D. Sotomayor-Ramírez, L. Pérez-Alegría Enzyme activities as affected by soil properties and land use in a tropical watershed Appl. Soil Ecol., 35 (2007), pp. 35–45

2. Acosta-Martínez et al., 2008 V. Acosta-Martínez, D. Acosta-Mercado, D. Sotomayor-Ramírez, L. Cruz-Rodríguez Microbial communities and enzymatic activities under different management in semiarid soils Appl. Soil Ecol., 38 (2008), pp. 249–260

3. Ajwa et al., 1999 H.A. Ajwa, C.J. Dell, C.W. Rice Changes in enzyme activities and microbial biomass of tallgrass prairie soil as related to burning and nitrogen fertilization Soil Biol. Biochem., 31 (1999), pp. 769–777

4. Allison et al., 2007 V.J. Allison, L.M. Condron, D.A. Peltzer, S.J. Richardson, B.L. Turner Changes in enzyme activities and soil microbial community composition along carbon and nutrient gradients at the Franz Josef chronosequence, New Zealand Soil Biol. Biochem., 39 (2007), pp. 1770–1781

5. Allison et al., 2011 S.D. Allison, M.N. Weintraub, T.B. Gartner, M.P. Waldrop Evolutionary-economic principles as regulators of soil enzyme production and ecosystem function G. Shukla, A. Varma (Eds.), Soil Enzymology, Springer, Berlin (2011), pp. 229–243

6. Aranda et al., 2011 V. Aranda, M.J. Ayora-Cañada, A. Domínguez-Vidal, J.M. Martín-García, J. Calero, R. Delgado, T. Verdejo, F.J. González-Vila Effect of soil type and management (organic vs. conventional) on soil organic matter quality in olive groves in a semi-arid environment in Sierra Mágina Natural Park (S Spain) Geoderma, 164 (2011), pp. 54–63

7. Bandick and Dick, 1999 A.K. Bandick, R.P. Dick Field management effects on soil enzyme activities Soil Biol. Biochem., 31 (1999), pp. 1471–1479

8. Bernard et al., 2012 E. Bernard, R.P. Larkin, S. Tavantzis, M.S. Erich, A. Alyokhin, G. Sewell, A. Lannan, S.D. Gross Compost, rapeseed rotation, and biocontrol agents significantly impact soil microbial communities in organic and conventional potato production systems Appl. Soil Ecol., 52 (2012), pp. 29–41

9. Blanchet et al., 2008 F. Blanchet, P. Legendre, D. Borcard Forward selection of explanatory variables Ecology, 89 (2008), pp. 2623–2632

10. Borcard et al., 1992 D. Borcard, P. Legendre, P. Drapeau Partialling out the spatial component of ecological variation Ecology, 73 (1992), pp. 1045–1055

11. Borcard et al., 2011 D. Borcard, F. Gillet, P. Legendre Canonical ordination Numerical Ecology with R, Springer, New York (2011), pp. 153–225

12. Bossio et al., 1998 D. Bossio, K.M. Scow, N. Gunapala, K. Graham Determinants of soil microbial communities: effects of agricultural management, season, and soil type on phospholipid fatty acid profiles Microb. Ecol., 36 (1998), pp. 1–12

13. Burger and Jackson, 2003 M. Burger, L.E. Jackson Microbial immobilization of ammonium and nitrate in relation to ammonification and nitrification rates in organic and conventional cropping systems Soil Biol. Biochem., 35 (2003), pp. 29–36

14. Burger et al., 2005 M. Burger, L.E. Jackson, E. Lundquist, D.T. Louie, R.L. Miller, D.E. Rolston, K.M. Scow Microbial responses and nitrous oxide emissions during wetting and drying of organically and conventionally managed soil under tomatoes Biol. Fertil. Soils, 42 (2005), pp. 109–118

15. Burns et al., 2013 R.G. Burns, J.L. DeForest, J. Marxsen, R.L. Sinsabaugh, M.E. Stromberger, M.D. Wallenstein, M.N. Weintraub, A. Zoppini Soil enzymes in a changing environment: current knowledge and future directions Soil Biol. Biochem., 58 (2013), pp. 216–234

16. Cabrera and Beare, 1993 M. Cabrera, M. Beare Alkaline persulfate oxidation for determining total nitrogen in microbial biomass extracts Soil Sci. Soc. Am. J., 57 (1993), pp. 1007–1012

17. Calderón et al., 2000 F.J. Calderón, L.E. Jackson, K.M. Scow, D.E. Rolston Microbial responses to simulated tillage in cultivated and uncultivated soils Soil Biol. Biochem., 32 (2000), pp. 1547–1559

18. Carrera et al., 2007 L.M. Carrera, J.S. Buyer, B. Vinyard, A.A. Abdul-Baki, L.J. Sikora, J.R. Teasdale Effects of cover crops, compost, and manure amendments on soil microbial community structure in tomato production systems Appl. Soil Ecol., 37 (2007), pp. 247–255

19. Cavigelli et al., 2013 M.A. Cavigelli, S.B. Mirsky, J.R. Teasdale, J.T. Spargo, J. Doran Organic grain cropping systems to enhance ecosystem services Renew. Agric. Food Syst., 28 (2013), pp. 145–159

20. CCOF, 2011 CCOF Organic Directory (2011) Available at: http://www.ccof.org/members/advanced (accessed 12.01.11.)

21. CIMIS, 2013 CIMIS California Irrigation Management Information System Department of Water Resources. Office of Water Use Efficiency (2013) Available at: http://wwwcimis.water.ca.gov/cimis/welcome.jsp (accessed 14.05.13)

22. Clark et al., 1998 M.S. Clark, W.R. Horwath, C. Shennan, K.M. Scow Changes in soil chemical properties resulting from organic and low-input farming practices Agron. J., 90 (1998), pp. 662–671

23. Cleveland and Liptzin, 2007 C.C. Cleveland, D. Liptzin C: N:P stoichiometry in soil: is there a "redfield ratio" for the microbial biomass? Biogeochemistry, 85 (2007), pp. 235–252

24. Culman et al., 2010 S.W. Culman, A. Young-Mathews, A.D. Hollander, H. Ferris, S. Sánchez-Moreno, A.T. O'Geen, L.E. Jackson Biodiversity is associated with indicators of soil ecosystem functions over a landscape gradient of agricultural intensification Landsc. Ecol., 25 (2010), pp. 1333–1348

25. Darnhofer et al., 2005 I. Darnhofer, W. Schneeberger, B. Freyer Converting or not converting to organic farming in Austria: farmer types and their rationale Agric. Hum. Values, 22 (2005), pp. 39–52

26. Darnhofer et al., 2010 I. Darnhofer, T. Lindenthal, R. Bartel-Kratochvil, W. Zollitsch Conventionalisation of organic farming practices: from structural criteria towards an assessment based on organic principles. A review Agron. Sustain. Develop., 30 (2010), pp. 67–81

27. de Vries and Bardgett, 2012 F.T. de Vries, R.D. Bardgett Plant–microbial linkages and ecosystem nitrogen retention: lessons for sustainable agriculture Front. Ecol. Environ., 10 (2012), pp. 425–432

28. de Vries et al., 2012 F.T. de Vries, P. Manning, J.R.B. Tallowin, S.R. Mortimer, E.S. Pilgrim, K.A. Harrison, P.J. Hobbs, H. Quirk, B. Shipley, J.H.C. Cornelissen, J. Kattge, R.D. Bardgett Abiotic drivers and plant traits explain landscape-scale patterns in soil microbial communities Ecol. Lett., 15 (2012), pp. 1230–1239

29. Dick et al., 1988 R.P. Dick, P.E. Rasmussen, E.A. Kerle Influence of long-term residue management on soil enzyme activities in relation to soil chemical properties of a wheat-fallow system Biol. Fertil. Soils, 6 (1988), pp. 159–164

30. Drinkwater and Wagoner, 1998 L.E. Drinkwater, P. Wagoner Legume-based cropping systems have reduced carbon and nitrogen losses Nature, 396 (1998), pp. 262–265

31. Drinkwater et al., 1995 L.E. Drinkwater, D. Letourneau, F. Workneh, A. van Bruggen, C. Shennan Fundamental differences between conventional and organic tomato agroecosystems in California Ecol. Appl., 5 (1995), pp. 1098–1112

32. Eshel et al., 2004 G. Eshel, G. Levy, U. Mingelgrin, M. Singer Critical evaluation of the use of laser diffraction for particle-size distribution analysis Soil Sci. Soc. Am. J., 68 (2004), pp. 736–743

33. Fierer et al., 2009 N. Fierer, M.S. Strickland, D. Liptzin, M.A. Bradford, C.C. Cleveland Global patterns in belowground communities Ecol. Lett., 12 (2009), pp. 1238–1249

34. Foster, 1995 J.C. Foster Soil nitrogen K. Alef, P. Nannipieri (Eds.), Methods in Applied Soil Microbiology and Biochemistry, Academic Press, San Diego (1995), pp. 79–87

35. Frankenberger and Tabatabai, 1991 W.T. Frankenberger, M.A. Tabatabai Factors affecting L-asparaginase activity in soils Biol. Fertil. Soils, 11 (1991), pp. 1–5

36. Fraterrigo et al., 2006 J.M. Fraterrigo, T.C. Balser, M.G. Turner Microbial community variation and its relationship with nitrogen mineralization in historically altered forests Ecology, 87 (2006), pp. 570–579

37. Frossard et al., 2012 A. Frossard, L. Gerull, M. Mutz, M.O. Gessner Disconnect of microbial structure and function: enzyme activities and bacterial communities in nascent stream corridors ISME J., 6 (2012), pp. 680–691

38. García-Ruiz et al., 2008 R. García-Ruiz, V. Ochoa, M.B. Hinojosa, J.A. Carreira Suitability of enzyme activities for the monitoring of soil quality improvement in organic agricultural systems Soil Biol. Biochem., 40 (2008), pp. 2137–2145

39. Gardner et al., 2011 T. Gardner, V. Acosta-Martínez, Z. Senwo, S.E. Dowd Soil rhizosphere microbial communities and enzyme activities under organic farming in Alabama Diversity, 3 (2011), pp. 308–328

40. Gattinger et al., 2012 A. Gattinger, A. Muller, M. Haeni, C. Skinner, A. Fliessbach, N. Buchmann, P. Mäder, M. Stolze, P. Smith, N.E.-H. Scialabba, U. Niggli Enhanced top soil carbon stocks under organic farming Proc. Natl. Acad. Sci. U. S. A., 109 (2012), pp. 18226–18231

41. Giacometti et al., 2013 C. Giacometti, M.S. Demyan, L. Cavani, C. Marzadori, C. Ciavatta, E. Kandeler Chemical and microbiological soil quality indicators and their potential to differentiate fertilization regimes in temperate agroecosystems Appl. Soil Ecol., 64 (2013), pp. 32–48

42. Guthman, 2000 J. Guthman Raising organic: an agro-ecological assessment of grower practices in California Agric. Hum. Values, 17 (2000), pp. 257–266

43. Guthman, 2004 J. Guthman Agrarian Dreams: the Paradox of Organic Farming in California University of California Press, Berkeley (2004)

44. Jackson et al., 2011 L.E. Jackson, S.M. Wheeler, A.D. Hollander, A.T. O'Geen, B.S. Orlove, J. Six, D.A. Sumner, F. Santos-Martin, J.B. Kramer, W.R. Horwath, R.E.

Howitt, T.P. Tomich Case study on potential agricultural responses to climate change in a California landscape Climat. Change, 109 (2011), pp. 407–427

45. Jackson et al., 2012 L.E. Jackson, T.M. Bowles, A.K. Hodson, C. Lazcano Soil microbial-root and microbial–rhizosphere processes to increase nitrogen availability and retention in agroecosystems Curr. Opin. Environ. Sustain., 4 (2012), pp. 517–522

46. Kallenbach and Grandy, 2011 C. Kallenbach, A.S. Grandy Controls over soil microbial biomass responses to carbon amendments in agricultural systems: a meta-analysis Agric. Ecosyst. Environ., 144 (2011), pp. 241–252

47. Kong et al., 2005 A.Y.Y. Kong, J. Six, D.C. Bryant, R.F. Denison, C. van Kessel The relationship between carbon input, aggregation, and soil organic carbon stabilization in sustainable cropping systems Soil Sci. Soc. Am. J., 69 (2005), pp. 1078–1085

48. Kong et al., 2011 A.Y.Y. Kong, K.M. Scow, A.L. Córdova-Kreylos, W.E. Holmes, J. Six Microbial community composition and carbon cycling within soil microenvironments of conventional, low-input, and organic cropping systems Soil Biol. Biochem., 43 (2011), pp. 20–30

49. Kramer et al., 2006 S.B. Kramer, J.P. Reganold, J.D. Glover, B.J.M. Bohannan, H.A. Mooney Reduced nitrate leaching and enhanced denitrifier activity and efficiency in organically fertilized soils Proc. Natl. Acad. Sci. U. S. A., 103 (2006), pp. 4522–4527

50. Lauber et al., 2008 C.L. Lauber, M.S. Strickland, M.A. Bradford, N. Fierer The influence of soil properties on the structure of bacterial and fungal communities across land-use types Soil Biol. Biochem., 40 (2008), pp. 2407–2415

51. Lazcano et al., 2008 C. Lazcano, M. Gómez-Brandón, J. Domínguez Comparison of the effectiveness of composting and vermicomposting for the biological stabilization of cattle manure Chemosphere, 72 (2008), pp. 1013–1019

52. Mäder et al., 2002 P.D.A. Mäder, A. Fließbach, D. Dubois, L. Gunst, P. Fried, U. Niggli Soil fertility and biodiversity in organic farming Science, 296 (2002), pp. 1694–1697

53. Marriott and Wander, 2006 E.E. Marriott, M. Wander Qualitative and quantitative differences in particulate organic matter fractions in organic and conventional farming systems Soil Biol. Biochem., 38 (2006), pp. 1527–1536

54. Marschner et al., 2003 P. Marschner, E. Kandeler, B. Marschner Structure and function of the soil microbial community in a long-term fertilizer experiment Soil Biol. Biochem., 35 (2003), pp. 453–461

55. Minoshima et al., 2007 H. Minoshima, L.E. Jackson, T.R. Cavagnaro, S. Sánchez-Moreno, H. Ferris, S.R. Temple, S. Goyal, J.P. Mitchell Soil food webs and carbon dynamics in response to conservation tillage in California Soil Sci. Soc. Am. J., 71 (2007), p. 952

56. Miranda et al., 2001 K.M. Miranda, M.G. Espey, D.A. Wink A rapid, simple spectrophotometric method for simultaneous detection of nitrate and nitrite Nitric Oxide: Biol. Chem., 5 (2001), pp. 62–71

57. Moeskops et al., 2010 B. Moeskops, D. Buchan, S. Sleutel, L. Herawaty, E. Husen, R. Saraswati, D. Setyorini, S. De Neve Soil microbial communities and activities under intensive organic and conventional vegetable farming in West Java, Indonesia Appl. Soil Ecol., 45 (2010), pp. 112–120

58. Moeskops et al., 2012 B. Moeskops, D. Buchan, S. Van Beneden, V. Fievez, S. Sleutel, M.S. Gasper, T. D'Hose, S. De Neve The impact of exogenous organic matter on SOM contents and microbial soil quality Pedobiologia, 55 (2012), pp. 175–184

59. Oksanen et al., 2012 J. Oksanen, F.G. Blanchet, R. Kindt, P. Legendre, P.R. Minchin, R.B. O'Hara, G.L. Simpson, P. Solymos, M.H.H. Stevens, H. Wagner Vegan: Community Ecology Package. R Package Version 2.0-5 (2012) http://CRAN.R-project.org/package=vegan

60. Olander and Vitousek, 2000 L. Olander, P. Vitousek Regulation of soil phosphatase and chitinase activity by N and P availability Biogeochemistry, 49 (2000), pp. 175–190

61. Olsen and Sommers, 1982 S.R. Olsen, L.E. Sommers Phosphorus,in: A.L. Page (Ed.), Methods of Soil Analysis, Part 2, Agron. Monogr. (second ed.), vol. 9, ASA and SSSA, Madison, WI (1982), pp. 403–430

62. Parham and Deng, 2000 J. Parham, S. Deng Detection, quantification and characterization of β-glucosaminidase activity in soil Soil Biol. Biochem., 32 (2000), pp. 1183–1190

63. Paul and Williams, 2005 J. Paul, B. Williams Contribution of α-amino N to extractable organic nitrogen (DON) in three soil types from the Scottish uplands Soil Biol. Biochem., 37 (2005), pp. 801–803

64. Peres-Neto et al., 2006 P.R. Peres-Neto, P. Legendre, S. Dray, D. Borcard Variation partitioning of species data matrices: estimation and comparison of fractions Ecology, 87 (2006), pp. 2614–2625

65. Piotrowska and Wilczewski, 2012 A. Piotrowska, E. Wilczewski Effects of catch crops cultivated for green manure and mineral nitrogen fertilization on soil enzyme activities and chemical properties Geoderma, 189–190 (2012), pp. 72–80

66. R Development Core Team, 2012 R Development Core Team R: a Language and Environment for Statistical Computing R Foundation for Statistical Computing, Vienna, Austria (2012)

67. Reed and Martiny, 2013 H.E. Reed, J.B.H. Martiny Microbial composition affects the functioning of estuarine sediments ISME J., 7 (2013), pp. 868–879

68. Ros et al., 2009 G.H. Ros, E. Hoffland, C. van Kessel, E. Temminghoff Extractable and dissolved soil organic nitrogen – a quantitative assessment Soil Biol. Biochem., 41 (2009), pp. 1029–1039

69. Saiya-Cork et al., 2002 K. Saiya-Cork, R.L. Sinsabaugh, D. Zak The effects of long term nitrogen deposition on extracellular enzyme activity in an Acer saccharum forest soil Soil Biol. Biochem., 34 (2002), pp. 1309–1315

70. Schafer and Singer, 1976 W. Schafer, M. Singer Influence of physical and mineralogical properties on swelling of soils in Yolo County, California Soil Sci. Soc. Am. J., 40 (1976), pp. 557–562

71. Schimel and Bennett, 2004 J.P. Schimel, J. Bennett Nitrogen mineralization: challenges of a changing paradigm Ecology, 85 (2004), pp. 591–602

72. Schipanski and Drinkwater, 2012 M.E. Schipanski, L.E. Drinkwater Nitrogen fixation in annual and perennial legume-grass mixtures across a fertility gradient Plant Soil, 357 (2012), pp. 147–159

73. Schutter and Dick, 2000 M.E. Schutter, R.P. Dick Comparison of fatty acid methyl ester (FAME) methods for characterizing microbial communities Soil Sci. Soc. Am. J., 64 (2000), pp. 1659–1668

74. Senwo and Tabatabai, 1996 Z.N. Senwo, M.A. Tabatabai Aspartase activity of soils Soil Sci. Soc. Am. J., 60 (1996), pp. 1416–1422

75. Seufert et al., 2012 V. Seufert, N. Ramankutty, J.A. Foley Comparing the yields of organic and conventional agriculture Nature, 485 (2012), pp. 229–232

76. Sinsabaugh et al., 2005 R.L. Sinsabaugh, M.E. Gallo, C. Lauber, M.P. Waldrop, D.R. Zak Extracellular enzyme activities and soil organic matter dynamics for northern hardwood forests receiving simulated nitrogen deposition Biogeochemistry, 75 (2005), pp. 201–215

77. Sinsabaugh et al., 2008 R.L. Sinsabaugh, C.L. Lauber, M.N. Weintraub, B. Ahmed, S.D. Allison, C. Crenshaw, A.R. Contosta, D. Cusack, S. Frey, M.E. Gallo, T.B. Gartner, S.E. Hobbie, K. Holland, B.L. Keeler, J.S. Powers, M. Stursova, C. Takacs-Vesbach, M.P. Waldrop, M.D. Wallenstein, D.R. Zak, L.H. Zeglin Stoichiometry of soil enzyme activity at global scale Ecol. Lett., 11 (2008), pp. 1252–1264

78. Smukler et al., 2008 S.M. Smukler, L.E. Jackson, L. Murphree, R. Yokota, S.T. Koike, R.F. Smith Transition to large-scale organic vegetable production in the Salinas Valley, California Agric. Ecosyst. Environ., 126 (2008), pp. 168–188

79. Smukler et al., 2010 S.M. Smukler, S. Sánchez-Moreno, S.J. Fonte, H. Ferris, K. Klonsky, A.T. O'Geen, K.M. Scow, K.L. Steenwerth, L.E. Jackson Biodiversity and multiple ecosystem functions in an organic farmscape Agric. Ecosyst. Environ., 139 (2010), pp. 80–97

80. Soil Survey Staff, 2011 Soil Survey Staff, Natural Resources Conservation Service, USDA Soil Survey Geographic (SSURGO) Database for Yolo Co., California (2011) Available at: http://soildatamart.nrcs.usda.gov (accessed 06.03.11.)

81. Stromberger et al., 2012 M.E. Stromberger, A.M. Keith, O. Schimdt Distinct microbial and faunal communities and translocated carbon in Lumbricus terrestris drilospheres Soil Biol. Biochem., 46 (2012), pp. 155–162

82. Štursová and Baldrian, 2010 M. Štursová, P. Baldrian Effects of soil properties and management on the activity of soil organic matter transforming enzymes and the quantification of soil-bound and free activity Plant and Soil, 338 (2010), pp. 99–110

83. Syswerda et al., 2012 S.P. Syswerda, B. Basso, S.K. Hamilton, J.B. Tausig, G.P. Robertson Long-term nitrate loss along an agricultural intensity gradient in the Upper Midwest USA Agric. Ecosyst. Environ., 149 (2012), pp. 10–19

84. Tabatabai, 1994 M. Tabatabai Soil enzymes R. Weaver, J. Angle, P. Bottomley (Eds.), Methods of Soil Analysis, Part 2: Microbiological and Biochemical Properties, Soil Science Society of America, Madison, WI (1994), pp. 775–833

85. Tiemann and Billings, 2010 L.K. Tiemann, S.A. Billings Indirect effects of nitrogen amendments on organic substrate quality increase enzymatic activity driving decomposition in a mesic grassland Ecosystems, 14 (2010), pp. 234–247

86. Vance et al., 1987 E. Vance, P. Brookes, D. Jenkinson An extraction method for measuring soil microbial biomass C Soil Biol. Biochem., 19 (1987), pp. 703–707

87. Vasseur et al., 2013 C. Vasseur, A. Joannon, S. Aviron, F. Burel, J.-M. Meynard, J. Baudry The cropping systems mosaic: how does the hidden heterogeneity of agricul-

tural landscapes drive arthropod populations? Agric. Ecosyst. Environ., 166 (2013), pp. 3–14

88. Wickham, 2009 H. Wickham ggplot2: Elegant Graphics for Data Analysis Springer, New York (2009)

89. Williams and Hedlund, 2013 A. Williams, K. Hedlund Indicators of soil ecosystem services in conventional and organic arable fields along a gradient of landscape heterogeneity in southern Sweden Appl. Soil Ecol., 65 (2013), pp. 1–7

90. Wu et al., 1990 J. Wu, R.G. Joergensen, B. Pommerening, R. Chaussod, P.C. Brookes Measurement of soil microbial biomass C by fumigation–extraction: an automated procedure Soil Biol. Biochem., 22 (1990), pp. 1167–1169

91. Young-Mathews et al., 2010 A. Young-Mathews, S.W. Culman, S. Sánchez-Moreno, A. Toby O'Geen, H. Ferris, A.D. Hollander, L.E. Jackson Plant–soil biodiversity relationships and nutrient retention in agricultural riparian zones of the Sacramento Valley, California Agrofor. Syst., 80 (2010), pp. 41–60

92. Yu et al., 2002 Z. Yu, Q. Zhang, T. Kraus, R. Dahlgren, C. Anastasio, R.J. Zasoski Contribution of amino compounds to dissolved organic nitrogen in forest soils Biogeochemistry, 61 (2002), pp. 173–198

93. Zelles, 1996 L. Zelles Fatty acid patterns of microbial phospholipids and lipopolysaccharides F. Schinner, R. Öhlinger, E. Kandeler, R. Margesin (Eds.), Methods in Soil Biology, Springer, Berlin (1996), pp. 80–93

94. Zelles, 1999 L. Zelles Fatty acid patterns of phospholipids and lipopolysaccharides in the characterisation of microbial communities in soil: a review Biol. Fertil. Soils, 29 (1999), pp. 111–129

There are several supplemental files that are not available in this version of the article. To view this additional information, please use the citation information cited on the first page of this chapter.

CHAPTER 14

NITRATE LEACHING FROM INTENSIVE ORGANIC FARMS TO GROUNDWATER

O. DAHAN, A. BABAD, N. LAZAROVITCH, E. E. RUSSAK, AND D. KURTZMAN

14.1 INTRODUCTION

Developing efficient productive agriculture, while preserving groundwater quality, is one of the most important challenges in water resource sustainability. On the one hand, developing agriculture is straightforward wherever agricultural input, such as water and nutrients, is unlimited. On the other hand, productive agriculture must inherently include the leaching of excess lower quality water below the root zone to the unsaturated zone and ultimately to the groundwater (Shani et al., 2007; Dudley et al., 2008). As such, maintaining the delicate balance between productive agriculture and groundwater quality requires a broad perspective over different time and dimensional scales. While agricultural productivity is measured on a timescale of seasons (several months to several years), its final impact on

This chapter was originally published under the Creative Commons Attribution License. Dahan O, Babad A, Lazarovitch N, Russak EE, and Kurtzman D. Nitrate Leaching From Intensive Organic Farms to Groundwater. Hydrology and Earth System Sciences *18 (2014). doi:10.5194/hess-18-333-2014.*

groundwater is a long-term cumulative process with a timescale of years to decades.

Public awareness of healthy food products that are free of chemical additives, along with a worldwide demand to reduce industrial pollution, has led, in recent years, to the development of organic farming (http//www.organiccenterewales.org.uk/). Although numerous studies have questioned organic agriculture's efficiency (Seufert et al., 2012), sustainability (Trewavas, 2001) and health (Jensen et al., 2012) aspects, organic food markets seem to be thriving in developed countries, as their output is perceived by the public to be healthier for both consumers and the environment. This type of agriculture depends mainly on fertilizers from biological sources, such as composted animal manure. Nevertheless, modern agriculture, whether practiced with conventional or organic methods, needs to reach the goals of mass production, i.e., large quantities and high quality, to satisfy market demand while maintaining economic standards of profitability. This goal is usually achieved through intensive agriculture in greenhouses where irrigation water and fertilizers are implemented in excess to satisfy crop demand and maximize productivity. In arid and semi-arid regions, where the climate is warm enough, intensive agriculture in greenhouses operates year round. Moreover, in many of these areas the agriculture is heavily dependent on groundwater resources for irrigation and therefore its quality is of great importance.

Mass production through intensive organic farming is very similar to conventional agriculture in its use of agricultural machinery and modern irrigation techniques; the main differences between the two approaches lie in fertilization and pest-control methodologies (http://www.epa.gov/oecaagct/torg.html; EPA, 2013). Unfortunately, the development of intensive agriculture is often associated with the long-term deterioration of groundwater quality, which is expressed mainly in elevated concentrations of nitrate and salinity (Vitousek et al., 2009; Burow et al., 2010; Kurtzman and Scanlon, 2011; Melo et al., 2012; Morari et al., 2012). Groundwater pollution is usually attributed to a very large array of chemicals. Nevertheless, on a global scale the main cause for drinking-water well shutdowns is a high nitrate concentration in the aquifer water (Osenbruck et al., 2006; Kourakos et al., 2012; Liao et al., 2012; Kurtzman et al., 2013).

The potential for groundwater contamination by nitrate from intensive agriculture is well known (Oren et al., 2004; Vazquez et al., 2006; Thompson et al., 2007). Two main approaches are often used for characterizing nitrate leaching from agricultural fields: (1) characterization of the chemical composition of the soil pore water in shallow depths under the root zone, as may be obtained by application suction lysimeters or sediment samples (Feaga et al., 2010), and (2) determining the cumulative long-term impact on groundwater as may be obtained from the chemical composition of well water (Harter et al., 2002). Apparently, the chemical characteristics of the root zone pore water may vary dramatically in timescales of days to seasons, according to irrigation patterns, fertilizer applications and crop-growing phases. However, the cumulative impact on groundwater develops in timescales of years to decades. The time lag between the initiations of a contamination event near the land surface to its detection in the aquifer water depends on the mechanisms controlling flow and transport in the vadose zone. The ability to characterize flow and transport processes in the vadose zone was recently improved following the development of a vadose zone monitoring system (VMS) that provides realtime, in situ information on the hydraulic and chemical state of the percolating water across the entire vadose zone. Up to now, the VMS has been implemented in a variety of hydrological setups, including (1) flood water percolation (Dahan et al., 2007, 2008; Amiaz et al., 2011), (2) rain water percolation through thick unsaturated sand and clay formations (Rimon et al., 2007, 2011b; Baram et al., 2012a), and (3) solute transport in the vadose zone (Dahan et al., 2009; Rimon et al., 2011a; Baram et al., 2012b).

A newly established agricultural area that has recently been modified from non-intensive open field agriculture to intensive organic and conventional agriculture in greenhouses provided a unique opportunity to investigate the contamination potential of these two agricultural regimes. As such, the main objective of the study was to compare the groundwater pollution potential of organic versus conventional greenhouses as it is expressed through the downleaching of nitrate through the vadose zone underlying these farms. The study was conducted using VMSs that allowed in situ monitoring of the unsaturated zone under selected organic and conventional greenhouses.

14.2 METHOD

14.2.1 STUDY AREA

The study area consists of 100 ha of new greenhouses that were constructed during 2008–2009 on land that had previously been cultivated for 3126 decades under a non-intensive growing regime (mostly rain-fed open field crops). The site is located on the Mediterranean coastal plain, south of the city of Ashkelon, Israel (3123–6 km from the sea shoreline). Most of the greenhouses in this area produce high-quality vegetables year round through organic methods (80 %), while the rest practice conventional methods.

Underneath these agricultural fields (31215–30m below the surface) lies a phreatic sandy aquifer that is characterized by high water quality (Cl$^-$ <200 mg L^{-1}; NO$_3^-$ <40 mg L^{-1}). This part of the aquifer is an important water source for the region. Its water is used for domestic and agricultural purposes through a large number of pumping wells. In addition, the aquifer in the region serves as an underground storage area through the artificial infiltration of flood water from the Shikma ephemeral stream and the water surplus from the national water carrier.

The climate in the area is Mediterranean with an annual average precipitation of 458mm (Israeli Water Authority, 2013). The majority of the rain events take place during the winter, between December and February. The annual average temperature is 20.2 C. The coldest month is January with average maximum and minimum temperatures of 17.2 C and 8.1 C, respectively. The warmest month is August with average maximum and minimum temperatures of 31.1 and 21.4 C, respectively (Israel Meteorological Service, 2013). The pan evaporation rate changes from an average of 2.2mm per day in January to 7.5mm per day in July (Israel Meteorological Service, 2013). Note that the studied sites are all located inside large greenhouses (3121 ha each) that are not influenced by the rain pattern, and their interior climate is warmer and more humid than external natural conditions. Most of the cultivated areas are located in large interdune valleys, surrounded by sandy dunes with sparse vegetation. The stratigraphic cross section of both sites is characterized by interchanges of sandy clay loam to clay loam layers. Yet the vadose zone under the organic farm includes more presence of clay loam layers.

TABLE 1: Vertical distribution of monitoring units across the vadose zone of the organic and conventional farms.

Farm	Probe	Depth to monitoring unit center (m)									Depth to water table (m)
Organic	FTDR	0.2	0.4	0.9	1.9	2.9	4.9	7.8	10	12.8	15
	VSP	0.2	0.4	1.3	2.3	3.3	5.3	8.3	10.5	13.3	
Conven-tional	FTDR	0.2	0.4	0.7	1.7	2.7	5.7	9.6	14.6	19.6	26
	VSP	0.2	0.4	1	2	3	6	10	15	20	

14.2.2 MONITORING SETUP

Two representative greenhouses, organic and conventional, that specialize in growing vegetables, such as cherry tomatoes, peppers and zucchini, were selected and instrumented with VMSs (Fig. 1). Technical descriptions of VMS structure, performance and installation procedures have been previously presented in other publications (Rimon et al., 2007; Dahan et al., 2008, 2009; Rimon et al., 2011a). To avoid overloading this manuscript with technical information, only a brief description will be presented here. The VMS is composed of a flexible sleeve installed in uncased slanted (35) boreholes hosting multiple monitoring units at various depths. Each monitoring unit has a flexible time domain reflectometry sensor (FTDR), for continuous measurements of sediment water content, and vadose zone sampling ports (VSP) for frequent collection of pore-water samples from the entire unsaturated zone. Slanted installation ensures that each monitoring unit faces an undisturbed sediment column that extends from land surface to the probe depth. In the borehole, the flexible sleeve is filled with high-density solidifying material to ensure sleeve expansion for proper attachment of the monitoring units to the borehole's irregular walls, sealing its entire void and preventing potential crosscontamination by preferential flow along the borehole. Each VMS included nine monitoring units distributed vertically and laterally along the entire vadose zone cross section (Table 1). Installation orientation was aligned with the growing rows. The representativeness of measurements made by the VMS, with respect to vadose zone sedimentological heterogeneity, has already been discussed in previous publications (Dahan et al., 2007; Rimon et al., 2007).

The VMS provided continuous information from 18 fixed points from the vadose zone underlying the two selected sites throughout a sampling period of 18 months. Each monitoring unit provided high-resolution information from multiple points that are located away from each other vertically and horizontally. Therefore the integrated data from the VMS should be regarded as representative of a wider zone rather than a vertical profile (Fig. 1). To overcome potential bias due to local heterogeneity, the VMS results were compared with the chemical composition of sediment samples that

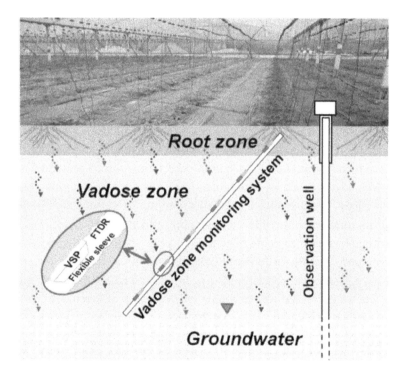

FIGURE 1: Schematic illustration of a Vadose zone Monitoring System (VMS) installed under the studied greenhouse. The VMS includes a flexible sleeve hosting multiple monitoring units. The monitoring units are composed of water content sensors (FTDR) and pore water sampling ports (VSP) that are distributed across the entire vadose zone from land surface to groundwater.

were obtained from additional locations in each greenhouse, and from additional greenhouses owned and cultivated by different farmers. In addition to the VMS that monitored the vadose zone, each site was instrumented with an observation well that penetrated the upper phreatic groundwater with screens to 5m below the water table (Fig. 1). In order to validate the results obtained by the VMS, nitrate profiles in the vadose zone pore water were compared with the nitrate concentrations in sediment samples from three additional boreholes in each greenhouse. Samples were collected at a 0.5m depth resolution from the top 3m of the profile using a standard hand auger.

14.2.3 CHEMICAL AND ISOTOPIC ANALYSIS

Nitrate concentrations in the water samples were determined using ion chromatography (DIONEX, 4500I). Soil samples were extracted by KCL and analyzed for ammonium by the Nesslerization method (APHA, 1989), for nitrite using the colorimetric method, and for nitrate using the second-derivative method (APHA, 2005). Total nitrogen in the soil was analyzed following the Kjeldahl method (Benton, 1999). The isotopic composition of ^{15}N and ^{18}O of nitrate in the water samples was determined through nitrate reduction to nitrogen dioxide, which was then analyzed using a gas mass spectrometer (McIlvin and Altabet, 2005).

Throughout the study, the following data were collected: (1) crop type and growing cycle, (2) irrigation quantity and quality, (3) fertilization regime, (4) temporal variation of the vadose zone water-content profile, (5) chemical composition of the vadose zone and underlying groundwater, and (5) isotopic composition (^{15}N–^{18}O) of nitrate obtained from the vadose zone pore water.

14.3 RESULTS AND DISCUSSION

14.3.1 AGROTECHNICAL REGIME

A comparison of agrotechnical regimes implemented in the organic and conventional greenhouses showed very similar general inputs. For example, during a single growing season for cherry tomato, which extended for 183 and 190 days in the organic and conventional greenhouses, respectively, the total irrigation amounts (applied through drip irrigation) were 3440 and 3570m^3 ha^{-1}, respectively (Fig. 2). Throughout the growing season, irrigation in the conventional farm gradually increased with the plant growing phase until the middle of the season, while on the organic farm irrigation is kept relatively constant throughout the entire season. In both greenhouses, the growing season began with an establishment irrigation of 200–240m^3 ha^{-1}. In general, establishment irrigation is implemented to prepare the upper soil for the new planting and down-leaching of salts such as sodium chloride that have accumulated in the root zone during the previous season (Yin et al., 2007; Ben-Gal et al., 2008). In addition, in the organic regime, the establishment irrigation enhances the mineralization of nutrients from the compost in the soil for plant uptake.

Total N-fertilizer implementation in both greenhouses from their establishment date (3164 yr) was very similar, 3800 and 3700 kgNha^{-1} for the organic and conventional greenhouses, respectively. In the organic greenhouse, 98% of the N was applied as compost (processed from dairy farms and poultry manure) that was mixed with the soil between the growing seasons, and guano (seabird excrement) that was embedded in the top soil by the plant stems during the growing season. In the conventional greenhouse, on the other hand, only 45% of the N was applied as compost, mainly as a soil amendment in the early stages after the establishment of the greenhouse, while the rest was provided through the drip irrigation system, as mineral liquid fertilizer from industrial sources.

14.3.2 NITRATE LEACHING IN THE VADOSE ZONE

Water samples from the vadose zone and groundwater were collected at both sites every 3 to 6 weeks for 19 months (total of 262 water samples). A comparison of nitrate concentrations in the vadose zone pore water at the two sites throughout all sampling campaigns exhibited a striking difference (Fig. 3). Nitrate concentrations below the root zone (>1 m) under the organic greenhouse exhibited a very high average concentration of 357 mg L^{-1}, with a peak average concentration of 724 mg L^{-1} at a depth

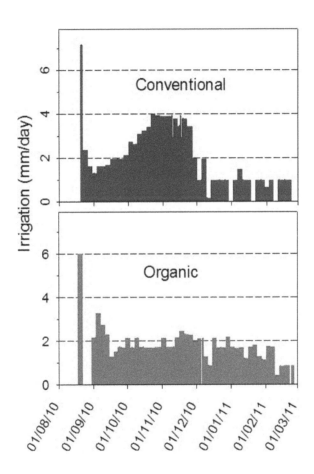

FIGURE 2: Daily irrigation during a single growing season of tomato in the organic and conventional greenhouses.

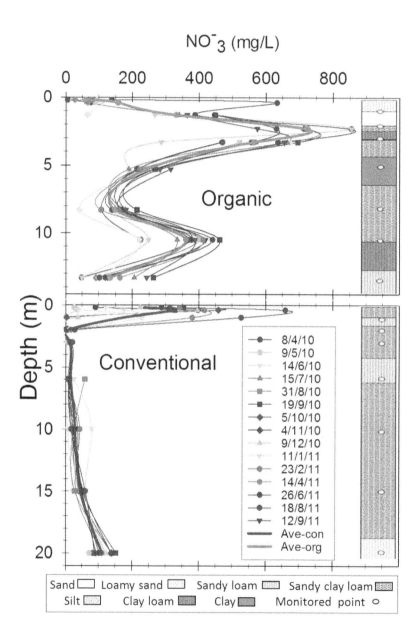

FIGURE 3: Nitrate concentrations in the water samples collected by the VMS from the vadose zone underlying the organic and conventional greenhouses along with the lithological cross section of each site. Nitrate profiles were established through 15 sampling campaigns over an 18 month period.

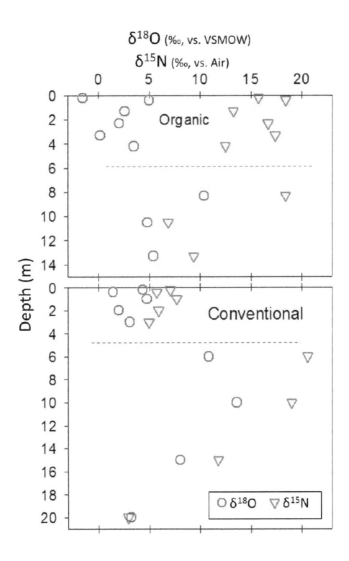

FIGURE 4: $\partial^{18}O$ and $\partial^{15}N$ profiles of nitrate in the water samples obtained from the vadose zone underlying the organic and conventional greenhouses (VSMOW refers to Vienna Standard Mean Ocean Water).

of 2.5 m. Nitrate concentrations below the root zone of the conventional greenhouse were much lower, with an average value of only 37.5 mg L^{-1}. A closer look at the upper part of the unsaturated zone of both sites exhibited an opposite concentration pattern. Nitrate concentrations under the conventional greenhouse showed a high average concentration of 270 mg L^{-1} in the root zone (<1 m) that quickly reduced in the deeper part of the vadose zone. This pattern is preferable for both agronomic and environmental reasons as nitrogen is available for root uptake in the shallow zone with minimal down migration of nitrate to the deeper parts of the vadose zone. On the other hand, under the organic greenhouse, the root zone suffered from a relative shortage of nitrate with an average concentration of a similar amount of N-fertilizers and use a similar amount of water. Nevertheless, the nitrate concentration profiles appear to be very different (Fig. 3). Identification of the nitrate sources in the vadose zone might be examined through the isotopic composition of $\partial^{15}N$ and $\partial^{18}O$ in the nitrate molecules. An isotopic analysis of nitrate in the water samples from the vadose zone exhibited significant differences in isotopic composition in the upper ~5m of each site (Fig. 4). $\partial^{15}N$ values in the upper part of the vadose zone underlying the organic greenhouse are rather heavier (average $\partial^{15}N=$ 15.6 ‰; STD = 2.12) in comparison with the depleted $\partial^{15}N$ values observed under the conventional greenhouse (average $\partial^{15}N=$ 6.26 ‰; STD = 0.98). The differences in isotopic composition in the upper part of the vadose zone correspond well with the expected isotopic values of processed manure versus synthetic fertilizer and natural soil nitrogen (Fig. 5) (Kendall and McDonnell, 1999).

One direct implication of the isotopic composition is the accurate identification of the penetration depth of the agricultural leachates. Though the nitrate concentration profiles are very different (Fig. 3), the isotopic fingerprint shows that after four years cultivation in the new established greenhouses agricultural leachates penetrated to a depth of ~5–7m in both greenhouses (Fig. 4). Although the impacted depth in both greenhouses is similar, the concentration pattern beneath the root zone in the conventional greenhouse shows significantly lower concentrations as compared with those of the organic greenhouse.

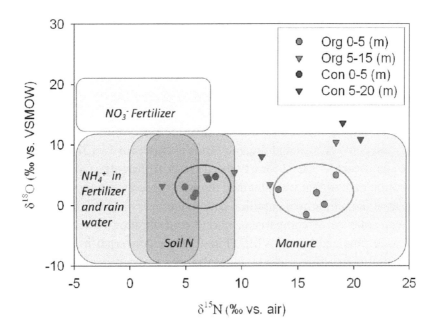

FIGURE 5: Isotopic composition of $\partial^{15}N$ vs. $\partial^{18}O$ of nitrate in the water samples from the vadose zone underlying organic (Org) and conventional (Con) greenhouses with respect to typical values in common N sources (VSMOWrefers to Vienna Standard Mean OceanWater).

14.3.4 RESULTS VALIDATION

To validate the representativeness of the results that were obtained by the VMS a sediment sampling campaign that included three boreholes to the upper vadose zone was carried in each greenhouse. Note that while the VMS collects the sediment mobile pore water, the sediment sample extract provides the entire solute capacity of the sediment, which includes both the mobile and immobile phases (Rimon et al., 2011a). Comparing nitrate concentrations that were obtained by the VMS with those obtained from the sediment extract shows that nitrate concentrations in sediment samples from the upper part of the vadose zone under the organic greenhouse gradually increase from an average value of 8.8 mg kg^{-1} in the root zone to 85.8 mg kg^{-1} at a depth of 3m (Fig. 6). However, nitrate concentrations in sediment samples obtained from the conventional greenhouse were significantly lower, with an average value of 14.7 mg kg^{-1} over the entire measured profile. The difference between the mean nitrate concentration under the conventional and organic farms at depths of 1 to 3m was found to be significant (P = 0.03 for a two-tail t test). Apparently, the nitrate concentration pattern that was obtained through the sediment samples is very similar to the nitrate concentration pattern that was observed by the VMS. For convenience of comparison, nitrate concentration in water samples that were obtained by the VMS (Fig. 3) were converted in Fig. 6 from concentrations in pore water to concentrations in dry soil.

To extend representativeness of the findings, the survey was extended to six greenhouses, three organic and three conventional. All six greenhouses were established at the same time, in the same area (within a distance of up to 5 km from each other). To avoid potential bias due to farmers' specific working methods, the extended survey included farms that were all owned and operated by different farmers. The top 3m of the sediment in each greenhouse was sampled in two boreholes with a depth sampling resolution of 0.5m (total of 72 sediment samples). The nitrate concentrations in the sediment samples from the extended survey showed ambiguous results (Fig. 7). Two organic greenhouses exhibited nitrate concentrations that increased with depth, from 27.1 mg kg^{-1} at 0.5m to 133 mg kg^{-1} at 3 m. On the other hand, the nitrate concentration patterns under all the conventional greenhouses exhibited lower average concentrations, ranging between 34.2 to 44.1 mg kg^{-1}. This difference in nitrate profile is very similar to the pattern that was observed by the VMS over a long period of continuous monitoring (Fig. 2). Nevertheless, nitrate concentration profiles under one of the organic greenhouses exhibited a rather low concentration profile with an average value of 29.9 mg kg^{-1} (ranging between 27.4 to 32.9 mg kg^{-1}). These results, however, did not match the results obtained from the other organic greenhouses. A closer examination of the fertilization methods practiced by the farmers showed that in the two greenhouses where nitrate concentrations increased with depth, the N-fertilizer source had been solid organic matter (compost and guano) mixed into the soil, a method that is commonly practiced in organic farming. On the other hand, in the organic greenhouse that exhibited low nitrate concentrations, N-fertilization relied mainly on the water extracts of guano excrement that had been applied directly through the drip irrigation throughout the growing season, and not as solid compost mixed with the soil prior to the growing season. Fertilization with the irrigation water (fertigation) throughout the growing season in response to the plant nutrient demand is a common practice in conventional agriculture.

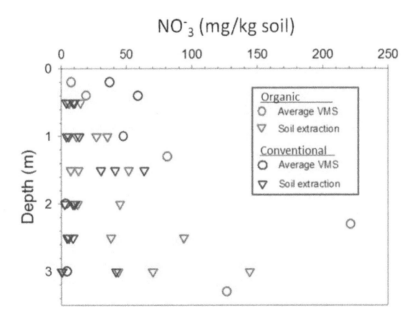

FIGURE 6: Nitrate concentrations in sediment samples (soil extraction) and pore water (average VMS) in the vadose zone underlying organic and conventional greenhouses.

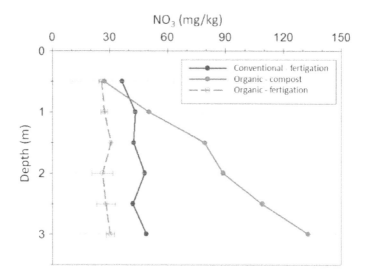

FIGURE 7: Average nitrate concentration in sediment samples obtained from the upper part of the vadose zone underlying: (1) organic greenhouses that implement compost as the sole fertilizer, (2) conventional greenhouses that rely on fertigation, and (3) organic greenhouses that use fertigation rather than solid compost.

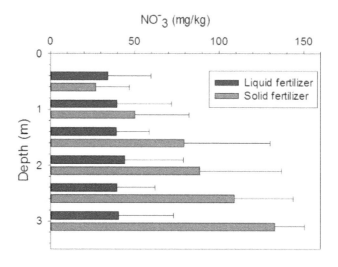

FIGURE 8: Nitrate concentrations in the vadose zone underlying greenhouses fertilized by solid fertilizers as compost and guano and liquid fertilizers.

14.3.5 FERTILIZATION METHOD AND NITRATE LEACHING

Reexamination of the nitrate concentration profiles in the vadose zone under the greenhouses suggests categorizing the potential down-leaching of nitrate according to the fertilization method rather than the general agrotechnical regime of organic versus conventional agriculture. As such the differences in nitrate profiles may be attributed to solid versus liquid fertilizer application. Solid fertilization refers to the application of organic matter such as compost directly to the soil, primarily before planting, as commonly practiced in organic agriculture. On the other hand, liquid fertilization refers to application of fertilizers via the irrigation system throughout the growing season (fertigation), as is commonly practiced in conventional agriculture. Under these two categories, the differences in nitrate concentrations in deep sections (>1 m) of the vadose zone were much more significant ($P = 0.0002$), showing a tremendous increase in nitrate concentration with depth in farms relying on solid fertilizer application (Fig. 8). Examination of the total nitrogen (TN) amount that was observed in the sediment samples from the upper section of the vadose zone under all of the greenhouses, organic and conventional, showed very similar values. Note that among the main nitrogen forms in the soil (nitrate, ammonium and organic-N), nitrate is much more mobile compared with other forms. Moreover, under aerobic conditions, which typically prevail in the vadose zone of semi-arid regions, nitrate is considered stable, and therefore, its transport across the vadose zone to groundwater persists. Ammonium and nitrite concentration in pore-water samples were either negligible or below detection limit and therefore not discussed here.

The observations of enhanced down-leaching of nitrate under intensive farms that rely on solid fertilizers may be attributed to unsynchronized nutrient release from the compost to the soil with respect to the nutrient uptake capacity of the plants. During the early stages of the growing season, an excess of irrigation water, with respect to the plant water and the nutrient uptake, creates a nutrient surplus in the soil (Pang and Letey, 2000). At this stage, the soil water content remains high as does the nutrient concentration, which is released from the compost to the wet soil. However, in the early stages of the growing season water and nutrient consumption by the undeveloped roots of the young plants is limited.

Therefore down-leaching of water enriched with nitrate to the deeper parts of the vadose zone is unavoidable. On the other hand, accurate fertigation methods that synchronize the fertilizer's implementation with the nutrient uptake capacity of the plants, as is commonly practiced in conventional agriculture, dramatically reduce the amount of nitrate leaching through the vadose zone to the groundwater.

14.4 SUMMARY AND CONCLUSIONS

Comparison of the groundwater pollution potential of greenhouses that grow year round vegetables under intensive regimes shows that commercial farms that rely on compost as the main fertilizer source, as commonly practiced in organic agriculture, result in substantial down-leaching of nitrate compared with farms that rely on fertigation methods, as commonly practiced in conventional agriculture.

The study implemented vadose zone monitoring technology that allowed frequent sampling of the sediment pore water at multiple points across the entire vadose cross section from the root zone to the water table over a long time period. Nitrate concentration profiles under the farms that rely on solid fertilizers revealed an increased concentration pattern with a depth to average of 724 mg L^{-1}. On the other hand, concentration profiles in farms that rely on implementation of liquid fertilizers through the irrigation systems during the growing season exhibited a reducing concentration with a depth to average concentration of only 37.5 mg L^{-1}, immediately under the root zone. Isotopic composition $\partial^{15}N$ and $\partial^{18}O$ in nitrate from the vadose zone under the sites confirmed that the high nitrate concentration under the studied organic farms is likely to have originated from composted manure, while the nitrate under the studied conventional greenhouse, though in low concentrations, is likely to be from industrial or natural soil sources.

All studied greenhouses, organic and conventional, were established at the same time, grow similar vegetables, use similar amounts of water and total N-fertilizers, and share most agro-technical practices. Nevertheless, the main difference between the greenhouses is related to the fertilization regime. While organic agriculture in greenhouses relies mostly on solid fertilizers such as compost that is mixed with the top soil prior to plantation, the conventional agriculture usually relies on fertigation methods where liquid fertilizer is implemented through the irrigation system during the growing season. As such, establishment irrigation in intensive organic farms, which is implemented in the early stages of the growing season, resulted in high down-leaching of nitrate that is produced in the wet top soil. This down-leaching is unavoidable since the soil is wet while plants roots are undeveloped and incapable of significant water and nutrient uptake. On the other hand, in typical conventional agriculture fertilizers are implemented along the growing season with the irrigation system according to the plant demand. As such nutrient uptake by the plant is more efficient, and down-leaching of nitrate is minimal. This concept has been approved on one of the organic farms that practice fertigation methods rather than application of solid compost in the soil. In this farm liquid fertilizers that were produced from guano extracts were implemented through the irrigation system. As a result, the vadose zone nitrate profile characteristics on this organic farm were similar to those observed in conventional farms and significantly lower than those observed in organic farms that rely on compost as the main fertilizer.

Long-term decision making on groundwater resource management requires substantial data on the fate of pollutant transport from their sources near land surface across the vadose zone to the groundwater. Data on the link between land use and groundwater quality may be obtained through monitoring technologies that are designed to provide real-time information on the quality of the percolating water across the vadose zone. In this study we have demonstrated how implementation of vadose zone monitoring systems provided valuable information on potential pollution threat to groundwater long before groundwater pollution became evident.

REFERENCES

1. Amiaz, Y., Sorek, S., Enzel, Y., and Dahan, O.: Solute transport in the vadose zone and groundwater during flash floods, Water Resour. Res., 47, W10513, doi:10.1029/2011WR010747, 2011.
2. APHA: Standard Methods for the Examination of Water and Wastewater, 17th Edn., American Public Health Association, Washington, D.C., 1989.

3. APHA: Standard Methods for the Examination of Water and Wastewater, 21st Edn., American Public Health Association, Washington, D.C., 2005.
4. Baram, S., Kurtzman, D., and Dahan, O.:Water percolation through a clayey vadose zone, J. Hydrol., 424, 165–171, 2012a.
5. Baram, S., Arnon, S., Ronen, Z., Kurtzman, D., and Dahan, O.: Infiltration mechanism controls nitrification and denitrification processes under dairy waste lagoon, J. Environ. Qual., 41, 1623– 1632, 2012b.
6. Ben-Gal, A., Ityel, E., Dudley, L., Cohen, S., Yermiyahu, U., Presnov, E., Zigmond, L., and Shani, U.: Effect of irrigation water salinity on transpiration and on leaching requirements: a case study for bell peppers, Agr. Water Manage., 95, 587–597, 2008.
7. Benton, J.: Soil Analysis Handbook of Reference Methods: Soil and Plant Analysis Counsil, Inc. CRC Press LLC, Boca Raton, FL, 1999.
8. Burow, K. R., Nolan, B. T., Rupert, M. G., and Dubrovsky, N. M.: Nitrate in groundwater of the United States, 1991–2003, Environ. Sci. Technol., 44, 4988–4997, 2010.
9. Dahan, O., Shani, Y., Enzel, Y., Yechieli, Y., and Yakirevich, A.: Direct measurements of floodwater infiltration into shallow alluvial aquifers, J. Hydrol., 344, 157–170, 2007.
10. Dahan, O., Tatarsky, B., Enzel, Y., Kulls, C., Seely, M., and Benito, G.: Dynamics of flood water infiltration and ground water recharge in hyperarid desert, Groundwater, 46, 450–461, 2008.
11. Dahan, O., Talby, R., Yechieli, Y., Adar, E., Lazarovitch, N., and Enzel, Y.: In situ monitoring of water percolation and solute transport using a vadose zone monitoring system, Vadose Zone J., 8, 916–925, 2009.
12. Dudley, L. M., Ben-Gal, A., and Lazarovitch, N.: Drainage water reuse: biological, physical, and technological considerations for system management, J. Environ. Qual., 37, 25–35, 2008. EPA: http://www.epa.gov/oecaagct/torg.html, last access: 1 December 2013.
13. Feaga, J. B., Selker, J. S., Dick, R. P., and Hemphill, D. D.: Longterm nitrate leaching under vegetable production with cover crops in the Pacific Northwest, Soil Sci. Soc. Am. J., 74, 186– 195, 2010.
14. Harter, T., Davis, H., Mathews, M. C., and Meyer, R. D.: Shallow groundwater quality on dairy farms with irrigated forage crops, J. Contam. Hydrol., 55, 287–315, 2002.
15. Israel Meteorological Service: http://ims.gov.il, last access: 2 December 2013.
16. Israeli water authority: http://www.water.gov.il/Hebrew/ ProfessionalInfoAndData/ Data-Hidrologeime/DocLib/PerennialData-Coastal02.pdf, last access: 2 December 2013.
17. Jensen, M. M., Jorgensen, H., Halekoh, U., Watzl, B., Thorup-Kristensen, K., and Lauridsen, C.: Health biomarkers in a rat model after intake of organically grown carrots, J. Sci. Food Agr., 92, 2936–2943, 2012.
18. Kendall, C. and McDonnell, J. J.: Isotope Tracers in Catchment Hydrology, Elsevier, Amsterdam, the Netherlands, 1999.
19. Kourakos, G., Klein, F., Cortis, A., and Harter, T.: A groundwater nonpoint source pollution modeling framework to evaluate longterm dynamics of pollutant exceedance probabilities in wells and other discharge locations, Water Resour. Res., 48, W00L13, doi:10.1029/2011WR010813, 2012.

20. Kurtzman, D. and Scanlon, B. R.: Groundwater recharge through vertisols: irrigated cropland vs. natural land, Israel, Vadose Zone J., 10, 662–674, 2011.
21. Kurtzman, D., Shapira, R. H., Bar-Tal, A., Fine, P., and Russo, D.: Nitrate fluxes to groundwater under citrus orchards in a Mediterranean climate: observations, calibrated models, simulations and agro-hydrological conclusions. J. Contam. Hydrol., 151, 93–104, 2013.
22. Liao, L., Green, C. T., Bekins, B. A., and Bohlke, J. K.: Factors controlling nitrate fluxes in groundwater in agricultural areas, Water Resour. Res., 48, W00L09, doi:10.1029/2011WR011008, 2012.
23. McIlvin, M. R. and Altabet, M. A.: Chemical conversion of nitrate and nitrite to nitrous oxide for nitrogen and oxygen isotopic analysis in freshwater and seawater, Anal. Chem., 77, 5589–5595, 2005.
24. Melo, A., Pinto, E., Aguiar, A., Mansilha, C., Pinho, O., and Ferreira, I.: Impact of intensive horticulture practices on groundwater content of nitrates, sodium, potassium, and pesticides, Environ. Monit. Assess., 184, 4539–4551, 2012.
25. Morari, F., Lugato, E., Polese, R., Berti, A., and Giardini, L.: Nitrate concentrations in groundwater under contrasting agricultural management practices in the low plains of Italy, Agr. Ecosys. Environ., 147, 47–56, 2012.
26. Oren, O., Yechieli, Y., Bohlke, J. K., and Dody, A.: Contamination of groundwater under cultivated fields in an arid environment, central Arava Valley, Israel, J. Hydrol., 290, 312–328, 2004.
27. Organic Centre Wales: http://www.organiccentrewales.org.uk/, last access: 1 December 2013.
28. Osenbruck, K., Fiedler, S., Knoller, K., Weise, S. M., Sultenfuss, J., Oster, H., and Pang, X. P., and Letey, J.: Organic farming: challenge of timing nitrogen availability to crop nitrogen requirements, Soil Sci. Soc. Am. J., 64, 247–253, 2000.
29. Rimon, Y., Dahan, O., Nativ, R., and Geyer, S.: Water percolation through the deep vadose zone and groundwater recharge: preliminary results based on a new vadose zone monitoring system, Water Resour. Res., 43, W05402, doi:10.1029/2006WR004855, 2007.
30. Rimon, Y., Nativ, R., and Dahan, O.: Physical and chemical evidence for pore-scale dual-domain flow in the vadose zone, Vadose Zone J., 10, 322–331, 2011a.
31. Rimon, Y., Nativ, R., and Dahan, O.: Vadose zone water pressure variation during infiltration events, Vadose Zone J., 10, 1105–1112, 2011b.
32. Seufert, V., Ramankutty, N., and Foley, J. A.: Comparing the yields of organic and conventional agriculture, Nature, 485, 229–232, 2012.
33. Shani, U., Ben-Gal, A., Tripler, E., and Dudley, L. M.: Plant response to the soil environment: an analytical model integrating yield, water, soil type, and salinity, Water Resour. Res., 43, W08418, doi:10.1029/2006WR005313, 2007.
34. Strauch, G.: Timescales and development of groundwater pollution by nitrate in drinking water wells of the Jahna- Aue, Saxonia, Germany, Water Resour. Res., 42, W121416, doi:10.1029/2006WR004977, 2006.
35. Thompson, R. B., Martinez-Gaitan, C., Gallardo, M., Gimenez, C., and Fernandez, M. D.: Identification of irrigation and N management practices that contribute to nitrate leaching loss from an intensive vegetable production system by use of a comprehensive survey, Agr. Water Manage., 89, 261–274, 2007.

36. Trewavas, A.: Urban myths of organic farming, Nature, 410, 409– 410, 2001.
37. Vazquez, N., Pardo, A., Suso, M. L., and Quemada, M.: Drainage and nitrate leaching under processing tomato growth with drip irrigation and plastic mulching, Agr. Ecosyst. Environ., 112, 313– 323, 2006.
38. Vitousek, P. M., Naylor, R., Crews, T., David, M. B., Drinkwater, L. E., Holland, E., Johnes, P. J., Katzenberger, J., Martinelli, L. A., Matson, P. A., Nziguheba, G., Ojima, D., Palm, C. A., Robertson, G. P., Sanchez, P. A., Townsend, A. R., and Zhang, F. S.: Nutrient imbalances in agricultural development, Science, 324, 1519–1520, 2009.
39. Yin, F., Fu, B. J., and Mao, R. Z.: Effects of nitrogen fertilizer application rates on nitrate nitrogen distribution in saline soil in the Hai River Basin, China, J. Soil. Sediment., 7, 136–142, 2007.

CHAPTER 15

IMPROVING NITROGEN USE EFFICIENCY IN CROPS FOR SUSTAINABLE AGRICULTURE

BERTRAND HIREL, THIERRY TÉTU, PETER J. LEA, AND FRÉDÉRIC DUBOIS

15.1 INTRODUCTION: SOCIOECONOMIC AND ENVIRONMENTAL STAKES

Today, the main method to maintain or restore soil nutrients and increase crop yields is the application of mineral fertilizers such as nitrogen (N). The N used in commercial fertilizers is particularly soluble for easy uptake and assimilation by plants. Because of the simplicity of its storage and handling, N can easily be applied when plants need it most. Mineral fertilizers are now the main source of nutrients applied to soils, even if the contribution of animal manure remains important, especially when there is densely populated livestock nearby. After World War II, N fertilizers have been used extensively to increase crop yield. The use of synthetic N fertilizers has eliminated a major elemental constraint with respect to enriching the soil stock of organic C and N originally managed by organic manure

This chapter was originally published under the Creative Commons Attribution License. Hirel B, Tétu T, Lea PJ and Dubois F. Improving Nitrogen Use Efficiency in Crops for Sustainable Agriculture. Sustainability 3,9 (2011). doi:10.3390/su3091452.

amendments, leguminous cultures and fallow periods. The formation of ammonia and thus synthetic N fertilizers by the Haber–Bosch process was one of the most important inventions of the 20th century, thus allowing the production of food for nearly half of the world population [1,2]. Consequently, a dramatic escalation has occurred in global consumption of synthetic N, from 11.6 million tonnes (Tg) in 1961 to 104 Tg in 2006 [3,4]. Over 40 years, the amount of mineral N fertilizers applied to agricultural crops increased by 7.4 fold, whereas the overall yield increase was only 2.4 fold [5]. This means that N use efficiency, (NUE) which may be defined as the yield obtained per unit of available N in the soil (supplied by the soil + N fertilizer) has declined sharply. This obviously implies that NUE is higher at reduced levels of crop production when the use of N fertilization is much lower. NUE is the product of absorption efficiency (amount of absorbed N/quantity of available N) and the utilization efficiency (yield/absorbed N). For a large number of crops, there is a genetic variability for both N absorption efficiency and for N utilization efficiency [6]. Moreover, the occurrence of interactions between the genotype and the level of N led to the conclusion that the best performing crop varieties at high N fertilization input are not necessarily the best ones when the supply of N is lower [7]. This is mainly because breeding for most crops has been conducted over the last 50 years in the presence of high mineral fertilization inputs, thus missing the opportunity to exploit genetic differences under a low level of mineral or organic N fertilization conditions [8].

In most intensive agricultural production systems, over 50% and up to 75% of the N applied to the field is not used by the plant and is lost by leaching into the soil [9-11]. Some microorganisms are able to improve soil fertility by metabolizing the N that is not absorbed by plants. It is however a lengthy process which involves a major risk because mineral N, especially nitrate (NO_3^-) and urea $\{CO(NH_2)_2\}$ are very soluble and can run off into the surface water or flow into the groundwater. Water contaminated by nitrate is not potable and at high concentrations can be a serious risk for human health [12,13]. Moreover, the water industry must bear additional costs to remove nitrates from groundwater sources [14,15].

The detrimental impacts of nitrate loss from the soil have toxicological implications for animals and humans [16] and also on the environment leading to the eutrophication of freshwater [17] and marine ecosystems

[18]. This phenomenon is manifested by a proliferation of green algae, reduced infiltration of light, oxygen depletion in surface water, disappearance of benthic invertebrates and the production of toxins harmful to fish, livestock and humans. Soils are also at risk from eutrophication, as excessive amounts of nutrients can cause oxygen depletion in the soil and thus prevent the proper functioning of natural microorganisms. This, in turn, affects soil fertility. Moreover, it has been reported that synthetic N fertilizers can promote microbial C utilization depleting both soil and sub-soil organic N content [4]. Eutrophic soils are the source for the emission of N_2O (nitrous oxide), which can react with the stratospheric ozone [19], thus increasing the greenhouse effect and also the emission of toxic ammonia (NH_3) into the atmosphere that can contribute to acidification [20-22]. The process of gaseous ammonia loss from plant foliage can range from 2 to 15kg N/ha/year released, depending on the crop examined or the location [23,24]. Additionally, when the plant does not take up urea fertilizers applied to the soil, up to 40% can also be lost in the form of ammonia [25,26].

Mineral N fertilizers produced by the Haber–Bosch process are very costly in energy production [1,27] and represent nowadays up to 50% of the operational cost for the farmer depending on the cultivated crop [28]. Thus, NUE and energy input are seen as important indicators for the environmental impact of the production of conventional crops but also of energy crops, since they have a large capacity to produce biomass with the minimal amount of N fertilizer [29]. Comparatively, the net energy cost of N_2 fixation in leguminous species is lower than that necessary for an equivalent production of synthetic N fertilizers [30,31]. Therefore, it will be advantageous to the farmer to include more legumes both in crop rotations and in cover crops, whether the main cultivated crop is grown for grain or biomass.

Biological dinitrogen (N_2) fixation is one of the most important sources of N in agricultural system, since it has been estimated to be around 122 Tg per year. The most important N-fixing agents are the symbiotic associations between crop and forage/fodder legumes and bacteria of the genus *Rhizobia* [31,32]. There are accurate estimations of annual inputs of symbiotically fixed N by legume crops. However, the amount of N fixed by other agricultural production systems involving non-symbiotic N_2 fixing associations, such as rice, sugar cane and cereals is much more difficult to estimate (see [30,33,34] for reviews).

To feed the world population in 2050, which will probably reach 9 billion people, it will be necessary to increase agricultural production by 1.7 fold [6]. It is clear that even if this increase in production must be realized in developing countries that need it most, other countries that use intensive agriculture do not consider reducing their production of N fertilizers. As such, they will continue to produce as much or more mineral fertilizers, while at the same time protecting the environment will be essential to preserve the equilibrium of most earth ecosystems. The detrimental impact of the overuse of N fertilizers on the environment can be minimized if it is accompanied by sustainable agricultural practices, such as fertilizer use rationalization, crop rotation, establishment of ground cover and burial of crop residues. Rational fertilization means that the application of fertilizers both organic and inorganic is performed under the proper conditions required to prevent runoff at the appropriate growth stages of the plant and in the correct doses [6]. For example, fractionating N fertilization is currently being performed to grow wheat and other crops such as rice and oilseed rape. Such fertilization strategies have in 15–20 years decreased by 15–20%, the amount of N fertilizer applied to crops in the field [35]. Alternatively, cropping systems using carefully designed species mixtures may be a way to lower N fertilization input, while maintaining economic profitability [36].

Other strategies to improve NUE are to use genetic modification or to breed for new varieties that take up more organic or inorganic N from the soil N and utilize the absorbed N more efficiently [6,37].

Additionally, breeding for more efficient symbioses with *Rhizobia* and arbuscular micorrhizal (AM) fungi can be an interesting alternative for increasing plant productivity using the same amount of synthetic N fertilizer [38,39]. Conservation tillage using no till and continuous cover cropping cultures are also known to increase significantly the potentiality and diversity of plant colonization by AM fungi in comparison to conventional tillage [40-43]. Thus, these new alternative farming techniques could also be an attractive way to increase NUE for a number of crops through the beneficial action of AM.

Lastly, the occurrence of plant growth promoting bacteria (PGPB) and its relationship with the improvement of N nutrition needs to be considered. Through the release of hormones PGPB, can stimulate root development thus increasing nutrient acquisition including N (see [44,45] for reviews).

15.2 NITROGEN FERTILIZATION IN AGRICULTURE

In the developed countries, mineral fertilizers are the main source of N applied to crops [46], followed closely by livestock manure [47]. There are also other sources of N to the soil: the major one being symbiotic N_2 fixation in legume nodules and in the rhizosphere of a range of plants [34,45]. Minor ones include N deposition from the atmosphere, in the form of ammonia and various nitrogen oxides, and the recycling of sewage sludge, which can be applied to cultivated land despite the presence of toxic compounds [48,49]. The importance of these varies from one country to another [50,51].

The mineral commercial fertilizers commonly applied to cultivated soils are anhydrous ammonia, urea, ammonium sulfate and ammonium nitrate. They are particularly soluble for easy assimilation by crops. Both urea and ammonia are converted to nitrate at different rates depending on the nature of the soil and of the climatic conditions, thus leading to various loss mechanisms either by volatilization for ammonia or runoff for nitrate or urea after heavy rainfall and leaching into groundwater [52,53]. However, it appears that the functional diversity of the autotrophic nitrifiers, the ecology (abundance and bacterial community structure) and the nitrification kinetics performed by bacterial ammonia-oxidizers, leading to nitrite (NO_2^-) production and its further oxidation to nitrate by nitrite-oxidizing microorganisms are affected by tillage practices or cover cropping systems [54-56]. Therefore, the final inorganic N budget is strongly affected by the nitrification process occurring in the soil via the action of root-associated or free living microbes that alter rates of nutrient supply and the partitioning of resources between the crop and the soil flora [57].

Manures are the second in nutrient inputs to agricultural land. The nutrient content of manure varies from one country to another and from one region to another within the same country. It depends on the type of farming, grazing systems and nutrient content of different foods and fodder for livestock. There is evidence that at least 50% of manure is lost in storage and transport and another 25% of manure is lost after application [58,59]. An incubation study with composted poultry manure showed a gradual release of inorganic N, mineralizing 0.4 to 5.8% of the total N over 56 days compared to 25.4–39.8% of the total N in uncomposted poultry manure [60].

The application of manure with different level of humification, (i.e composted), has frequently been shown to increase soil fertility [61] and to stimulate soil microbial activity through the improvement of soil structure [62]. Additionally, it has been demonstrated that humic substances have auxin-like activity and positive effects on plant physiology by influencing nutrient uptake and root architecture [63,64]. Simultaneously, it has been shown that through the use of flow-through colorimetry that there is and adsorption of nitrate on to humic substances, thus improving N availability to the plant [65].

From information on N inputs to agricultural soils and estimates of N uptake by crops and grass, a calculation of the excess amounts of N applied to agricultural land can be established. This method of calculating the excess N is known as N balance at the surface [66]. The surface balance can be used as an indicator that highlights areas potentially threatened by N pollution under various environmental scenarios [67,68]. In addition, monitoring the evolution of these surpluses over several years can be used to evaluate the effectiveness of agri-environmental measures to avoid pollution by nitrates. The calculation of the surplus cannot however be immediately interpreted as an indicator of N loss in water. The balance between inputs and outputs for a system includes all potential losses described in the above sections, and inventory changes of N, mainly in the soil.

15.3 NITROGEN FERTILIZATION USING GREEN MANURE AND COVER CROPS

Green manure fertilization (see [69] for a review) aims to improve soil fertility and quality by incorporation into the soil of any field or forage crop while the cultivated plant is still at the green vegetative stage, or just after the flowering stage. Green manure can also be crushed or rolled before no-till seeding (Figure 1).

A cover crop is any crop grown to provide soil cover, regardless of whether it is later incorporated into the soil. Cover crops are grown primarily to prevent soil erosion by wind and water. Cover crops and green manures can be annual, biennial, or perennial herbaceous plants grown in a pure or mixed stand during all or part of the year (Figure 1). In addition to providing ground cover and, in the case of a symbiotic N_2-fixing legume, they provide substantial amounts of N. They also help suppress weeds [70] via allelopathic legume cover and mulching species [71] and reduce insect pests and diseases [72-74]. When cover crops are planted to reduce nutrient leaching (N in particular) following a main crop, they are often termed "catch crops." [75,76]. Moreover, growing green manures on site is a way to prevent the often inhibitive handling and transportation costs of other organic inputs [69]. There are a large variety of cover crop species that are appropriate for a farmer and a particular region. Details on the use of catch crops to prevent N leaching losses during the winter period and of N fertilization using green manures (including N fixing legumes), can be found in the review by Thorup-Kristensen et al. [77] and in the handbook: *Managing Cover Crop Profitability* [78].

Legumes are widely used as cover crops since there is a large choice of different species suited to a particular environment (Figure 1). Legumes are defined by their unique flower structure, their pod, and the ability of 88% of the species examined so far to form atmospheric N_2 fixing nodules [79,80]. Legumes are only of second importance after grasses to humans, by contributing significantly to grain, pasture and forage, and forestry production [33,81]. Since legumes are able to fix symbiotically atmospheric N_2, they require minimal or even no inputs of N fertilizers. If part of this "free" N is made available to a following cultivated crop, the use of legumes in a rotation can allow a significant reduction in the use of N fertilizers. Additionally, legumes can also enhance both the colonization of crop roots by mycorrhizae [82] and the tripartite symbiosis between the host plant AM fungi and N-fixing bacteria thus finally affecting N uptake by the host plant [83]. The legumes used as cover crops or green manure can be classified into two categories: tropical and temperate.

FIGURE 1: Example showing utilization of a mixture of legume and non-legume cover crops for green fertilization. (A) Autumn wheat: no till direct sowing onto a cover crop mixture of radish with berseem (*Trifolium alexandrinum*), simultaneously with a frontal crushing of the two cover crops. Note that on top of providing N for wheat growth, the use of crushing simultaneously to sowing, avoids the utilization of herbicides that are often used in direct seeding culture systems to remove the cover crop. (B) Close-up view of the cover crop mixture composed of radish (r) and berseem (b). (C) Close-up view of the radish root system used as a cover crop. (D) View of the wheat culture in winter after direct sowing and simultaneous crushing of the cover crops. (E) Close-up view of the wheat culture showing the presence of residual crushed cover crop that provides organic N to the soil, thus avoiding the requirement for additional mineral N fertilization.

Warmer climates or warmer winter temperatures allow temperate species to persist during the winter, and tropical species are more adapted to the summer months. It is the intra- and inter-specific genetic variability that partly explains why some legumes grow more and accumulate more N than others. However, it is mainly the soil and climatic conditions that are the predominant factors that restrict the selection of the best performing legumes species. For example, Brandsaeter et al. [84] showed in a recent study that the biochemical quality of the plants differed between species and dates of harvesting, and that this was reflected in the dynamics of net N mineralization. A number of reviews have focused on selection criteria, breeding methods and genetic modification approaches and have covered future improvements in legume crops that will be beneficial not only to the environment and farmers but also to consumers in both developed and developing countries [85-87]. Studies using quantitative genetics approaches to improve NUE in legumes are scarce. However, it seems that both root and nodule traits are important for efficient N assimilation for further translocation to the seeds [88].

N production from legumes is a key benefit of growing cover crops and green manures. The amount of N available from legumes depends on the species of legume grown, the total biomass produced, and the percentage of N in the plant tissue. Cultural and environmental conditions that limit legume growth, such as a delayed planting date, poor stand establishment, and drought will reduce the amount of N produced. Conditions that encourage good N production include getting a good stand, optimum soil nutrient levels and soil pH, good nodulation, and adequate soil moisture. The portion of green-manure N available to a following crop is usually about 40% to 60% of the total amount contained in the legume [76]. Interestingly, it has been demonstrated that leguminous cover crops were also able to replace 60% of the chemical N fertilization for cotton production, although the quantity of available N derived from the cover crop was not synchronized with the requirements of the cotton plant [89]. In turn, one has to consider that NUE is strongly affected by the organic residues remaining from the preceding crop and the application rate of both synthetic N or organic fertilizers applied to the next crop [90].

Both raw and composted manures are useful in organic crop production (for a review see [91]). Used properly, with attention to balancing soil fertility, manures can supplant all or most needs for purchased N fertilizer, especially when combined with a whole system fertility plan that includes crop rotation and cover cropping with N-fixing legumes. However, there is often a lack of synchronization between the timing of N mineralization originating from the catch crop and the N requirement of the main crop, thus leading to a loss of part of the N initially saved by the catch crop. It is therefore necessary to improve estimates of the longer-term N effects of catch crops and to optimize crop sequences in order to estimate accurately the turnover of N retained in the soil by the nitrate catch crops [92,93]. Thus, the grower needs to monitor nutrients in the soil via soil testing, and learn the characteristics of the manure and/or compost to be used. The grower should adjust the rates and select additional fertilizers and amendments accordingly. Finally, development of viable green manure-based alternatives leading to applied crop synergisms will probably not occur without refinement of whole-systems approaches within which green manure secure multiple ecosystemic services [94], utilizing and conserving functional agro-biodiversity services [95].

In addition to legumes, commonly used cover crops include annual cereals (rye, wheat, barley oats), annual or perennial forage grasses such as ryegrass, warm season grasses such as sorgum-sundangrass hybrids and brassicas (Figure 1) including mustard (see [78], for details on their benefits and management).

If organic farming needs to use both classical and green manure to replace chemical N fertilization, it appears that plant genetic adaptations and breeding for these alternative farming techniques are needed to increase crop NUE, for example in wheat [96-99]. Additionally, the development of biomarkers for determining the potential of NUE and optimization of N inputs in crop plants under organic farming cultivation conditions will be required [100].

15.4 NITROGEN ASSIMILATION BY PLANTS

Nitrate is the principal N source for most wild and crop species, whatever the source of inorganic or organic N provided to the plant [101,102]. It is taken up by means of specific high and low affinity transporters located in the root cell membrane [103,104]. Nitrates are then reduced to nitrite through the reaction catalysed by the enzyme nitrate reductase (NR; EC 1.6.6.1), [105] followed by the reduction of nitrite to ammonia catalysed by the enzyme nitrite reductase (NiR; EC 1.7.7.1), [106]. Under particular environments, root ammonia transporters [107] can allow a direct uptake of ammonia when available in the soil, in rice paddy fields or in acidic forest habitats [101,108]. Ammonia can be generated inside the plant by a variety of metabolic pathways such as photorespiration, phenylpropanoid metabolism, utilization of N transport compounds and amino acids catabolism. Symbiotically fixed N is also an important source of ammonia readily available to herbaceous plants or woody species that are able to form a symbiotic relationship with N fixing microorganisms [87,109]; (Figure 2).

Several studies have shown that a wide variety of plant species are able to take up organic N compounds, especially under low N conditions [10,102,110-113]. However, the importance of this N source and the methods used to evaluate its contribution to plant N requirements has been questioned. A few studies have been done on the uptake of organic N by commercial crops: e.g., corn [114], agricultural grasses including species of clover [112] and wheat [96]. Despite these limited studies, they demonstrate the ability of plants to directly take up organic N, but have not established the importance and significance of organic N as a source of crop N, for example when they are grown under organic farming conditions.

In line with the finding that plants can take organic N up directly, there is also an interesting report in which it has been shown that herbaceous species can use protein as a N source without the assistance of other organisms. This indicates that the spectrum of N compounds that can be taken up by the roots is quite diverse, indicating that the relationships existing between the soil fauna and the plant for N capture is more complex than originally thought [115].

Urea is a low molecular weight organic molecule containing N that exists in natural systems and is also applied as a synthetic fertilizer in conventional agriculture. It is well known that urea is absorbed as an intact molecule by plant leaves and roots [116] by means of specific root transporters [117,118]. Although the use of urea is mainly as a source of N fertilizer, the contribution of plant urea uptake and metabolism in a physiological and agricultural context is still not investigated. However, plants possess urea transporters, and can hydrolyse and use urea very efficiently [119].

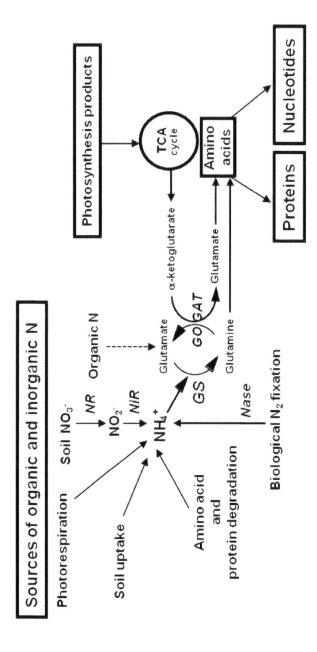

FIGURE 2: Main reactions involved in nitrogen assimilation in higher plants. NO_3^- = nitrate; NO_2^- = nitrite; NH_4^+ = ammonium, N_2 = atmospheric dinitogen. The main enzymes involved in nitrate reduction and ammonia assimilation are indicated in italics: NR = nitrate reductase; NiR = nitrite reductase; Nase = nitrogenase; GS = glutamine synthetase; GOGAT = glutamate synthase. The ultimate source of inorganic N available to the plant is ammonium, which is incorporated into organic molecules in the form of Glutamine and Glutamate through the combined action of the two enzymes GS and GOGAT. Carbon originating from photosynthesis through the tricarboxylic acid cycle (TCA cycle) provides the α–ketoglutarate needed for the reaction catalyzed by the enzyme GOGAT. Amino acids are further used for the synthesis of proteins, nucleotides and all N-containing molecules.

The importance of AM fungi for nutrient uptake by plants is well documented [120-122]. Several studies have shown that AM fungi-infected plants can take up organic N compounds [10,111,112]. Thus, AM fungi can be used as a source of biological fertilization, since they are able to develop symbiotic associations with most terrestrial plants. They are able to alleviate the effects of different stresses both on growth and yield, by significantly increasing the uptake of water and nutrients (including N) by the host plant [123-128]. In particular, it has been reported that the hyphae of AM are able to use inorganic N more efficiently, thus enabling the host plant to indirectly have access to soil N through its fungal partner [129]. However the quantitative contribution of AM fungi to the direct uptake of organic N by plants is still not well established [128], even though recent progress have been made in this field of research. Nevertheless, Tian et al. [130] showed that AM fungi were able to absorb both organic and inorganic N and synthesize organic N molecules such as arginine that are further released by the fungal hyphae and then absorbed by the host plant. Interestingly, the occurrence of a transfer of symbiotically fixed N to a crop such as maize via vesicular-AM hyphae has been demonstrated [131-133], indicating that associated or continuous cover cropping systems could be an alternative way to rationalize plant N nutrition by optimizing field conditions favourable to mycorrhizal colonization.

Ammonia, which is the ultimate form of inorganic N available to the plant, is then incorporated into the amino acid glutamate through the action of two enzymes. The first reaction catalyzed by enzyme glutamine synthetase (GS; EC 6.3.1.2) [134] is considered to be the major route facilitating the incorporation of inorganic N into organic molecules in conjunction with the second enzyme glutamate synthase (GOGAT; EC 1.4.7.1) [135], which recycles glutamate and incorporates C skeletons as a form of 2-oxoglutarate into the cycle. The amino acids glutamine and glutamate are then further used as amino group donors to all the other N-containing molecules notably other amino acids used for storage, transport and protein synthesis and to nucleotides used as basic molecules for RNA and DNA synthesis [134-136].

The two enzymes GS and GOGAT are present in the plant in several isoenzymic forms located in different cellular compartments and differentially expressed in a particular organ or cell type according to the developmental stage. The GS enzyme exists as a cytosolic form (GS1) present in a variety of organ and tissues such as roots, leaves, phloem cells and a plastidic form (GS2) localized in the chloroplasts of photosynthetic tissues and the plastids of roots and etiolated tissues. It has also been proposed that GS2 is located in the mitochondria [137]. However, in numerous previous studies using immunocytolocalization techniques, the presence of the enzyme in the mitochondria has never been reported [138]. The relative proportions of GS1 and GS2 vary within the organs of the same plant and between plant species, each GS isoform playing a specific role in a given metabolic process, such as photorespiratory ammonia assimilation, nitrate reduction, N translocation and recycling [134,139]. The enzyme GOGAT also exists as two forms that have specific roles during primary N assimilation or N recycling. A ferredoxin-dependent iseoenzyme (Fd-GOGAT) is mainly involved, in conjunction with GS2, in the reassimilation of photorespiratory ammonia and a pyridine nucleotide-dependent isoenzyme (NADH-GOGAT; EC 1.4.1.14) involved in the synthesis of glutamate both in photosynthetic and non-photosynthetic organs or tissues to sustain plant growth and development [134,136]. Moreover, by virtue of their differential mode of expression regulated either at the transcriptional and post transcriptional levels, both GS and GOGAT isoenzymes have been shown to play a specific role at particular stages of the plant life cycle and under particular environmental conditions related mainly to the mode of N nutrition [134,135,139].

The reversible reaction catalyzed by the enzyme glutamate dehydrogenase (GDH; EC 1.4.1.2) [134], which has theoretically the capacity to incorporate ammonia into 2-oxoglutarate to form glutamate, was originally thought to be the main enzyme involved in inorganic N assimilation in plants. Later on, a number of experiments using ^{15}N labeling techniques and mutants deficient in GS and GOGAT have demonstrated that over 95% of the ammonia made available to the plant is assimilated via the GS/GOGAT pathway [134,140]. A number of ^{15}N labeling experiments followed by GCMS or NMR-spectroscopy analysis have shown that GDH operates in the direction of glutamate deamination to provide organic acids notably when the cell is C-limited [141,142]. The finding that under certain physiological conditions GDH is able to assimilate ammonia also needs to be taken into consideration, although the rate of glutamate synthesis is probably far lower than that formed through the GS/GOGAT pathway [143]. Recently the hypothesis that GDH plays an important role in controlling glutamate homeostasis has been put forward [142]. This function, which may have a signaling role at the interface of C and N metabolism, may be of importance under certain phases of plant growth and development when there is an important release or accumulation of ammonia [144-146].

Over the last two decades, our knowledge of the various pathways involved in the synthesis of the twenty amino acids that are used to build up proteins, particularly those derived from glutamate and glutamine, has been increased through the use of mutant and transgenic plants in which amino acid biosynthesis has been altered. There are excellent reviews describing extensively our current knowledge on plant amino acid biosynthesis and its regulation [136,143]. Therefore, we will not cover this complex aspect of N assimilation in this review, even though it is of major importance for plant growth and productivity. However, there are some examples of genetic modification in crops in which these pathways have been altered particularly to increase the content of lysine and methionine, which are often the most limiting for both humans and animal nutrition [147-149].

Significant progress has been made during the last few years on the regulation of inorganic N metabolism and the relationships with C metabolism, both at the cellular and organ levels. In particular, attempts to integrate large transcriptomic and physiological data sets at the whole plant level have increased our understanding of the regulation of N assimilation not only under controlled growth conditions but also under the constantly changing environmental constraints usually occurring in field situations [6]. This integration is required, because in addition to regulating a range of cellular processes including N assimilation itself through the co-ordination of nitrate or ammonia uptake and use, nitrate and N metabolite levels in the cell can regulate directly or indirectly a number of closely related metabolic and developmental processes [150,151]. These processes, which may also be regulated through the action of hormones [152], include the synthesis and accumulation of amino acids and organic acids and the modification of plant development including the extent and form of root growth and the timing of flower induction. All these processes, acting either individually or synergistically, condition N allocation in newly developing tissues or in storage organs to finally ensure plant vegetative or sexual reproduction.

15.5 IMPROVEMENT OF NITROGEN UTILIZATION USING GENETICALLY MODIFIED CROPS

Nitrate reduction is rarely limiting for optimal grain yield or biomass production. In contrast, this is not the case for the ammonia assimilatory pathway [153]. For example the work of Fuentes et al. [154] showed that, in tobacco, overexpression of a gene encoding cytosolic glutamine synthetase (GS1) from alfalfa, causes an increase in photosynthesis and growth under a low N fertilization regime. These results suggest that the transgenic tobacco plants overexpresing GS1 are able to utilize N more efficiently under N stress conditions. Interestingly, Oliveira et al. [155] also showed that in tobacco, the overexpression of a gene encoding a pea GS1 lead to increased biomass production both under limiting and non-limiting N feeding conditions.

By overexpressing a pine GS1 gene in poplar, Jing et al. [156] and Man et al. [157] observed that the transgenic trees, which were older than five years exhibited a 41%, increase in growth rate, whereas the other phenotypic characteristics of the genetically modified plants remained similar.

In wheat, the overexpression of a gene for GS1 from French bean led to an increase in grain yield (grain weight in particular) and therefore of NUE, which was estimated to be about 20% [158]. However, to our knowledge there has been no further development of this interesting study, either because of the difficulty of field testing in Europe or because this testing is currently being performed in the private sector. Similar work was conducted in maize consisting in the overexpression of a native gene encoding GS1 (*Gln1-3*) of maize. Grain yield (mainly grain number) of the maize transgenic plants grown under greenhouse conditions was increased by about 30%. However, grain N content and biomass production of the transgenic plants were not modified at maturity [159]. More recently, transgenic rice lines overexpressing GS1 showed improved harvest index, N harvest index and N utilization efficiency. However, these lines did not exhibit higher NUE under N-limiting conditions compared to non-limiting N conditions [160].

In other species, the overexpression of GS1 had a rather negative impact on growth and yield of the plant. For example, overexpression of a GS1gene from tobacco in the legume birds foot trefoil (*Lotus corniculatus* L.) grown on nitrate led to an acceleration of senescence, which was apparently detrimental to the overall plant developmental process [161]. When the transgenic *L. corniculatus* plants were grown under symbiotic N-fixing conditions an increase in plant biomass production was unexpectedly observed. However, the physiological mechanisms involved in this increase remain unknown [86].

In rape (canola), the overexpression of a gene encoding the enzyme alanine aminotransferase (AlaAT) from barley, directed by a rape root-specific promoter, led to a dramatic increase in biomass production and seed yield [162]. Improvement of plant productivity was only observed under low N fertilization conditions and was attributed to a higher flux of nitrate, associated or induced by a decrease in the content of glutamine and glutamate in the stem. In the field when the applied N fertilizer rate was reduced by 40%, the agronomic performance of the transgenic rapeseed plants overexpressing AlaAT was similar to that of untransformed control plants grown under higher optimal N fertilizer rates.

Overexpression of the same gene in rice led to increased biomass production and N content of stems [163]. Unlike in rapeseed, there was an increase of glutamine and asparagine content both in the stems and in the roots. The genetically modified rice plants had a finer, denser and more branched root system, which was presumably more favorable for the absorption of N. This result indicates that genetic modification targeted to improve N utilization efficiency also had an impact on plant development, although the effect of AlaAT overexpression was variable from one species to another in terms of both plant growth and metabolic activity.

There are a few other examples of successful genetic modification of N metabolism using either structural or putative regulatory genes. When the bacterial enzyme glutamate dehydrogenase (GDH A) from *E. coli* was constitutively overexpressed in tobacco, biomass production of the transgenic plants was increased by about 10–15%. In addition to the increase in biomass production GDHA overexpressors had more leaves and their free amino acid content was higher, suggesting that both N metabolism and C metabolism were modified [164]. The transgenic tobacco plants were also more tolerant to water stress.

In rice, overexpression of a gene of unknown function OsENOD93-1, a N-responsive gene identified following genome-wide gene expression profiling, led to an increase in grain yield, of 13–14% and 19–23% under limiting and non-limiting N nutrition conditions respectively [165]. When a gene encoding NAD(H)-dependent GOGAT from alfalfa was constitutively expressed in tobacco, a significant increase in biomass production was observed [166]. Overexpression of the native NAD(H)-dependent GOGAT in rice led to an increase in grain weight [167,168]. These results suggest that the GOGAT enzyme plays a major role with respect to organic N management and is used either for growth or for grain production depending on the species examined.

There are fewer studies in which the importance of regulatory genes has been clearly demonstrated [169]. When a *Dof1* gene encoding a transcription factor from maize was overexpressed in Arabidopsis (*Arabidopsis thaliana* L.), an increase in amino acid content and of N uptake was observed, especially when plants were grown at a low level of N supply. In addition, the transgenic plants produced more biomass under low N supply and they did not exhibit symptoms of N deficiency in comparison to the untransformed control plants, which developed much earlier symptoms of senescence. When the *Dof 1* gene was overexpressed in potato, transgenic plants accumulated more amino acids especially glutamine and glutamate [169]. These two sets of experiments suggest that this gene could be used to improve the uptake and utilization of N in several species. Thus, overexpressing regulatory genes rather than structural genes, such as genes encoding GS, GOGAT or AlaAT appears to be an interesting alternative to improve plant NUE and overall plant growth and development in a more stable and balanced way across species.

When vegetable crops such as lettuce or spinach are grown under greenhouse conditions they can accumulate substantial amounts of nitrate in the leaf cell vacuoles. The threshold of nitrate accumulation often exceeds the limits permitted by law, even when N fertilization is reduced because mineralization of soil organic matter always provides a surplus of nitrate to the plant [170]. In human food, when nitrate is absorbed in excess, its reduction to nitrite during digestion can oxidize hemoglobin, causing a kind of anemia. Moreover, nitrites can be converted to carcinogenic nitrosamines [12,13]. Conventional methods of selection have led to the development of varieties able to reduce the absorbed nitrate more efficiently instead of storing it, but these varieties are not able to completely eliminate any risk of toxic accumulation. Studies were therefore undertaken to limit nitrate accumulation by increasing the capacity of a plant to reduce nitrate by increasing nitrate reductase (NR) activity in genetically modified plants, by overexpressing a gene that allows the deregulation of the synthesis of the enzyme [171]. In tobacco a 50% reduction in leaf nitrate content was observed after introduction of the native structural NR gene (*Nia2*) placed under the control of the 35S strong constitutive promoter. Using the same approach, encouraging results were obtained in a variety of potato [172] that showed a 95% decrease in the amount nitrate in the tubers. In another variety of potato, the transgenic plants showed a marked improvement in biomass production, especially in tubers, with still lower amounts of nitrate. The more effective reduction of nitrate probably allowed a better allocation of N to the photosynthetic apparatus and to enzymes involved in C metabolism, which was demonstrated by higher leaf chlorophyll content in the transgenic potato plants [173].

In lettuce transformed with the same 35S-*Nia2* construct, a problem of post-transcriptional regulation of the NR enzyme was encountered [174]. The transgenic lettuce accumulated 21% less nitrate after 22 days. However, the nitrate content was only 4% lower in 84 days-old transgenic plants. The hypothesis that the strength of the 35S promoter decreases during plant ageing was put forward, suggesting that a way to maintain NR activity at a high level regardless of plant age needs to be found. Such a strategy to reduce the nitrate content in vegetable crops requires further research before the use of the *Nia2* transgene can be efficiently mastered.

Although we do not have any clear information from the private sector about the recent development and commercialization of transgenic plants modified for NUE, it seems to be likely that crops overexpressing the enzymes AlaAT and GS1 will be commercially released within the next five years, following extensive validation of their function under different field trial conditions and using different genetic backgrounds.

15.6 DECIPHERING THE GENETIC BASIS OF NITROGEN USE EFFICIENCY IN CROPS

There have been an increasing number of studies only performed on the model species Arabidopsis, in an attempt to link plant physiology to whole genome expression in order to obtain an integrated view on how the expression of genes can affect overall plant functioning [151]. When a structural or regulatory gene putatively involved in the control of a metabolic pathway or a developmental process or both is identified, information can then be obtained by producing overexpressors or selecting deficient mutants of the gene in question. By studying the impact of the genetic modification or the mutation on the phenotype or the physiology of the plant, it is often possible to determine whether the expression of this specific gene is a limiting step in the development of a particular organ or of a metabolic pathway. In general, this targeted approach, which allows the identification of a single limiting reaction, or a co-limiting/non-limiting reaction does not adequately take into account the variation in complex traits such as those controlling NUE, which involves multiple genes and thus multiple enzyme reactions and regulatory factors.

Over the last ten years, quantitative genetics, through the detection of quantitative trait loci (QTL), has become an important approach for identifying key regulatory or structural genes involved in the expression of complex physiological and agronomic traits in an integrated manner and for the study of plant responses to environmental constraints [175]. When QTLs for agronomic and phenotypic traits are located on a genetic map, it is possible to look for their genetic significance by establishing the co-location of QTLs for physiological or biochemical traits with genes putatively involved in the control of the trait of interest (candidate genes). Validation of candidate genes can then be undertaken using transgenic technologies (forward genetics) or mutagenesis (reverse genetics) or by studying the relationship between allelic polymorphism and the trait of interest (association genetics; Figure 3) either at a single gene or genome-wide level [176]. Positional cloning is another alternative strategy that can be used to focus on the chromosomal region controlling the trait of interest and that ultimately allows access to a single gene [177].

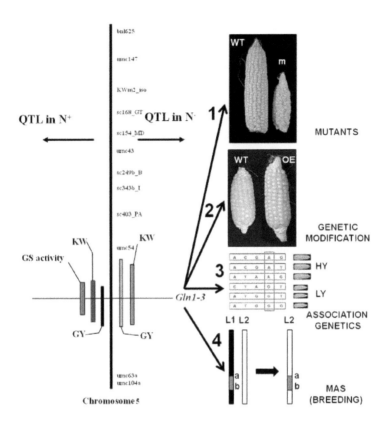

FIGURE 3: Example of identification and validation of a candidate gene involved in the control of NUE and yield in maize. On the left is shown a chromosomal colocation of QTLs for different yield traits (KW = kernel weight and GY = grain yield) and for glutamine synthetase (GS) activity at the level of the Gln1-3 locus (encoding a cytosolic GS involved in ammonia assimilation; see paragraph 4 and Figure 2). N+ means with high N fertilization, N- with low N fertilization. Such a result shows that the Gln1-3 gene is a good candidate gene for explaining variation in NUE. Validation of the candidate gene Gln1.3 was then performed using: (1) mutants {reduction of grain yield in the mutant (m) compared to the wild type (WT)}; (2) genetic modification by overexpressing the Gln1.3 in transgenic maize plants {increase in grain yield in the trangenics (OE) compared to the untransformed plant (WT); see [159]; (3) association genetics linking Gln1.3 gene nucleotide polymorphism to the increase in yield (HY = high yield, LY = low yield) to identify the best performing Gln1.3 allele among a population covering maize genetic diversity; (4) marker assisted selection (MAS) can be then undertaken by breeders where a trait of interest (yield associated to the presence of the Gln1.3 locus) is selected not based on the trait itself, but on a marker or markers linked (marker a and b)to it and introduced in the desired elite line (L2) from the donor line (L1) containing the best performing Gln1.3 allele in terms of yield.

Therefore, quantitative genetic approaches were developed first in maize for which recombinant inbred lines (RIL) populations were used to build-up genetic maps and then study QTLs. The aim of such studies was to identify chromosomal regions involved in the control of yield and its components at high or low N fertilization input, and to determine whether or not some of these regions were specific for one of the two nutrition regimes. In one study, a limited number of QTLs for yield was detected only at low N-input [178]. In another study, it was found that most of the chromosomal regions for grain composition and traits related to NUE detected at low N-input, corresponded to QTLs detected at high N-input [179]. These contrasting results suggest that depending on the RIL population, the response of yield to various levels of N fertilization could be different and thus controlled by a different set of genes.

In a more detailed investigation by Bertin and Gallais [179] using maize RILs, agronomic traits, NUE and physiological traits were associated with DNA markers [180,181]. Interestingly, coincidences were detected between QTLs for yield and two genes encoding cytosolic GS (*Gln1-3* and *Gln1-4*) and whole leaf enzyme activity. As a result of which, the hypothesis that cytosolic GS activity could be a major element controlling grain yield was put forward. [180]. Since a QTL for a thousand kernel weight was coincident with the *Gln1-4* locus and QTLs for a thousand kernel weight and yield were coincident with the *Gln1-3* locus (Figure 3), further work was undertaken to validate the function of these two putative candidate genes. In another study also performed in maize, fine QTL mapping of C and N metabolism enzymes activities was performed on a different RIL population. These QTLs did not colocalize with those reported by other authors [180], which indicates that there are large differences in diversity traits in maize [182].

The impact of the knockout mutations *gln1-3* and *gln1-4* on kernel yield and its components were examined in plants grown under controlled conditions [159]. The phenotype of the two mutant lines was characterized by a reduction of kernel size in the *gln1-4* mutant and by a reduction of kernel number in the *gln1-3* mutant. In the *gln1-3/1-4* double mutant, a cumulative effect of the two mutations was observed. In transgenic plants overexpressing *Gln1-3* constitutively in the leaves, there was an increase in kernel number, thus providing further evidence that the cytosolic GS isoenzyme GS1-3 plays a major role in controlling kernel yield [159]; Figure 3). The hypothesis that GS is one of the key steps in the control of cereal productivity was strengthened by a study performed on rice, in which a co-localization of a QTL for the *GS1;1* locus and a QTL for one-spikelet weight was identified [183]. As a confirmation, a strong reduction in growth rate and grain yield was observed in rice GS1;1 deficient mutants [184].

The role of the GS enzyme and other N-related physiological traits in the control of agronomic performance in wheat still remains to be clearly established. Using a quantitative genetics approach, Fontaine et al. [185] found only a co-localization between a QTL for GS activity and *GSe*, a structural gene encoding cytosolic GS, but no obvious colocalization with a QTL for yield, in agreement with previous work published by Habash et al. [158]. In contrast, in recent work, physical mapping, sequencing, annotation and candidate gene validation of an NUE QTL on wheat chromosome 3B suggested that the NADH-dependent GOGAT enzymes contribute to NUE in wheat and other cereals [186] in agreement with work previously performed on rice [167].

Interestingly, in a woody species such as maritime pine that is far away from cereals on an evolutionary point of view, a protein QTL for GS co-localized with a GS gene and a QTL for biomass [187]. Functional validation of the pine GS gene in transgenic poplars (see above), which can be considered as a crop for wood production, shows once again that quantitative genetics represent one of the most powerful approaches for identifying NUE candidate genes that may be involved in the control of plant productivity.

To date, there are only a few reports reporting specific breeding for organic input systems and especially N [188]. A question that could be addressed is whether the genetic control of NUE under organic or conventional fertilization conditions is similar or if there are specific genes or combinations of genes that are more adapted to one mode of fertilization compared to the other, taking into account that organic material can be directly taken up by the plant [189]. Moreover its appears that using appropriate selection environments is important for breeding crops adapted to organic farming systems [190].

Further work is necessary to identify whether other root and shoot enzymes or regulatory proteins could play a specific role under low or high

N availability, whatever the type of N fertilization conditions (organic or mineral). Such proteins include those directly involved in N metabolism or those positioned at the interface between C and N metabolism during plant growth and development [150,191,192]. It will be necessary therefore to identify new N-responsive genes through detailed analyses of transcriptomic data sets [189], including using systems biology approaches [109]. The analyses will be targeted specifically to N uptake, assimilation and recycling in vegetative [165 and reproductive organs [193] at various stages of plant development, using plants grown under different levels of N fertilization. Systems biology consists in taking advantage of various 'omics' data sets including transcriptomics, proteomics and metabolomics that can be further analysed in an integrated manner through the utilization of various mathematical, bioinformatic and computational tools [192]. Ultimately, such integrated analyses may allow the identification of the key individual or common regulatory elements involved in the control of a given biological process [157]. Such an approach, originally developed for the model plant Arabidopsis by virtue of the wealth of information available at the transcriptome level, when transferred to crops, may help in identifying key master genes involved in the control of NUE. In parallel, metabolomic studies are becoming more and more extensively used for the high throughput phenotyping necessary for large scale molecular and quantitative genetic studies aimed at identifying new candidate genes involved in the control of plant productivity [194,195]. This has prompted a number of groups, to focus their research efforts on developing data integration tools for metabolic reactions that complement gene expression studies. Encouragingly, on the modeling side, an increasing number of genome-scale metabolic models of plants have recently been released [196,197]. Such metabolic models should help to unravel key reactions and thus limit the steps required for the control of NUE, taking into account both tissue-specificities and environmental constraints.

Using the knowledge gained from these various systems biology approaches, it should then be possible to map the newly identified genes encoding regulatory proteins or enzymes, taking advantage of the recent progress in crop genomics through the availability of both physical and genetic high density maps and QTL or Meta-QTL genetic map positions generated by the plant science community [186,198]. Comparative genomics and synteny approaches similar to those of Quraishi et al. [186] can complete such analyses by linking the genetic maps of maize, rice, barley and wheat harboring N related QTLs, thus allowing the reinforcement of the weight of selected putative candidate genes.

Ultimately, following the functional validation of candidate genes using all the available approaches offered by mutagenesis, genetic modification and association genetics, marker-assisted selection (MAS) can be then undertaken (Figure 3). However, there are still a number of technical and scientific challenges that remain to be resolved before MAS can be routinely used in breeding for complex traits such as NUE. This is mainly due to the number of interactions that govern the expression of such traits both at the genetic and environmental levels [199], whether we are dealing with conventional or organic farming growth conditions.

15.7 CONCLUSION AND PERSPECTIVES

A large number of studies have been carried out over the last two decades to identify by means of agronomic, physiological and genetic studies, the rate limiting steps of NUE both in model and crop species, as a function of environmental conditions. For abiotic stress improvement in crops, NUE has become the second priority after drought both in the private and in the public sector. To decipher the genetic and physiological basis of NUE, many tools are available for most crops and for cereals in particular. They include mutant collections, wide genetic diversity, recombinant inbred lines (RILs) or Doubled Haploid Line populations (DHLs), straightforward transformation protocols and physiological, biochemical and genomic data for systems biology development [6,200]. In addition, the commercial crop research effort is paralleled by research in the public sector, notably with the release of the genome sequences for rice [201] and maize [202] and the current development of sequencing projects for wheat [203], barley [204] and a number of other crops.

Cereal grains such as rice, wheat and maize provide 60% of the world's nutrition, the rest being represented by barley, coarse grains of legumes along with root crops. These crops account for the majority of end products used for human diets [205] and it is likely that they will still contribute either directly in the human diet or indirectly as animal feed in the next century [200]. Thus, considering both the economical and environmental challenge represented by reducing both the cost and application of N fertilizers, all major maize seed breeding companies such as Monsanto, DuPont-Pioneer and Syngenta are investing in genomic research for improving NUE. Moreover, improvement in yield for most crops over the last 50 years has been estimated to be 40%, due to improvements in cultural practices and 60% due to genetic gains, thus indicating that breeding for improved NUE is still possible [206]. However, to our knowledge, improving NUE either through genetic engineering or marker assisted breeding is still at the stage of proof of concept. Therefore, very little information is currently released from both the private and public sector in consideration of the potential economic value of crop NUE improvement.

However, both on the genetic and physiological side, the identification of key steps involved in the control of NUE from gene expression to metabolic activity remains incomplete. It is likely because the regulatory mechanisms involved in the control of the two components of NUE (N uptake and utilization efficiencies) are species-specific [6]. Moreover, they are subjected to changes or adaptation in a constantly changing soil and aerial environment during plant growth and development that require the taking into account the various genotypic/environment interactions [207].

NUE is controlled by a complex array of physiological, developmental and environmental interactions that are organ and tissue-specific and which are specific to the genotype of a given species. It is therefore essential that a much more extensive survey of a wide range of genotypes covering the genetic diversity of a crop should be performed. This can be achieved using the various available "omics" techniques, combined with agronomic and physiological approaches in order to identify both common and specific elements controlling NUE and plant productivity of plants grown in the field under organic or mineral N fertilizer conditions [208].

Over the last two decades, the construction of cereals that can fix atmospheric N has always been a challenge for plant scientists, in order to reduce the need for mineral N fertilization. Although, the signaling pathway for recognition of N-fixing bacteria is present in cereals, complex genetic modification will be necessary to allow bacterial colonization and nodule organogenesis [209].

At the field level, only agronomic predictive models using the appropriate biogical and environmental parameters [210] should be able to take into account interactions between plants and their environment to obtain an integrated view of the various inputs or outputs, influencing crop NUE [211,212]. One of the main challenges in the future will be to develop reliable decision support systems with the help of sensors [213,214] and biological diagnostic tools in precision agriculture, in order to optimize the application of N under organic or conventional conditions in a more sustainable manner. Moreover, the establishment of such models will need to be scaled up at the ecological level [44], in order to obtain a better understanding as to how N cycling is occurring from organisms to the whole ecosystem [57].

A proposed strategy for integrating multidisciplinary approaches for improving crop NUE is summarized in Figure 4. This strategy highlights the necessity to develop an integrated approach between the public and private sectors to improve our understanding and control of the biological and agronomic basis of NUE in crops of major economical importance. However, the nature of an agronomic trait such as NUE is complex, due to the intervention of multiple elements interacting with each other as a function of both plant development and environmental constraints. Moreover, the interaction between these elements appears to be not only species-specific but also specific to a given genetic background. Therefore, improvement of this understanding will require the development of a multi-disciplinary approach, integrating expertise from fundamental and more applied studies in crop developmental biology, physiology, genomics, genetics, physiology, modeling, agronomy and breeding [212]. In addition, taking advantage of the genetic variability that already exists or that can be created, will provide a valuable contribution to the genetic and physiological dissection of NUE under mineral and organic N nutrition conditions and an evaluation of the genes or group of genes involved. The major breakthrough expected from this multidisciplinary approach will be to provide 1) useful alleles or gene-based markers to breeders for the production of genetically modified plants or for marker assisted selection (MAS) ; 2) predictive biological markers for breeders to improve selection for higher NUE by conventional breeding; 3) tools for farmers to monitor and adjust mineral and/or organic N fertilization for obtaining optimal yields compatible with a strategy for sustainability of the agricultural practices needed to feed the world population, while preserving the environment.

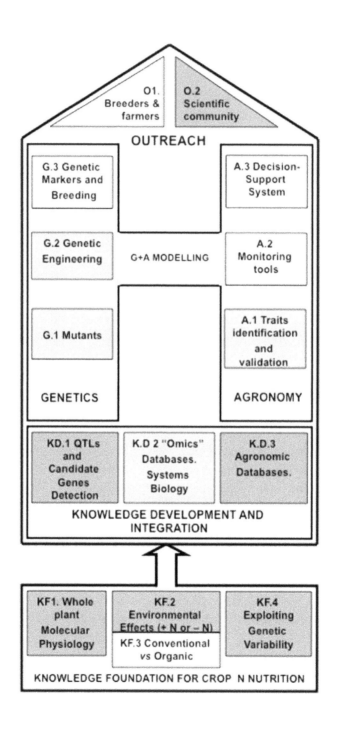

FIGURE 4: Proposed strategy for improving N use efficiency in crops. This strategy is built around two main agronomic and genetic studies conducted in parallel. Each of these two main studies is divided into a subset of approaches strongly interacting with each other within and across them. It will be necessary to integrate current knowledge in agronomy, molecular physiology, eco-physiology and genetics to guide, develop and integrate novel methods and concepts for improving NUE in crops. This knowledge development and integration can be performed through the use of quantitative genetics for QTL and candidate gene detection (KD.1), through the exploitation of all the 'omics' databases using a systems biology approach (KD.2) and through the use of agronomic databases gathering all the information concerning plant performance under various environmental scenarios (KD3). The basis of this knowledge is represented by: (1) the numerous whole plant physiology studies performed over the last two decades on both model and crop species (KF.1) ; (2) the studies aimed at identifying the influence of N fertilization on crop growth and development and its physiology either under organic (KF2) or organic N nutrition (KF3) ; (3) through the exploitation of genetic variability of a given species using different modern and ancient genotypes, landraces, lines, hybrids originating from different parts of the world. The primary goal of the genetic studies is to provide breeders with markers genes or loci aimed at selecting varieties more efficient at utilizing N, identified through the use of quantitative genetics (KD.1), mutagenesis (G.1) and genetic engineering (G.2) for further commercialization by breeding companies (O.1). The aim of the agronomic studies is to provide tools for breeders and agronomists to create and evaluate new varieties in cropping systems under low and adequate N input in conventional or organic farming systems. To achieve this it will be necessary to identify key agronomic traits that can be use to predict plant performance under low or high N input and according to various environmental conditions (A.1). Plant performance could also be predicted and monitored through the use of monitoring tools or sensors (A.2 = metabolic, enzymatic and molecular markers for NUE; see [109] for details) and through the development of plant and crop modeling approaches integrating agronomic, physiological and molecular data (G+A) [213]. These monitoring tools and models will also help the farmers to rationalize N fertilization when integrated into decision support systems (A.3). In addition the knowledge gained from these complementary studies will be useful to the scientific community to improve our understanding of N assimilation by plants both at the whole plant and canopy levels (O.2). The boxes shaded in dark grey indicate where significant progress has been made in the area. Those in pale grey indicate that work is still currently being actively performed. Those in white indicate the research area for which results and data are scarce or missing.

REFERENCES

1. Erisman, J.W.; Sutton, M.A.; Galloway, J.N.; Klimont, Z.; Winiwarter, W. How a century of ammonia synthesis changed the world. Nat. Geosci. 2008, 1, 636–639, doi:10.1038/ngeo325.

2. Galloway, J.N.; Townsend, A.R.; Erisman, J.W.; Bekunda, M.; Cai, Z.; Freney, J.R.; Martinelli, L.A.; Seitzinger, S.P.; Sutton, M.A. Transformation of the nitrogen cycle: Recent trends, questions, and potential solutions. Science 2008, 320, 889–892, doi:10.1126/science.1136674.

3. Hoang, V.N.; Alauddin, M. Assessing the eco-environmental performance of agricultural production in OECD countries: The use of nitrogen flows and balance. Nutr. Cycl. Agroecosys. 2010, 87, 353–36, doi:10.1007/s10705-010-9343-y.

4. Mulvaney, R.L.; Khan, S.A.; Ellsworth, T.R. Synthetic nitrogen depleted soil nitrogen: A global dilemma for sustainable cereal production. J. Environ. Qual. 2009, 38, 2295–2314, doi:10.2134/jeq2008.0527.

5. Tilman, D.; Cassman, K.G.; Matson, P.A.; Naylor, R.; Polasky, S. Agricultural sustainability and intensive production pratices. Nature 2002, 418, 671–677, doi:10.1038/nature01014.

6. Hirel, B.; Le Gouis, J.; Ney, B.; Gallais, A. The challenge of improving nitrogen use efficiency in crop plants: Towards a more central role for genetic variability and quantitative genetics within integrated approaches. J. Exp. Bot. 2007, 58, 2369–2387, doi:10.1093/jxb/erm097.

7. Gallais, A.; Coque, M. Genetic variation and selection for nitrogen use efficiency in maize: A synthesis. Maydica 2005, 50, 531–537.

8. Ceccarelli, S. Adaptation to low/high input cultivation. Euphytica 1995, 92, 203–204.

9. Raun, W.R.; Johnson, G.V. Improving nitrogen use efficiency for cereal production. Agron. J. 1999, 91, 357–363, doi:10.2134/agronj1999.00021962009100030001x.

10. Hodge, A.; Robinson, D.; Fitter, A. Are microorganisms more effective than plants at competing for nitrogen? Trends Plant Sci. 2000, 5, 304–308, doi:10.1016/S1360-1385(00)01656-3.

11. Asghari, H.R.; Cavagnaro, T.R. Arbuscular mycorrhizas enhance plant interception of leached nutrients. Funct. Plant Biol. 2011, 38, 219–226, doi:10.1071/FP10180.

12. Al-Redhaiman, K.N. Nitrate accumulation in plants and hazards to man and livestock: A review. J. King Saud Univ. 2000, 12, 143–156.

13. Umar, A.S.; Iqbal, M. Nitrate accumulation in plants, factors affecting the process, and human health implications. A review. Agron. Sustain. Dev. 2007, 27, 45–57, doi:10.1051/agro:2006021.

14. Harris, R.C.; Skinner, A.C. Controlling diffuse pollution of groundwater from agriculture and industry. Water Environ. J. 1992, 6, 569–574, doi:10.1111/j.1747-6593.1992.tb00792.x.

15. Cameron, S.G.; Schipper, L.A. Nitrate removal and hydraulic performance of organic carbon for use in denitrification beds. Ecol. Eng. 2010, 36, 1588–1595, doi:10.1016/j.ecoleng.2010.03.010.

16. Camarguo, J.A.; Alonso, A. Ecological and toxicological effects of inorganic nitrogen pollution in aquatic ecosystems: A global assessment. Environ. Int. 2006, 32, 831–849, doi:10.1016/j.envint.2006.05.002.
17. London, J.G. Nitrogen study fertilizes fears of pollution. Nature 2005, 433, 791, doi:10.1038/433791a.
18. Beman, J.M.; Arrigo, K.; Matson, P.M. Agricultural runoff fuels large phytoplankton blooms in vulnerable areas of the ocean. Nature 2005, 434, 211–214, doi:10.1038/nature03370.
19. Sutton, M.; Howard, C.M.; Erisman, J.W.; Billen, G.; Bleeker, A.; Grennfelt, P.; van Grinsven, H.; Grizetti, B. Assessing our nitrogen inheritance. In The European Nitrogen Assessment. Sources, Effects and Policy Perspectives; Sutton, M.A., Howard, C.M., Erisman, J.W., Billen, G., Bleeker, A., Grennfelt, P., van Grinsven, H., Grizetti, B., Eds.; Cambridge University Press: Cambridge UK, 2011; pp. 1–6.
20. Ramos, C. Effect of agricultural practices on the nitrogen losses to the environment. Fertilizer Res. 1996, 43, 183–189, doi:10.1007/BF00747700.
21. Stulen, I.; Perez-Soba, M.; De Kok, L.J.; Van Der Eerden, L. Impact of gaseous nitrogen deposition on plant functioning. New Phytol. 1998, 139, 61–70, doi:10.1046/j.1469-8137.1998.00179.x.
22. David, M.; Loubet, B.; Cellier, P.; Mattson, M.; Schjoerring, J.K.; Nemitz, E.; Roche, R.; Riedo, M.; Sutton, M.A. Ammonia sources and sinks in an intensively managed grassland canopy. Biogeosciences 2009, 6, 1903–1915.
23. Ammonia emission from mineral fertilizers and fertilized crops. Adv. Agron. 2004, 82, 557–622.
24. Wang, L.; Xu, Y-C.; Schjoerring, J. K. Seasonal variation in ammonia compensation point and nitrogen pools in beech leaves (Fagus sylvatica). Plant Soil 2011, 343, 51–66, doi:10.1007/s11104-010-0693-7.
25. Fowler, D.B.; Brydon, J. No-till winter wheat production in the Canadian prairies: Timing of nitrogen fixation. Agron. J. 1989, 81, 817–825, doi:10.2134/agronj1989. 00021962008100050024x.
26. San Francisco, S.; Urrutia, O.; Martin, V.; Peristeropoulos, A.; Garcia-Mina, J.M. Efficiency of urease and nitrification inhibitors in reducing ammonia volatilization from diverse nitrogen fertilizers applied to different soil types and wheat straw mulching. J. Sci. Food. Agr. 2011, 91, 1569–1575, doi:10.1002/jsfa.4349.
27. Olson, R.A. Fertilizers for food production vs energy needs and environmental quality. Ecotox. Environ. Safe. 1977, 1, 311–26, doi:10.1016/0147-6513(77)90023-9.
28. Reganold, J.P.; Papendick, R.I.; Parr, F.F. Sustainable agriculture. Sci. Am. 1990, 262, 112–120.
29. Lewandowski, I.; Schmidt, U. Nitrogen, energy and land use efficiencies of miscanthus, reed canary grass and triticale as determined by the boundary line approach. Agr. Ecosyst. Environ. 2006, 112, 335–346, doi:10.1016/j.agee.2005.08.003.
30. Andrews, M.; Lea, P.J.; Raven, J.A.; Azevedo, R.A. Nitrogen use efficiency. 3. Nitrogen fixation: Genes and costs. Ann. Appl. Biol. 2009, 155, 1–13, doi:10.1111/j.1744-7348.2009.00338.x.
31. Fustec, J.; Lesuffleur, F.; Mahieu, S.; Cliquet, J.B. Nitrogen rhizodeposition of legumes. A review. Agron. Sustain. Dev. 2010, 30, 57–66, doi:10.1051/agro/2009003.

32. Liu, Y.Y.; Wu, L.H.; Baddeley, J.A.; Watson, C.A. Models of biological nitrogen fixation of legumes. A review. Agron. Sustain. Dev. 2011, 31, 155–172, doi:10.1051/agro/2010008.

33. Herridge, D.F.; People, M.B.; Boddey, R.M. Global inputs of biological nitrogen fixation in agricultural systems. Plant Soil 2008, 311, 1–18, doi:10.1007/s11104-008-9668-3.

34. Andrews, M.; Hodge, S.; Raven, J.A. Positive plant microbial reactions. Ann. Appl. Biol. 2010, 157, 317–320, doi:10.1111/j.1744-7348.2010.00440.x.

35. Meynard, J.; Sebillotte, M.M. L'élaboration du rendement du blé, base pour l'étude des autres céréales à talles. In Elaboration du Rendement des Principales Cultures Annuelles; Combe, L., Picard, D., Eds.; INRA: Paris, France, 1994; pp. 31–51.

36. Malézieux, E.; Crozat, Y.; Dupraz, C.; Laurans, M.; Makowski, D.; Ozier-Lafontaine, H.; Rapidel, B.; de Tourdonnet, S.; Valentin-Morison, M. Mixing plant species in cropping systems: Concepts, tools and models. A review. Agron. Sustain. Dev. 2009, 29, 43–62, doi:10.1051/agro:2007057.

37. Good, A.G.; Shrawat, A.K.; Muench, D.G. Can less yield more? Is reducing nutrient input into the environment compatible with maintaining crop production? Trends Plant Sci. 2004, 9, 597–605, doi:10.1016/j.tplants.2004.10.008.

38. Rengel, Z. Breeding for better symbiosis. Plant Soil 2002, 245, 147–162, doi:10.1023/A:1020692715291.

39. Raviv, M. The use of mycorrhiza in organically-grown crops under semi arid conditions: A review of benefits, constraints and future challenges. Symbiosis 2010, 52, 65–74, doi:10.1007/s13199-010-0089-8.

40. Kabir, Z.; O'Halloran, I.P.; Hamel, C. Seasonal changes of arbuscular mycorrhizal fungi as affected by tillage practices and fertilization. Plant Soil 1997, 192, 285–293, doi:10.1023/A:1004205828485.

41. Kabir, Z.; O'Halloran, I.P.; Hamel, C. Overwinter survival of arbuscular mycorrhizal hyphae is favored by attachment to roots but diminished by disturbance. Mycorrhiza 1997, 7, 197–200, doi:10.1007/s005720050181.

42. Kabir, Z.; Rhamoun, M.; Lazicki, P.; Horwath, W. Cover crops and conservation tillage increase mycorrhizal colonization of corn and tomato roots. Sustainable Agriculture Farming System Project. Volume 9, No. 1; University of California: Davis CA, USA, 2008. Available online: http://safs.ucdavis.edu/newsletter/v09n1/page3.htm (accessed on 23 August 2011).

43. Aggarwal, N.A.; Gaur, A.; Bhalla, E.; Gupta, S.R. Soil aggregate carbon and diversity of mycorrhiza as affected by tillage practices in a rice-wheat cropping system in northern India. Int. J. Ecol. Environ. Sci. 2010, 36, 233–243.

44. Kraiser, T.; Gras, D.; Gutièrrez, A.G.; Gonzalez, B.; Gutièrrez, A.R. A holistic view of nitrogen acquisition in plants. J. Exp. Bot. 2011, 62, 1455–1466, doi:10.1093/jxb/erq425.

45. Tikhonovich, I.A.; Provorov, N.A. Microbiology is the basis of sustainable agriculture: An opinion. Ann. Appl. Biol. 2011, 159, 155–168, doi:10.1111/j.1744-7348.2011.00489.x.

46. Robertson, G.P.; Vitousek, P.M. Nitrogen in Agriculture: Balancing the cost of an essential resource. Annu. Rev. Envir. Resour. 2009, 34, 97–125, doi:10.1146/annurev.environ.032108.105046.

47. Hooda, P.S.; Edwards, A.C.; Anderson, H.A.; Miller, A. A review of water quality concerns in livestock farming areas. Sci. Total. Environ. 2000, 250, 143–167, doi:10.1016/S0048-9697(00)00373-9.

48. Smith, S.R. Organic contaminants in sewage sludge (biosolids) and their significance for agricultural recycling. Phil. Trans. R. Soc. B. 2009, 367, 4005–4041.

49. Giller, K.E.; Witter, E.; McGrath, S.P. Heavy metals and soil microbes. Soil Biol. Biochem. 2009, 41, 2031–2037, doi:10.1016/j.soilbio.2009.04.026.

50. Billen, G.; Beusen, A.; Bouwman, L.; Garnier, J. Anthropogenic nitrogen autotrophy and heterotrophy of the world's watersheds: Past, present, and future trends. Global. Biogeochem. Cy. 2010, 24, GB0A11, doi:10.1029/2009GB003702.

51. Spiertz, J.H.J. Nitrogen, sustainable agriculture and food security. A review. Agron. Sustain. Develop. 2010, 30, 43–55, doi:10.1051/agro:2008064.

52. Vitosh, M. L.; Johnson, J. W.; Mengel, D. B. Tri-State Fertilizer Recommendations for Corn, Soybean, Wheat and Alfalfa. Extension Bulletin; Ohio State University: Columbus Ohio, USA, 1995; p. E-2567. Available online http://ohioline.osu.edu/e2567/ (accessed 23 August 2011).

53. Jarvis, S.; Hutchings, N.; Brentrup, F.; Olesen, J.E.; van de Hoek, K.W. Nitrogen flows in farming systems across Europe. In The European Nitrogen Assessment. Sources, Effects and Policy Perspectives; Sutton, M.A., Howard, C.M., Erisman, J.W., Billen, G., Bleeker, A., Grennfelt, P., van Grinsven, H., Grizetti, B., Eds.; Cambridge University Press: Cambridge, UK, 2011; pp. 21–28.

54. Webster, G.; Embley, T.M.; Freitag, T.E.; Smith, Z.; Prosser, J.I. Links between ammonia oxidizer species composition, functional diversity and nitrification kinetics in grassland soils. Environ. Microbiol. 2005, 7, 676–684, doi:10.1111/j.1462-2920.2005.00740.x.

55. Le Roux, X.; Poly, F.; Currey, P.; Commeaux, C.; Hai, B.; Nicol, G.W.; Prosser, I.; Schloter, M.; Attard, E.; Klumpp, K. Effect of aboveground grazing on coupling among nitrifier activity, abundance and community structure. ISME J. 2008, 2, 221–232, doi:10.1038/ismej.2007.109.

56. Attard, E.; Poly, F.; Commeaux, C.; Laurent, F.; Terada, A.; Smets, B.F.; Recous, S.; Le Roux, X. Shifts between Nitrospira- and Nitrobacter-like nitrite oxidizers underlie the response of soil potential nitrite oxidation to changes in tillage practices. Environ. Microbiol. 2010, 12, 315–326, doi:10.1111/j.1462-2920.2009.02070.x.

57. Van der Heidjen, M.G.A.; Bardgett, R.D.; van Straalen, N.M. The unseen majority: soil microbes as drivers of plant diversity and productivity in terrestrial ecosystems. Ecol. Lett. 2008, 11, 296–310, doi:10.1111/j.1461-0248.2007.01139.x.

58. Bouldin, D.R.; Klausner, S.D.; Reid, S.D. Use of nitrogen from manure. In Nitrogen in Crop Production; Hauck, R.D., Ed.; ASA-CSSA-SSSA: Madison, WI, USA, 1984; pp. 221–245.

59. Maguire, R.O.; Kleinman, P.J.A.; Beegle, D.B. Novel manure management technologies in no-till and forage systems: Introduction to the special series. J. Environ. Qual. 2011, 40, 287–291, doi:10.2134/jeq2010.0396.

60. Tyson, S.C.; Cabrera, M.L. Nitrogen mineralization in soils amended with composted and uncomposted poultry litter. Commun. Soil Sci. Plant Anal. 1993, 24, 2361–2374, doi:10.1080/00103629309368961.

61. Glaser, B.; Lehmann, J.; Zech, W. Ameliorating physical and chemical properties of highly watered soils in the tropics with charcoal—A review. Biol. Fert. Soils 2002, 35, 219–230, doi:10.1007/s00374-002-0466-4.

62. Watson, C.A.; Atkinson, D.; Gosling, P.; Jackson, L.R.; Rayns, F.W. Managing soil fertility in organic farming systems. Soil Use Manag. 2002, 18, 239–247, doi:10.1079/SUM2002131.

63. Canellas, L.P.; Teixeira Junior, L.R.; Teixeira Junior, L.R.L.; Dobbss, L.B.; Silva, C.A.; Medici, L.O.; Zandonadi, D.B.; Façanha, A.R. Humic acid crossinteractions with root and organic acids. Ann. Appl. Biol. 2008, 153, 157–166.

64. Trevisan, S.; Francioso, O.; Quaggiotti, S.; Nardi, S. Humic substances biological activity at the plant-soil interface; From environmental aspects to molecular factors. Plant Signal. Behav. 2010, 5, 635–643, doi:10.4161/psb.5.6.11211.

65. Klucakova, M. Adsorption of nitrate on humic acids studied by flow-through coulometry. Environ. Chem. Lett. 2010, 8, 145–148, doi:10.1007/s10311-009-0201-6.

66. Goodwin, D.C.; Singh, U. Nitrogen balance and crop response to nitrogen in upland and lowland cropping systems. In Understanding Options for Agricultural Production; Tsuji, G.Y., Hoogenboom, G., Thornton, P.K., Eds.; Kluwer: Dordrecht, The Netherlands, 1998; Volume 7, pp. 55–78.

67. Eulenstein, F.; Werner, A.; Willms, M.; Juszczak, R.; Schlindwein, S.L.; Chijjnicki, B.H.; Olenik, J. Model based scenario studies to optimize the regional nitrogen balance and reduce leaching of nitrate and sulfate of an agricultural water catchment. Nutr. Cycl. Agroecosys. 2008, 82, 33–49, doi:10.1007/s10705-008-9167-1.

68. Bechini, L.; Castoldi, N. Calculating the soil surface nitrogen balance at regional scale: Example application and critical evaluation of tools and data. Ital. J. Agron. 2006, 1, 665–676.

69. Cherr, C.M.; Scholberg, J.M.S.; McSorley, R. Green manure approaches to crop production: A synthesis. Agron. J. 2006, 98, 302–319, doi:10.2134/agronj2005.0035.

70. Burgos, N.R.; Talberg, T.E. Weed control and sweet corn (Zea mays, var. Rugosa) response in no-till system with cover crops. Weed Sci. 1996, 44, 355–361.

71. Caamal-Maldonado, J.A.; Jimenez-Osornio, J.J.; Torres-Barragan, A.; Anaya, A.L. The use of allelopathic legume cover and mulch species for weed control in cropping systems. Agron. J. 2001, 93, 27–36, doi:10.2134/agronj2001.93127x.

72. Caswell, E.P.; Defranck, J.; Apt, W.J.; Tang, C.S. Influence of non-host plants on population decline of Rotylenchus reniformis. J. Nematol. 1991, 23, 91–98.

73. Stevenson, F.C.; Van Kessel, C. The nitrogen and non-nitrogen rotation benefits of pea to succeeding crops. Can. J. Plant Sci. 1996, 76, 735–734, doi:10.4141/cjps96-126.

74. Unkovich, M.J.; Pate, J.S.; Sanford, P. Nitrogen fixation by annual legumes in Australian Mediterranean agriculture. Aust. J. Agr. Res. 1997, 48, 267–293, doi:10.1071/A96099.

75. Sullivan, P.S. Overview of cover crops and green manures. National Center for Appropriate Technology Sustainable Agricultural Project; National Center for Appropriate Technology: Butte, MT, USA, 2003; p. IP024. Available online: https://attra.ncat.org/attra-pub/summaries/summary.php?pub=288 (accessed on 23 August 2011).

76. Rinnofner, T.; Friedel, J.K.; de Kruiff, R.; Freyer, G.P. Effect of catch crops on N dynamics and following crops in organic farming. Agron. Sustain. Dev. 2008, 28, 551–558, doi:10.1051/agro:2008028.

77. Thorup-Kristensen, K.; Magrid, J.; Stoumann Jesen, L. Catch crops and green manures as biological tools in nitrogen management in temperate zone. Adv. Agron. 2003, 79, 227–302.

78. Managing Cover Crops Profitability., Third Ed. ed.; Handbook Series 9; Sustainable Agriculture Research and Education: Beltsville, MD, USA, 2007; p. 224.

79. De Faria, S.M.; Lewis, G.P.; Sprent, J.I.; Sutherland, J.M. Occurrence of nodulation in the leguminosae. New Phytol. 1989, 111, 607–619, doi:10.1111/j.1469-8137.1989. tb02354.x.

80. Novak, K. On the efficiency of legume supernodulating mutants. Ann. Appl. Biol. 2010, 157, 321–342, doi:10.1111/j.1744-7348.2010.00431.x.

81. Graham, P.H.; Vance, C.P. Legumes: Importance and constraints to greater use. Plant Physiol. 2003, 131, 872–877, doi:10.1104/pp.017004.

82. Hoffman, C.A.; Carroll, C.R. Can we sustain the biological basis of agriculture? Annu. Rev. Ecol. Syst. 1995, 26, 69–92, doi:10.1146/annurev.es.26.110195.000441.

83. Wang, X.; Pan, Q.; Chen, F.; Yan, X.; Liao, H. Effects of co-inoculation with arbuscular mycorrhizal fungi and rhizobia on soybean growth as related to root architecture and availability of N and P. Mycorrhiza 2011, 21, 173–181, doi:10.1007/ s00572-010-0319-1.

84. Brandsaeter, L.O.; Heggen, H.; Riley, H.; Stubhaug, E.; Henriksen, T.M. Winter survival, biomass accumulation and N mineralization of winter annual and biennial legumes sown at various times of the year in northern temperate regions. Eur. J. Agron. 2008, 28, 437–448, doi:10.1016/j.eja.2007.11.013.

85. Ranalli, P. Breeding methodologies for the improvement of grain legumes. In Improvement strategies for Leguminosae Biotechnology; Jaiwal, P.K., Singh, R.P., Eds.; Kluwer: Dordrecht, The Netherlands, 2003; pp. 3–21.

86. Hirel, B.; Harrison, J.; Limami, A. Improvement of Nitrogen Utilization. In Improvement strategies for Leguminosae Biotechnology; Jaiwal, P.K., Singh, R.P., Eds.; Kluwer: Dordecht, The Netherlands, 2003; pp. 201–220.

87. Valentine, A.J.; Vagner, A.; Benedito, A.; Kandy, Y. Legume nitrogen and soil abiotic stress: From physiology to genomics and beyond. Valentine. In Annual Plant Reviews, Nitrogen Metabolism in Plants in the Post-genomic Era; Foyer, C.H., Zhang, H., Eds.; Wiley-Blackwell: Chichester, UK, 2011; Volume 42, pp. 207–248.

88. Bourion, V.; Hasan Risvi, S.M.; Fournier, S.; de Lambergue, H.; Galmiche, F.; Marget, P.; Duc, G.; Burstin, J. Genetic dissection of nitrogen nutrition in pea through a QTL approach of root, nodule, and shoot variability. Theor. Appl. Genet. 2010, 212, 71–86.

89. Zablotowicz, R.M.; Reddy, K.N.; Krutz, L.J.; Gordon, R.E.; Jackson, R.E.; Price, L.D. Can leguminous cover crops partially replace nitrogen fertilization in Mississipi delta cotton production? Int. J. Agron. 2011, doi:10.1155/2011/135097.

90. Rahimizadeh, M.; Kashani, A.; Zare-Feizabadi, A.; Koocheki, A.R.; Nassiri-Mahallati, M. Nitrogen use efficiency of wheat as affected by preceding crop, application rate of nitrogen and crop residues. Aust. J. Crop Sci. 2010, 4, 363–368.

91. Kuepper, G. Manures for organic crop production. National Center for Appropriate Technology Sustainable Agriculture Project; National Center for Appropriate Technology: Butte, MT, USA, 2003. Available online: http://attra.ncat.org/attra-pub/manures.html (accessed on 23August 2011).

92. Thomsen, I.K.; Christensen, B.T. Nitrogen conserving potential successive ryegrass catch crops in continuous spring barley. Soil Use Manage. 1999, 15, 195–200.

93. Möller, K.; Stinner, W.; Leithold, G. Growth, composition, biological N2 fixation and nutrient uptake of a leguminous cover crop mixture and the effect of their removal on field nitrogen balance and nitrate leaching risk. Nutr. Cycl. Agroecosys. 2008, 82, 233–249, doi:10.1007/s10705-008-9182-2.

94. Anderson, R.L. Synergism: A rotation effect of improved growth efficiency. Adv. Agron. 2011, 112, 205–226.

95. Jackson, L.E.; Pascual, U.; Hodgkin, T. Utilizing and conserving agrobiodiversity in agriculture landscapes. Agr. Ecosys. Environ. 2007, 121, 196–210, doi:10.1016/j.agee.2006.12.017.

96. Näsholm, T.; Huss-Danell, K.; Högberg, P. Uptake of glycine by field grown wheat. New Phytol. 2001, 150, 59–63, doi:10.1046/j.1469-8137.2001.00072.x.

97. Baresel, J.P.; Zimmerman, G.; Reents, H.J. Effect on genotype and environment on N uptake and N partition in organically grown winter wheat (Triticum aestivum L.) in Germany. Euphytica 2008, 163, 347–354, doi:10.1007/s10681-008-9718-1.

98. Loschenberger, F.; Fleck, A.; Grausgruber, H.; Hetzendorfer, H.; Hof, G.; Lafferty, J.; Marn, M.; Neumayer, A.; Pfaffinger, G.; Birschitzsky, J. Breeding for organic agriculture: the example of winter wheat in Austria. Euphytica 2008, 163, 469–480, doi:10.1007/s10681-008-9709-2.

99. Reeve, J.R.; Smith, J.L.; Carpenter-Boggs, L.; Reganold, J.P. Glycine, nitrate and ammonium uptake by classic and modern wheat varieties in a short-term microcosm study. Biol. Fertil. Soils 2009, 45, 723–732, doi:10.1007/s00374-009-0383-x.

100. Kumar, A.; Gupta, N.; Gupta, A.K.; Gaur, V.K. Identification of biomarkers for determining genotypic potential of nitrogen-use-efficiency and optimization of the nitrogen inputs in crop plants. J. Crop Sci. Biotech. 2009, 12, 183–194, doi:10.1007/s12892-009-0105-9.

101. Salsac, L.; Chaillou, S.; Morot-Gaudry, J.F.; Lesaint, C.; Jolivet, E. Nitrate and ammonium nutrition in plants. Plant Physiol. Biochem. 1987, 25, 805–812.

102. Näsholm, T.; Kielland, K.; Ganeteg, U. Uptake of organic nitrogen by plants. New Phytol. 2009, 182, 31–48, doi:10.1111/j.1469-8137.2008.02751.x.

103. Miller, A.J.; Fan, X.; Orsel, M.; Smith, S.J.; Wells, D.M. Nitrate transport and signaling. J. Exp. Bot. 2007, 58, 2297–2306, doi:10.1093/jxb/erm066.

104. Dechorgnat, J.; Nguyen, C.T.; Armengaud, P.; Jossier, M.J.; Diatloff, E.; Filleur, S.; Daniel-Vedele, F. From the soil to the seeds: The long journey of nitrate in plants. J. Exp. Bot. 2011, 62, 1349–1359, doi:10.1093/jxb/erq409.

105. Kaiser, W.M.; Planchet, E.; Rümer, S. Nitrate reductase and nitric oxide. In Annual Plant Reviews, Nitrogen Metabolism in Plants in the Post-genomic Era; Foyer, C.H., Zhang, H., Eds.; Wiley-Blackwell: Chichester, UK, 2011; Volume 42, pp. 127–146.

106. Sétif, P.; Hirasawa, M.; Cassan, N.; Lagoutte, B.; Tripathy, J.N.; Knaff, D.B. New insights into the catalytic cycle of plant nitrite reductase. Electron transfer kinetics and charge storage. Biochemistry 2009, 48, 2828–2838, doi:10.1021/bi802096f.

107. Ludewig, U.; Neuhäuser, B.; Dynowski, M. Molecular mechanisms of ammonium transport and accumulation in plants. FEBS Lett. 2007, 581, 2301–2308, doi:10.1016/j.febslet.2007.03.034.

108. Mae, T. Physiological nitrogen efficiency in rice: Nitrogen utilization, photosynthesis and yield. In Plant Nutrition for Sustainable Food Production and Environment; Ando, T., Fujita, K., Mae, T., Matsumoto, H., Mori, S., Sekiya, J., Eds.; Kluwer Academic Publishers: Dordrecht, The Netherlands, 1997; pp. 51–60.

109. Hirel, B.; Lea, P.J. The molecular genetics of nitrogen use efficiency in crops. In The Molecular and Physiological Basis of Nutrient Use Efficiency in Crops; Hawkesford, M.J., Barraclough, P.B., Eds.; Wiley-Blackwell: Chichester, UK, 2011; pp. 139–164.

110. Schimel, J.P.; Chapin, F.S. Tundra plant uptake of amino acid and NH4+ nitrogen in situ: Plants compete well for amino acid N. Ecology 1996, 77, 2142–2147, doi:10.2307/2265708.

111. Näsholm, T.; Ekblad, A.; Nordin, R.; Giesler, M.; Hogberg, M.; Hogberg, P. Boreal forest plants take up organic nitrogen. Nature 1998, 392, 914–916, doi:10.1038/31921.

112. Näsholm, T.; Huss-Danell, K.; Högberg, P. Uptake of organic nitrogen in the field by four agriculturally important plant species. Ecology 2000, 81, 1155–1161.

113. Harrison, K.A.; Bol, R.; Bardgett, R.D. Do plant species with different growth strategies vary in their ability to compete with soil microbes for chemical forms of nitrogen? Soil Biol. Biochem. 2008, 40, 228–237, doi:10.1016/j.soilbio.2007.08.004.

114. Biernath, C.; Fischer, H.; Kuzyakov, Y. Root uptake of N-containing and N-free low molecular weight organic substances by maize. A 14C/15N tracer study. Soil Biol. Biochem. 2008, 40, 2237–2245, doi:10.1016/j.soilbio.2008.04.019.

115. Paugfoo-Lonhienne, C.; Lonhienne, T.G.A.; Rentch, D.; Robinson, N.; Christie, M.; Webb, R.I.; Gamage, H.K.; Caroll, B.J.; Schenk, P.M.; Schmidt, S. Plants can use protein as a nitrogen source without assistance from other organisms. Proc. Natl. Acad. Sci. USA 2007, 105, 4524–4529.

116. Tan, X.W.; Ikeda, H.; Oda, M. The absorption, translocation, and assimilation of urea, nitrate or ammonium in tomato plants at different plant growth stages in hydroponic culture. Sci. Hortic. Amsterdam 2000, 84, 275–283, doi:10.1016/S0304-4238(99)00108-9.

117. Kojima, S.; Bohner, A.; von Wirén, N. Molecular mechanisms of urea transport in plants. J. Membrane. Biol. 2006, 212, 83–91, doi:10.1007/s00232-006-0868-6.

118. Kojima, S.; Bohner, A.; Gassert, B.; Yuan, L.; von Wirén, N. AtDUR3 represents the major transporter for high-affinity urea transport across the plasma membrane of nitrogen-deficient Arabidopsis roots. Plant J. 2007, 52, 30–40, doi:10.1111/j.1365-313X.2007.03223.x.

119. Witte, C.P. Urea metabolism in plants. Plant Sci. 2010, 180, 431–438.

120. Smith, S.E.; Read, D.J. Mycorrhizal Symbiosis, 3rd ed. ed.; Academic Press: London, UK, 2008; p. 800.

121. Hodge, A.; Helgason, T.; Fitter, A.H. Nutritional ecology of arbuscular mycorrhizal fungi. Fungal Ecol. 2010, 3, 267–273, doi:10.1016/j.funeco.2010.02.002.

122. Peay, K.G.; Bidartondo, M.I.; Arnold, A.E. Not every fungus is everywhere: Scaling to the biogeography of fungal-plant interactions across roots, shoots and ecosystems. New Phytol. 2010, 185, 878–882, doi:10.1111/j.1469-8137.2009.03158.x.

123. Tanaka, Y.; Yano, K. Nitrogen delivery to maize via myccorhizal hyphae depends on the form of N supplies. Plant Cell Environ. 2005, 28, 1247–1254, doi:10.1111/j.1365-3040.2005.01360.x.
124. Jackson, L.E.; Burger, M.; Cavagnaro, T.R. Nitrogen transformation and ecosystem services. Annu. Rev. Plant. Biol. 2008, 59, 341–363, doi:10.1146/annurev.arplant.59.032607.092932.
125. Miransari, M.; Bahrami, H.A.; Rejali, F.; Malakouti, M.J. Using arbuscular mycorrhiza to reduce the stressful effects of soil compaction on wheat (Triticum aestivum L.) growth. Soil. Biol. Biochem. 2008, 40, 1197–1206, doi:10.1016/j.soilbio.2007.12.014.
126. Miransari, M.; Rejali, F.; Bahrami, H.A.; Malakouti, M.J. Effect of soil compaction and arbuscular mycorrhiza on corn (Zea mays L.) nutrient uptake. Soil. Till. Res. 2009, 103, 282–290, doi:10.1016/j.still.2008.10.015.
127. Daei, G.; Ardakani, M.; Rejali, F.; Teimuri, S.; Miransari, M. Alleviation of salinity on wheat yield, yield components, and nutrient uptake using arbuscular myccorhizal fungi under field condition. J. Plant Physiol. 2009, 166, 617–625, doi:10.1016/j.jplph.2008.09.013.
128. Miransari, M. Arbuscular mycorrhizal fungi and nitrogen uptake. Arch. Microbiol. 2011, 193, 77–81, doi:10.1007/s00203-010-0657-6.
129. Tobar, R.; Azcon, R.; Barea, J.M. Improved nitrogen uptake and transport from 15N-labelled nitrate by external hyphae of arbuscular mycorrhiza under water stressed conditions. New Phytol. 1994, 126, 119–122, doi:10.1111/j.1469-8137.1994.tb07536.x.
130. Tian, C.; Kasiborski, B.; Koul, R.; Mammers, P.J.; Bucking, H.; Shachar-Hill, Y. Regulation of the nitrogen transfer pathway in the arbuscular mycorrhizal symbiosis: Gene characterization and the coordination of expression with nitrogen flux. Plant Physiol. 2010, 153, 1175–1187, doi:10.1104/pp.110.156430.
131. Frey, B.; Schüpp, H. Transfer of symbiotically fixed nitrogen from berseem (Trifolium alexandrinum to maize via vesicular arbuscular mychorrhizal hyphae. New Phytol. 1992, 122, 447–454, doi:10.1111/j.1469-8137.1992.tb00072.x.
132. Bonfante, P.; Anca, I.A. Plants, mycorrhizal fungi, and bacteria: A network of interactions. Annu. Rev. Microbiol. 2009, 63, 363–383, doi:10.1146/annurev.micro.091208.073504.
133. Bonfante, P.; Genre, A. Mechanisms underlying beneficial plant-fungus interactions in mycorrhizal symbiosis. Nat. Commun. 2010, 1, 1–11.
134. Lea, P.J.; Miflin, B.J. Nitrogen assimilation and its relevance to crop improvement. In Annual Plant Reviews, Nitrogen Metabolism in Plants in the Post-genomic Era; Foyer, C.H., Zhang, H., Eds.; Wiley-Blackwell: Chichester, UK, 2011; Volume 42, pp. 1–40.
135. Suzuki, A.; Knaff, D.B. Glutamate synthase: Structural, mechanistic and regulatory properties, and role in the amino acid metabolism. Photosynth. Res. 2005, 83, 191–217, doi:10.1007/s11120-004-3478-0.
136. Hirel, B.; Lea, P.J. Amino acid metabolism. In Plant Nitrogen; Lea, P.J., Morot-Gaudry, J.F., Eds.; INRA, Springer-Verlag: Berlin, Germany, 2001; pp. 79–99.
137. Taira, M.; Valtersson, U.; Burkhardt, B.; Ludwig, R.A. Arabidopsis thaliana GLN2-encoded glutamine synthetase is dual targeted to leaf mitochondria and chloroplasts. Plant Cell 2004, 16, 2048–2058, doi:10.1105/tpc.104.022046.

138. Dubois, F.; Brugière, N.; Sangwan, R.S.; Hirel, B. Localization of tobacco cytosolic glutamine synthetase enzymes and the corresponding transcripts shows organ- and cell-specific patterns of protein synthesis and gene expression. Plant Mol. Biol. 1996, 31, 803–817, doi:10.1007/BF00019468.

139. Cren, M.; Hirel, B. Glutamine synthetase in higher plants: Regulation of gene and protein expression from the organ to the cell. Plant Cell Physiol. 1999, 40, 1187–1193, doi:10.1093/oxfordjournals.pcp.a029506.

140. Lea, P.J.; Ireland, R.J. Nitrogen metabolism in higher plants. In Plant Amino Acids; Singh, B.K., Ed.; Dekker M.: New York, NY, USA, 1999; pp. 1–47.

141. Aubert, S.; Bligny, R.; Douce, R.; Ratcliffe, R.G.; Roberts, J.K.M. Contribution of glutamate dehydrogenase to mitochondrial metabolism studied by 13C and 31P nuclear magnetic resonance. J. Exp. Bot. 2001, 52, 37–45, doi:10.1093/jexbot/52.354.37.

142. Labboun, S.; Tercé-Laforgue, T.; Roscher, A.; Bedu, M.; Restivo, F.M.; Velanis, C.N.; Skopelitis, D.S.; Moshou, P.N.; Roubelakis-Angelakis, K.A.; Suzuki, A. Resolving the role of plant glutamate dehydrogenase: I. In vivo real time nuclear magnetic resonance spectroscopy experiments. Plant Cell Physiol. 2009, 50, 1761–1773, doi:10.1093/pcp/pcp118.

143. Skopelitis, D.S.; Paranychiankis, N.V.; Paschalidis, K.A.; Plianokis, E.D.; Delis, I.D.; Yakoumakis, D.I.; Kouvarakis, A.; Papadakis, E.D.; Stephanou, E.G.; Roubelakis-Angelakis, K.A. Abiotic stress generates ROS that signal expression of anionic glutamate dehydrogenase to form glutamate for proline synthesis in tobacco and grapevine. Plant Cell 2006, 18, 2767–2781, doi:10.1105/tpc.105.038323.

144. Masclaux, C.; Quilleré, I.; Gallais, A.; Hirel, B. The challenge of remobilisation in plant nitrogen economy. A survey of physio-agronomic and molecular approaches. Ann. Appl. Biol. 2001, 138, 69–81, doi:10.1111/j.1744-7348.2001.tb00086.x.

145. Stitt, M.; Müller, C.; Matt, P.; Gibon, Y.; Carillo, P.; Morcuende, R.; Sheible, W.R.; Krapp, A. Steps towards an integrated view of nitrogen metabolism. J. Exp. Bot. 2002, 53, 959–970, doi:10.1093/jexbot/53.370.959.

146. Tercé-Laforgue, T.; Dubois, F.; Ferrario-Mery, S.; Pou de Crecenzo, M.A.; Sangwan, R.; Hirel, B. Glutamate dehydrogenase of tobacco (Nicotiana tabacum L.) is mainly induced in the cytosol of phloem companion cells when ammonia is provided either externally or released during photorespiration. Plant Physiol. 2004, 136, 4308–4317, doi:10.1104/pp.104.047548.

147. Morot Gaudry, J.F.; Job, D.; Lea, P.J. Amino acid metabolism. In Plant Nitrogen; Lea, P.J., Morot Gaudry, J.F., Eds.; INRA Springer-Verlag: Berlin, Germany, 2001; pp. 167–211.

148. Lea, P.J.; Azevedo, R.A. Nitrogen use efficiency. 2. Amino acid metabolism. Ann. Appl. Biol. 2007, 151, 269–275, doi:10.1111/j.1744-7348.2007.00200.x.

149. Galili, S.; Amir, R.; Galili, G. Genetic engineering of amino acids in plants. Adv. Plant Biochem. Mol. Biol. 2008, 1, 49–80.

150. Gutiérrez, R.A.; Lejay, L.V.; Dean, A.; Chiaromonte, F.; Shasha, D.E.; Coruzzi, G.M. Qualitative network models and genome-wide expression data define carbon/nitrogen-responsive molecular machines in Arabidopsis. Genome Biol. 2007, 8, R7, doi:10.1186/gb-2007-8-1-r7.

151. Coruzzi, G.M.; Burga, A.R.; Katari, M.S.; Gutiérrez, R.A. Systems biology: Principles and applications in plant research. In Annual Plant Reviews, Plant Systems Biology; Coruzzi, G.M., Guttiérez, R.A., Eds.; Wiley-Blackwell: Chichester, UK, 2009; Volume 35, pp. 3–40.

152. Thum, K.E.; Shin, M.J.; Gutiérrez, R.A.; Mukherjee, I.; Katari, M.S.; Nero, D.; Shasha, D.; Coruzzi, G.M. An integrated genetic, genomic and systems approach defines gene networks regulated by the interaction of light and carbon signalling pathways in Arabidopsis. BMC Syst. Biol. 2008, 2, 31, doi:10.1186/1752-0509-2-31.

153. Andrews, M.; Lea, P.J.; Raven, J.A.; Lindsey, K. Can genetic manipulation of plant nitrogen assimilation enzymes result in increased crop yield and greater N-use efficiency? An assessment. Ann. Appl. Biol. 2004, 145, 25–40, doi:10.1111/j.1744-7348.2004.tb00356.x.

154. Fuentes, S.I.; Alen, D.J.; Ortiz-Lopez, A.; Hernandez, G. Overexpression of cytosolic glutamine synthetase increases photosynthesis and growth at low nitrogen concentrations. J. Exp. Bot. 2001, 52, 1071–1081, doi:10.1093/jexbot/52.358.1071.

155. Oliveira, I.C.; Brears, T.; Knight, T.J.; Clark, A.; Coruzzi, G.M. Overexpression of cytosolic glutamine synthetase. Relation to nitrogen, light, and photorespiration. Plant Physiol. 2002, 129, 1170–1180, doi:10.1104/pp.020013.

156. Jing, Z.P.; Gallardo, F.; Pascual, M.B.; Sampalo, R.; Romero, J.; Torres de Vavarra, A.; Canovas, F.M. Improved growth in a field trial of transgenic hybrid poplar overexpressing glutamine synthetase. New Phytol. 2004, 164, 137–145, doi:10.1111/j.1469-8137.2004.01173.x.

157. Man, H.M.; Boriel, R.; El-Khatib, R.; Kirby, E.G. Characterization of transgenic poplar with ectopic expression of pine cytosolic glutamine synthetase under conditions of varying nitrogen ability. New Phytol. 2005, 167, 31–39, doi:10.1111/j.1469-8137.2005.01461.x.

158. Habash, D.Z.; Massiah, A.J.; Rong, H.L.; Wallsgrove, R.M.; Leigh, R.A. The role of cytosolic glutamine synthetase in wheat. Ann. Appl. Biol. 2001, 138, 83–89, doi:10.1111/j.1744-7348.2001.tb00087.x.

159. Martin, A.; Lee, J.; Kichey, T.; Gerentes, D.; Zivy, M.; Tatou, C.; Balliau, T.; Valot, B.; Davanture, M.; Dubois, F.; et al. Two cytosolic glutamine synthetase isoforms of maize (Zea mays L.) are specifically involved in the control of grain production. Plant Cell 2006, 18, 3252–3274, doi:10.1105/tpc.106.042689.

160. Brauer, E.K.; Rochon, A.; Bi, Y.M.; Bozzo, G.G.; Rothstein, S.J.; Shelp, B. Reappraisal of nitrogen use efficiency in rice overexpressing glutamine synthetase 1. Physiol. Plantarum. 2011, 141, 361–372, doi:10.1111/j.1399-3054.2011.01443.x.

161. Vincent, R.; Fraisier, V.; Chaillou, S.; Limami, M.A.; Deléens, E.; Phillipson, B.; Douat, C.; Boutin, J.P.; Hirel, B. Overexpression of a soybean gene encoding cytosolic glutamine synthetase in shoots of transgenic Lotus corniculatus L. plants triggers changes in ammonium assimilation and plant development. Planta 1997, 201, 424–433, doi:10.1007/s004250050085.

162. Good, A.G.; Johnson, S.J.; De Pauw, M.; Carroll, R.T.; Savodiv, N.; Vidmar, J.; Lu, Z.; Taylor, G.; Stroeher, V. Engineering nitrogen use efficiency with alanine aminotransferase. Can. J. Bot. 2007, 85, 252–262, doi:10.1139/B07-019.

163. Shrawat, A.K.; Carroll, R.T.; DePauw, M.; Taylor, G.J.; Good, A.G. Genetic engineering of improved nitrogen use efficiency in rice by the tissue specific expression

of alanine amino-transferase. Plant Biotechnol. J. 2008, 6, 722–732, doi:10.1111/j.1467-7652.2008.00351.x.

164. Ameziane, R.; Bernhard, K.; Lightfoot, D. Expression of the bacterial gdhA gene encoding a NADPH-glutamate dehydrogenase in tobacco affects plant growth and development. Plant Soil 2000, 221, 47–57, doi:10.1023/A:1004794000267.

165. Bi, Y.M.; Kant, S.; Clark, J.; Gidda, S.; Ming, F.; Xu, J.; Rochon, A.; Shelp, B.J.; Hao, L.; Zhao, R.; et al. Increased nitrogen use efficiency in transgenic rice plants over-expressing a nitrogen-responsive early nodulin gene identified from rice expression profiling. Plant Cell Environ. 2009, 32, 1749–1760, doi:10.1111/j.1365-3040.2009.02032.x.

166. Chichkova, S.; Arellano, J.; Vance, C.P.; Hernandez, G. Transgenic tobacco plants that overexpress alfalfa NADH-glutamate synthase have higher carbon and nitrogen content. J. Exp. Bot. 2001, 52, 2079–2087.

167. Yamaya, T.; Obara, M.; Nakajima, H.; Sasaki, S.; Hayakawa, T.; Sato, T. Genetic manipulation and quantitative trait loci mapping for nitrogen recycling in rice. J. Exp. Bot. 2002, 53, 917–925, doi:10.1093/jexbot/53.370.917.

168. Tabuchi, M.; Abiko, T.; Yamaya, T. Assimilation of ammonium ions and reutilization of nitrogen in rice (O. sativa L.). J. Exp. Bot. 2007, 58, 2319–2327, doi:10.1093/jxb/erm016.

169. Yanagisawa, S.; Akiyama, A.; Kisaka, H.; Uchimiya, H.; Miwa, T. Metabolic engineering with Dof1 transcription factor in plants: Improved nitrogen assimilation and growth under low-nitrogen conditions. Proc. Natl. Acad. Sci. USA 2004, 101, 7833–7838, doi:10.1073/pnas.0402267101.

170. Fonder, N.; Heens, B.; Xanthoulis, D. Optimisation de la fertlisation azotée de cultures industrielles légumières sous irrigation. Biotechnol. Agron. Soc. Environ. 2010, 14, 103–111.

171. Quilléré, I.; Dufossé, C.; Roux, Y.; Foyer, C.H.; Caboche, M.; Morot-Gaudry, J.F. The effects of deregulation of NR gene expression on growth and nitrogen metabolism of Nicotiana plumbaginifolia plants. J. Exp. Bot. 1994, 278, 1205–1211.

172. Djennane, S.; Chauvin, J.E.; Quilléré, I.; Meyer, C.; Chupeau, Y. Introduction and expression of a deregulated tobacco nitrate reductase gene in potato lead to highly reduced nitrate levels in transgenic tubers. Transgenic Res. 2002, 11, 175–184, doi:10.1023/A:1015299711171.

173. Djennane, S.; Quilléré, I.; Leydecker, M.-T.; Meyer, C.; Chauvin, J.E. Expression of a deregulated tobacco nitrate reductase gene in potato increases biomass production and decreases nitrate concentration in all organs. Planta 2004, 219, 884–893.

174. Curtis, I.S.; Power, J.B.; de Laat, A.M.M.; Caboche, M.; Davey, M.R. Expression of a chimeric nitrate reductase gene in transgenic lettuce reduces nitrate in leaves. Plant Cell Rep. 1999, 18, 889–896, doi:10.1007/s002990050680.

175. Xu, Y. Quantitative trait loci: Separating, pyramiding, and cloning. Plant Breed. Rev. 1997, 15, 85–139.

176. Yu, J.; Buckler, E.S. Genetic association mapping and genome organization of maize. Cur. Opin. Biotech. 2006, 17, 155–160, doi:10.1016/j.copbio.2006.02.003.

177. Salvi, S.; Tuberosa, R. Cloning QTLs in Plants. In Genomics-Assisted Crop Improvement; Varshney, R.K., Tuberosa, R., Eds.; Springer: Dordrecht, The Netherlands, 2007; Volume 1, pp. 207–226.

178. Agrama, H.A.S.; Zacharia, A.G.; Said, F.B.; Tuinstra, M. Identification of quantitative trait loci for nitrogen use efficiency in maize. Mol. Breed. 1999, 5, 187–195, doi:10.1023/A:1009669507144.

179. Bertin, P.; Gallais, A. Physiological and genetic basis of nitrogen use efficiency in maize. II. QTL detection and coincidences. Maydica 2001, 46, 53–68.

180. Hirel, B.; Gallais, A.; Bertin, P.; Quillere, I.; Bourdoncle, W.; Attagan, C.; Dellay, C.; Gouy, A.; Cadiou, S.; Retaillau, C.; et al. Towards a better understanding of the genetic and physiological basis for nitrogen use efficiency in maize. Plant Physiol. 2001, 125, 1258–1270, doi:10.1104/pp.125.3.1258.

181. Gallais, A.; Hirel, B. An approach of the genetics of nitrogen use efficiency in maize. J. Exp. Bot. 2004, 55, 295–306, doi:10.1093/jxb/erh006.

182. Zhang, N.; Gibon, Y.; Gur, A.; Chen, C.; Lepak, N.; Höne, M.; Zhang, Z.; Kroon, D.; Tschoep, H.; Stitt, M.; et al. Fine quantitative trait loci mapping of carbon and nitrogen metabolism enzyme activities and seedling biomass in the maize IBM mapping population. Plant Physiol. 2010, 154, 1753–1765, doi:10.1104/pp.110.165787.

183. Obara, M.; Kajiura, M.; Fukuta, Y.; Yano, M.; Hayashi, M.; Yamaya, T.; Sato, T. Mapping of QTLs associated with cytosolic glutamine synthetase and NADH-glutamate synthase in rice (Oryza sativa L.). J. Exp. Bot. 2001, 52, 1209–1217, doi:10.1093/jexbot/52.359.1209.

184. Tabuchi, M.; Sugiyama, T.; Ishiyama, K.; Inoue, E.; Sato, T.; Takahashi, H.; Yamaya, T. Severe reduction in growth and grain filling of rice mutants lacking OsGS1;1, a cytosolic glutamine synthetase 1;1. Plant J. 2005, 42, 641–655, doi:10.1111/j.1365-313X.2005.02406.x.

185. Fontaine, J.X.; Ravel, C.; Pageau, K.; Heumez, E.; Dubois, F.; Hirel, B.; Le Gouis, J. A quantitative genetic study for elucidating the contribution of glutamine synthetase, glutamate dehydrogenase and other nitrogen-related physiological traits to the agronomic performance of common wheat. Theor. Appl. Genet. 2009, 119, 645–662, doi:10.1007/s00122-009-1076-4.

186. Quraishi, U.M.; Abrouk, M.; Murat, F.; Pont, C.; Foucrier, S.; Demaizieres, G.; Confolent, C.; Rivière, N.; Charmet, G.; Paux, E.; et al. Cross-genome map based dissection of a nitrogen use efficiency ortho-meta QTL in bread wheat unravels concerted cereal genome evolution. Plant J. 2011, 65, 745–756, doi:10.1111/j.1365-313X.2010.04461.x.

187. Plomion, C.; Bahrmann, N.; Costa, P.; Frigério, J.M.; Gerber, S.; Gion, J.M.; Lalanne, C.; Madur, D.; Pionneau, C. Proteomics for genetic and physiological studies in forest trees: Application in maritime pine. In Molecular Genetics and Breeding of Forest Trees; Kumar, S., Fladung, M., Eds.; Haworth Press: New York, NY, USA, 2004; pp. 53–80.

188. Lammerts van Buren, E.T.; Jones, S.S.; Tamn, L.; Murphy, K.M.; Myers, J.R.; Leifert, C.; Mesmer, M.M. The need to breed crop varieties suitable for organic farming, using wheat, tomato and broccoli as examples: A review. NJAS Wageningen J. Life Sci. 2010, doi:10.1016/j.njas.2010.04.001.

189. Hawkesford, M.J.; Howarth, J.R. Transcriptional profiling approaches for studying nitrogen use efficiency. In Annual Plant Reviews, Nitrogen Metabolism in Plants in the Post-genomic Era; Foyer, C.H., Zhang, H., Eds.; Wiley-Blackwell: Chichester, UK, 2011; Volume 42, pp. 41–62.

190. Messmer, M.M.; Burger, H.; Schmidt, W.; Geiger, H.H. Importance of appropriate selection environments for breeding maize adapted to organic farming systems. Tagung der Vereinigung der Pflanzenzücher und Saatgutkaufleute Österreichs 2009, 1–3.

191. Krapp, A.; Truong, H.N. C/N interaction in model plant species. In Enhancing the Efficiency of Nitrogen Utilisation in Plants; Goyal, S.S., Tischner, R., Basra, A.S., Eds.; Haworth Press: Binghampton, NY, USA, 2005; pp. 127–173.

192. Gutiérrez, R.A.; Gifford, M.L.; Poultney, C.; Wang, R.; Shasha, D.E.; Coruzzi, G.M.; Crawford, N.M. Insights into the genomic nitrate response using genetics and the sungear software system. J. Exp. Bot. 2007, 58, 2359–2367, doi:10.1093/jxb/erm079.

193. Cañas, R.A.; Quilleré, I.; Christ, A.; Hirel, B. Nitrogen metabolism in the developing ear of maize (Zea mays L.): Analysis of two lines contrasting in their mode of nitrogen managementX. J. Exp. Bot. 2009, 184, 340–352.

194. Meyer, R.C.; Steinfath, M.; Lisec, J.; Becher, M.; Witucha-Wall, H.; Törjék, O.; Fienh, O.; Eckardt, A.; Willmitzer, L.; Selbig, J.; et al. The metabolic signature related to high plant growth rate in Arabidopsis thaliana. Proc. Nat. Acad. Sci. USA 2007, 104, 4759–4664, doi:10.1073/pnas.0609709104.

195. Lisec, J.; Meyer, R.C.; Steinfath, M.; Redestig, H.; Becher, M.; Witucka-Wall, H.; Fienh, O.; Törjék, O.; Selbig, J.; Altman, T.; et al. Identification of metabolic and biomass QTL in Arabidopsis thaliana in a parallel analysis of RIL and Il populations. Plant J. 2008, 53, 960–972.

196. Radrich, K.; Tsuruoka, Y.; Dobson, P.; Gevorgyan, A.; Swaitson, N.; Schwartz, J.M. Integration of metabolic databases for the reconstruction of genome-scale metabolic networks. BMC Syst. Biol. 2010, 4, 1–16, doi:10.1186/1752-0509.

197. De Oliveira Dal'Molin, C.G.; Quek, L.E.; Palfreyman, R.W.; Brumbley, S.M.; Nielsen, L.K. AraGEM, a genome-scale reconstruction of the primary metabolic network in Arabidopsis. Plant Physiol. 2010, 152, 579–589, doi:10.1104/pp.109.148817.

198. Coque, M.; Martin, A.; Veyrieras, J.B.; Hirel, B.; Gallais, A. Genetic variation for N-remobilization and postsilking N-uptake in a set of maize recombinant inbred lines. 3. QTL detection and coincidences. Theor. Appl. Genet. 2008, 117, 729–747, doi:10.1007/s00122-008-0815-2.

199. Xu, Y.; Crouch, J.H. Marker-assisted selection in plant breeding: From publication to practice. Crop Sci. 2008, 48, 391–407, doi:10.2135/cropsci2007.04.0191.

200. Hirel, B.; Le Gouis, J.; Bernard, M.; Perez, P.; Falque, M.; Quétier, F.; Joets, J.; Montalent, P.; Rogwoski, P.; Murigneux, A.; et al. Genomics and plant breeding: Maize and wheat. In Functional Plant Genomics; Morot-Gaudry, J.-F., Lea, P.J., Briat, J.-F., Eds.; Science Publishers: Enfield, NH, USA, 2007; pp. 614–635.

201. Goff, S.A.; Ricke, D.; Lan, T.H.; Presting, G.; Wang, R.; Dunn, M.; Glazebrook, J.; Session, A.; Oeller, P.; Varma, H.; et al. A draft sequence of the rice genome (Oryza sativa L. ssp. japonica). Science 2002, 296, 92–100, doi:10.1126/science.1068275.

202. Schnable, P.S.; Ware, D.; Fulton, R.S.; Stein, J.C.; Wei, F.; Pasternak, S.; Liang, C.; Khang, J.; Fulton, L.; Graves, T.A.; et al. The B73 maize genome: Complexity, diversity and dynamics. Science 2009, 326, 1112–1115, doi:10.1126/science.1178534.

203. International Wheat Genome Organization. Available online: http://www.wheatgenome.org/ (accessed on 1 September 2011).
204. International Barley Sequencing Consortium. Available online: http://barleygenome.org/ (accessed on 1 September 2011).
205. Food and Agriculture Organization. Available online: http://www.fao.org/worldfoodsituation/wfs-home/csdb/en/ (accessed on 1 September 2011).
206. Edgerton, M.D. Increasing crop productivity to meet global needs for feed, food and fuel. Plant Physiol. 2009, 149, 7–13, doi:10.1104/pp.108.130195.
207. Pilbeam, D.J. The utilization of nitrogen by plants: A whole plant perspective. In Annual Plant Reviews, Nitrogen Metabolism in Plants in the Post-genomic Era; Foyer, C.H., Zhang, H., Eds.; Wiley-Blackwell: Chichester, UK, 2011; Volume 42, pp. 305–352.
208. Cañas, R.A.; Amiour, N.; Quilleré, I.; Hirel, B. An integrated statistical analysis of the genetic variability of nitrogen metabolism in the ear of three maize inbred lines (Zea mays L.). J. Exp. Bot. 2010, 62, 2309–2318.
209. Charpentier, M.; Oldroyd, G. How close are we to nitrogen-fixing cereals? Curr. Opin. Plant Biol. 2010, 13, 556–564, doi:10.1016/j.pbi.2010.08.003.
210. Tremblay, M.; Wallach, D. Comparison of parameter estimation methods for crop models. Agronomie 2004, 24, 351–365, doi:10.1051/agro:2004033.
211. McCown, R.L.; Hammer, G.L.; Hargreaves, J.N.G.; Holzworth, D.P.; Freebairn, D.M. APSIM: A novel software system for model development, model testing and simulation in agricultural systems research. Agr. Syst. 1996, 50, 255–271, doi:10.1016/0308-521X(94)00055-V.
212. Stark, C.H.; Richards, K.G. The continuing challenge of agricultural nitrogen loss to the environment in the context of global change and advancing research. Dyn. Soil. Dyn. Plant 2008, 2, 1–12.
213. Hammer, G.L.; Kropff, M.J.; Sinclair, T.R.; Porter, J.R. Future contributions of crop modeling—from heuristics and supporting decision-making to understanding genetic regulation and aiding crop improvement. Eur. J. Agron. 2002, 18, 15–31, doi:10.1016/S1161-0301(02)00093-X.
214. Samborski, S.M.; Tremblay, N.; Fallon, E. Strategies to make use of plant sensors-based diagnostic information for nitrogen recommendations. Agron. J. 2009, 101, 800–816, doi:10.2134/agronj2008.0162Rx.

AUTHOR NOTES

CHAPTER 1

Acknowledgments

This study was supported by the EU grant no. KBBE-245058-SOLIBAM. Special thanks go to the case farmer for providing the data and allowing us to perform research at his interesting farm. Further, we acknowledge the contribution from other SOLIBAM participants, specifically Elena Tavella at the University of Copenhagen, who calculated the revenues.

Author Contributions

The study was designed by Mads V. Markussen in collaboration with all co-authors. Data was collected by Mads V. Markussen and Michal Kulak with assistance from Laurence G. Smith. All data analysis related to the emergy evaluation was done by Mads V. Markussen, and all data analysis related to LCA was done by Michal Kulak. All authors contributed to combining the analyses and interpreting the results.

Conflicts of Interest

The authors declare no conflict of interest.

CHAPTER 4

Acknowledgments

The research was funded by the Centre of Potential for Excellence fund (CPE) of University Grand Commission, New Delhi, India and it is to be gratefully acknowledged.

CHAPTER 5

Acknowledgments

This work was supported in part by MEXT through Special Coordination Funds for Promoting Science and Technology, as part of the research project for "Sustainable agriculture practices to mitigate and adapt to global warming" undertaken by the Institute for Global Change Adaptation Science, Ibaraki University.

CHAPTER 6

Competing Interests

The authors declare that they have no competing interests.

Author Contributions

This work is part of the Ph.D. thesis of MRD where SNS supervised the thesis and MM participated in the statistical analysis and correction of the manuscript. All authors read and approved the final manuscript.

Acknowledgments

Authors are grateful to Dr. H.S. Guar, Dean and Joint Director (Education), IARI, New Delhi for his untiring help and support.

CHAPTER 9

Competing Interests

The authors declare that they have no competing interests.

Author Contributions

JP: Collected and reviewed the literature and drafted the manuscript. NS: Formulated the objectives, provided guidance and improved the quality of the manuscript. Both authors read and approved the final manuscript.

Acknowledgments

The financial support from Department of Biotechnology (DBT), New Delhi, India, and Department of Science and Technology (DST), New

Delhi, India, through Fund for Improvement of Science and Technology Infrastructure in Higher Educational Institutions (FIST), is gratefully acknowledged.

CHAPTER 11

Conflict of Interest
The authors declare no conflict of interest.

Acknowledgments
Part of the research leading to these results has received funding from the Specific Targeted Research Project of the European Commission 6th Framework Program Priority 8.1 SSP: Opportunities for farm seed conservation, breeding and production Proposal/Contract no.: SSP-CT-2006-04434, and the European Community's Seventh Framework Programme (FP7/ 2007–2013) under the Grant Agreement n245058-Solibam (Strategies for Organic and Low-Input Breeding and Management). JC Dawson was supported by an INRA postdoctoral grant and P Riviere was supported by a doctoral grant from the DIM ASTREA program of the Region Ile de France. The authors would like to thank the following people: Véronique Chable, Jérôme Enjalbert, Nicolas Schermann, Henri Ferté, Giandommenico Cortiana, Thomas Levillain, Silvio Pino, Aart Osman, Elie Guillo and all the farmers involved in the FSO and Croisement du Roc projects.

CHAPTER 12

Acknowledgments
We wish to thank the farmers involved in this study. We also thank the European Community for funding the Farm Seed Opportunity project (FP6 STREP, contract no. 044345, priority 8.1, "Specific Support to Policies").

Conflict of Interest
The authors declare no conflict of interest.

CHAPTER 13

Acknowledgments

We thank the eight growers in Yolo Co. for collaborating on this project, allowing us to sample their fields, and providing management information. Andrew Margenot, Felipe Barrios-Masias, Amanda K. Hodson, Cristina Lazcano, and Alia Tsang provided field assistance. Steve Culman provided statistical advice. This research was funded by the USDA NIFA Organic Agriculture Research and Education Initiative Award2009-01415, the UC Davis Graduate Group in Ecology Block Grants, and an NSF Graduate Research Fellowship to T.M. Bowles.

CHAPTER 14

Acknowledgments

The authors wish to express their gratitude to the farmers who allowed us to conduct this study in their fields. We would like to thank Sara Elchanani and the Division of Water Quality of the Israel Water Authority for supporting and funding the project. The project was partly supported by the Koshland foundation.

INDEX

D

Milton Keynes UK
Ingram Content Group UK Ltd.
UKHW022056141024
449569UK00031B/1653